LASER SYSTEMS FOR APPLICATIONS

Edited by **Krzysztof Jakubczak**

Laser Systems for Applications
Edited by Krzysztof Jakubczak

Published by InTech
Janeza Trdine 9, 51000 Rijeka, Croatia

Publishing Process Manager Petra Nenadic

Technical Editor Teodora Smiljanic

Cover Designer InTech Design Team

Additional hard copies can be obtained from orders@intechopen.com

Laser Systems for Applications, Edited by Krzysztof Jakubczak
p. cm.
ISBN 978-953-307-429-0

For EU product safety concerns:
IN TECH d.o.o., Prolaz Marije Krucifikse Kozulić 3, 51000 Rijeka, Croatia,
info@intechopen.com or visit our website at intechopen.com.

Meet the editor

Krzysztof Jakubczak is a graduate from Military University of Technology, Warsaw, Poland. He achieved his education in the fields of laser physics, optics, infrared technology, and light modulation and detection. He earned his MSc degree for research on laser-plasma source for lithography and micro- and nano-structures production with the lobster-eye optics assembly. He continued his education at Czech Technical University in Prague. He earned his PhD for construction, examination and applications of high-order harmonics source using titanium-sapphire femtosecond laser system and noble gases. Amongst others he expanded his professional skills to such areas like vacuum technology, spectroscopy, plasma physics, ultrafast laser systems, and optics. He started his professional career as R&D Engineer with the Academy of Sciences of the Czech Republic. Then he got involved in HiLASE Project, as Deputy Project Manager, and led the R&D team Research Program Leader working on development of thin-disk lasers. Finally, he decided to pursue an industrial career in medical devices (intraocular lenses). He works as Head of Engineering Department with Croma Polska sp. z o.o. and is now based in Warsaw, Poland. Krzysztof Jakubczak is a member of SPIE and Photonics Society of Poland.

Contents

Preface

Well known and established techniques, which allow obtaining intense laser pulses, have led to the development of advanced laser system used in science and various industries. Often at times, they combine Q-switching and mode-locking techniques in a single device to achieve unprecedented average pulse powers (e.g. HiLASE, HiPER, ELI, NIF, LiFE) for scientific applications, such as secondary sources of EUV, X-ray or particles, or as inertial confinement fusion (ICF) drivers. On the other hand, the great advent of industrial lasers has been observed throughout recent years. This resulted in a renaissance of CO_2 laser systems, development of micro-lasers, high-average power fiber systems, and thin-disk lasers. Many of these systems became commercially available either as "off-the-shelf" or customized solutions tailored to a client's specific requirements. The two trends present the generic flavor of the present laser market. This book gives a brief overview into how such systems are built, what physical phenomena they make use of, or even how to tune laser design for particular applications.

Dr. Krzysztof Jakubczak
Croma Polska Sp. z o.o.
Warsaw
Poland

Part 1

Laser and Terahertz Sources

1

Q-Switching with Single Crystal Photo-Elastic Modulators

F. Bammer[1], T. Schumi[1], J. R. Carballido Souto[1], J. Bachmair[1], D. Feitl[2], I. Gerschenson[2], M. Paul[2] and A. Nessmann[2]
[1]*Vienna University of Technology*
[2]*Gymnasium Stubenbastei, Vienna*
Austria

1. Introduction

Q-switching is a common technology to produce laser pulses. It is based on a fast optical switch in the laser resonator blocking laser light while pumping energy is stored in the gain medium. Usually electro- or acousto-optic Q-switches are used for this task. We introduce here Single Crystal Photo-Elastic Modulators (SCPEM) that combine features of both technologies, namely first the change of polarization in case of electro-optics, and second the use of the photo-elastic effect as in acousto-optics. These resonant devices allow running lasers on a constant pulse repetition frequency determined by the crystal size and excited eigenmode.

2. Photo-elastic modulators

We will give here an overview about the basics and theory of Photo-Elastic Modulators (PEM). This will include information about classical PEM excited, from the end, from the sides, and from two sides. Then we will describe single crystal photo-elastic modulators with focus on the material from the crystal symmetry group 3m and short discussion of other groups.

2.1 Conventional photo-elastic modulators

Conventional photo-elastic modulators (PEMs, Fig. 1, left) modulate the polarization of a light beam and are mainly used in ellipsometry in the form of Kemp-modulators (Kemp, 1969; Kemp, 1987; www.hindsinstruments.com). They are made of a piece of optical glass which is glued to a quartz-crystal. Both pieces are adjusted to have the same longitudinal resonance frequency. When the quartz-crystal is electrically excited with the proper frequency the system will start a strong resonant oscillation. Since damping is low the strain amplitudes become even at low voltage amplitudes so large that a significant artificial birefringence modulation due to the photo-elastic effect is caused. Polarized light passing the glass will experience a strong modulation of its polarization.
A more advanced modulator is shown in the middle of Fig. 1 which is used for the infrared, where due to the longer wavelengths strong retardation is required, usually with ZnSe or

GaAs as the optical material. Its special shape allows together with the two actuators a superposition of a vertical and horizontal longitudinal mode.
Further Canit & Badoz (1983) proposed an advanced design for a conventional PEM, where small actuators are glued on the sides of the glass piece (Fig. 1, right) to excite the 3rd harmonics of the longitudinal horizontal mode.

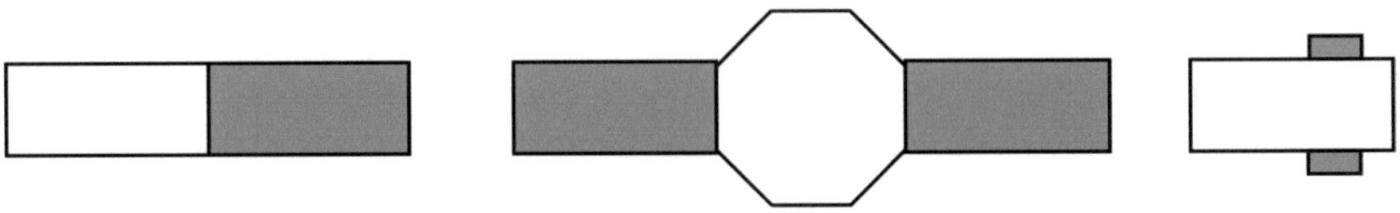

Fig. 1. Different shapes of classical photo-elastic modulators. The actuators are shown grey.

Disadvantages of conventional PEMs are:

- High precision is needed to adjust the system such that it oscillates with high merit.
- Furthermore the device is very large compared to its useful aperture.
- No superposition of frequency adjusted higher modes is possible
- the gluing process of the actuator(s) can lead to stresses in the glass and hence to an unwanted stray birefringence

2.2 Definition and description of a SCPEM

A Single Crystal Photo-Elastic Modulator (SCPEM, Fig. 1) is a piezoelectric optical transparent crystal that is electrically excited on one of its resonance frequencies. Many possible configurations are described by Bammer (2007).

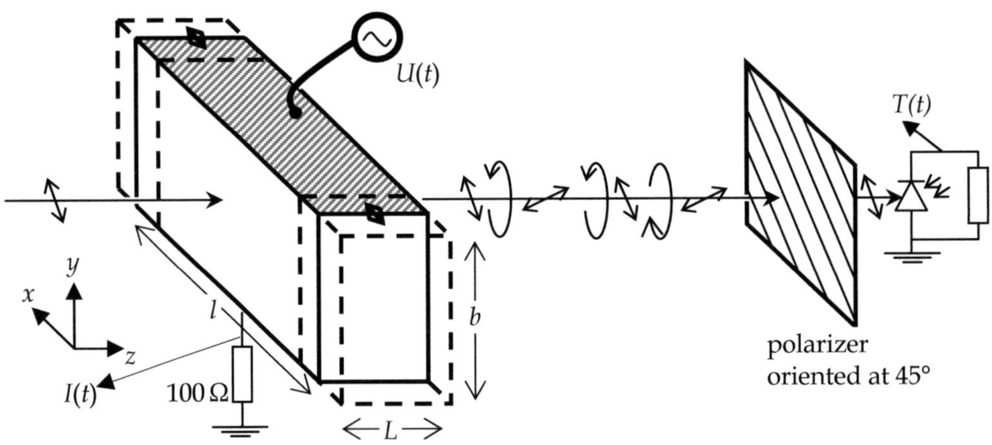

Fig. 2. Single Crystal Photo-Elastic Modulator made of a 3m-crystal (Bammer, 2009).

A necessary feature is that the polarization of light of any wavelength passing this crystal must not be changed when the crystal is at rest. This for example cannot be fulfilled with the quartz crystal of a conventional PEM, due to its optical activity (Nye, 1985), which would need to be compensated by a second reversely oriented quartz crystal as proposed in an early patent on PEMs (n.n., 1925).

SCPEMs for the MIR are based on 43m crystals and are experimentally and theoretically described by Weil & Halido (1974). Fig. 2 shows now one favourable configuration based on a crystal with symmetry 3m.
The light travels along the optical axis (= z-axis), the exciting electrical field points into the y-direction, and in most cases the longitudinal x-eigenmode is used. Two further important eigenmodes are the longitudinal y-eigenmode and the shear yz-eigenmode. Of course there exist infinitely many higher eigen-modes and frequencies.
In most cases a polarizer (analyzer) oriented at 45° is placed behind the modulator. Fig. 2 further shows a photo diode to get a transmission signal T and a resistor (here with 100 Ω) to generate a measure for the piezo electric current I generated by the crystal.

2.2.1 Use as a Q-switch

We define now for monochromatic light with the wavelength λ the retardation as

$$\delta = 2\pi \frac{L\,(n_x - n_y)}{\lambda} \tag{1}$$

where L is the z-dimension of the crystal and n_x, n_y are the refractive indices for x- and y-polarized light. They are calculated in chapter 2.2 for the photo-elastic effect and in chapter 2.6 for the electro-optic effect.

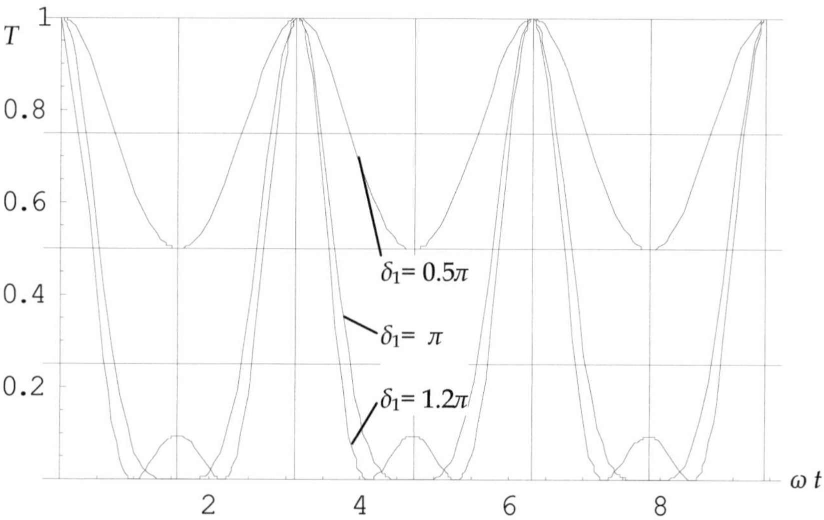

Fig. 3. Transmission curves for the configuration in Fig. 2.

The transmission $T(\delta)$ for the situation of Fig. 2 is

$$T(\delta) = T_{min} + (T_{max} - T_{min})\cos^2(\delta/2) = T_{min} + (T_{max} - T_{min})\,(1 + \cos\delta)/2\,. \tag{2}$$

The parameters T_{min} and T_{max} take into account that in reality no perfect optical "on" or "off" is possible. For the discussion in this chapter we set $T_{min} = 0$ and $T_{max} = 1$ and the Eq. 2 will reduce to the theoretical formulas found in elementary books like Bass, (1995) or Maldonado, (1995).
Based on Eq. (2) and with a retardation course

$$\delta(t) = \delta_1 \sin(\omega t) \tag{3}$$

transmission courses as depicted in Fig. 3 can be realized (ω... angular frequency of one resonance).

The choice $\delta_1 = \pi$ leads to quite sharp transmission peaks, while during the off-time a second peak with ~10% transmission evolve. When this is used for Q-switching it can be expected that the small off-transmission will not lead to an emission while the sharp transmission peaks will cause defined pulsing of the laser. This was first demonstrated for a low power fibre laser (Bammer & Petkovsek, 2007). For high gain lasers however this will not longer work, typical problems like pre-/post-lasing or spiking are encountered, and a second eigenmode must be added to get even sharper transmission peaks as discussed later.

2.2.2 Retardation control

Fig. 4 shows typical resonance curves of crystal-current and crystal-deformation (e.g. the movement of one x-facet or the angle of the z-facet in case of a shear-mode) at any (well working) resonance frequency.

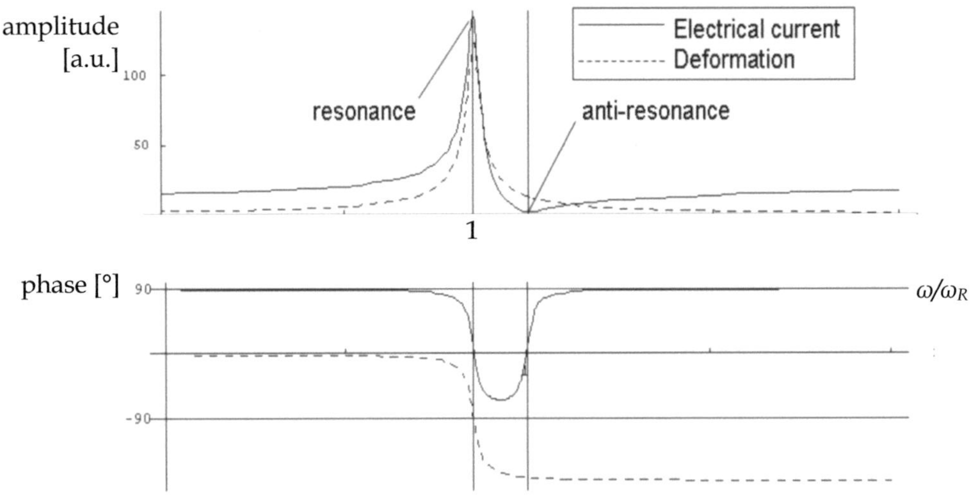

Fig. 4. Typical resonance curves of a piezo crystal for current and deformation

One important fact here is that the relation between deformation and current is approximately constant within the FWHM-resonance bandwidth (usually in the range 1/1000-1/10000 of the resonance frequency). In fact it changes linearly with the frequency, but since the frequency change is so small within the bandwidth and since the modulator is not useful outside of this bandwidth, a constant value can be assumed. For Q-switching this is sufficiently precise, for other applications, e.g. ellipsometry more accuracy is needed.

The reason why we point out this relation between current and deformation, which is directly connected to the retardation[1], is the following: The crystal frequency and the damping can change during operation, e.g. when the crystal heats up, expands and has then

[1] This is only true if the electro-optic effect is neglected, which can be safely done since the usual PEM-voltage-amplitudes in the order of ~10V are much lower than the kilo-Volts needed for a measureable electro-optic effect.

increased contact and damping forces with the contacts. Since this directly influences the relation between voltage amplitude and retardation amplitude, it is better to track and control the current amplitude during in order to keep the retardation amplitude on a desired value. (Bammer, 2007; Bammer & Petkovsek 2011) give formulas for the calculation of these relations with ~10% accuracy, such that practically one has to measure the real values anyway. Typically the retardation per current of the most used material for SCPEM, namely $LiTaO_3$, is around 50nm/mA corresponding to ~0.5rad/mA for $\lambda = 632$nm.
Regard that not al 3m-materials will show the x-oscillation discussed now. For BBO it seems that this x-oscillation cannot be excited (Bammer & Petkovsek 2011), where only a yz-shear-eigenmode seems to work (Fig. 6), due to its degenerated elasticity matrix.

2.2.3 Dual mode operation

Besides the x-eigenmode sketched in Fig. 2 a y-excitation can also excite longitudinal y-modes and shear yz-modes. In case of the y-modes different shapes are found corresponding to different acoustic waveguide modes (Dieulesaint & Royer, 2000). E.g. for a $LiTaO_3$-crystal with the dimensions 28x9.5x4mm Fig. 5 shows two different y-eigenmodes at the frequencies 258.8 kHz and 273.2 kHz. The higher mode shows additionally a strong xy-shear component.

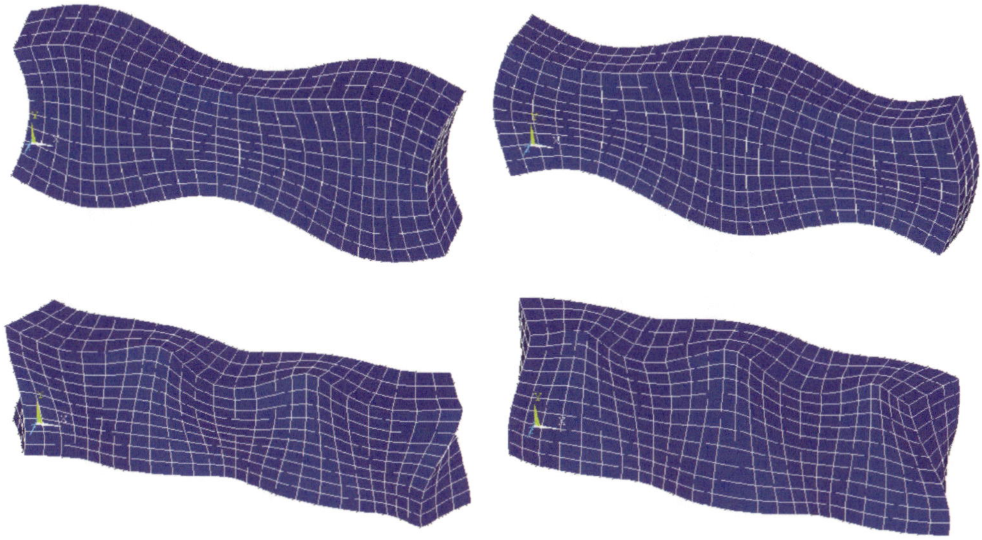

Fig. 5. ANSYS-simulation: y-eigenmode of a y-excited $LiTaO_3$-crystal with dimension 28.8x9.5x4mm with two knots at 258.8 kHz (upper pictures) and with four knots at 273.2 kHz (lower pictures). The real measured values were for one sample (out of five with all slightly different eigenfrequencies) 260.62 kHz and 275.86 kHz.

Eigenmodes as shown in Fig. 5 can now be superposed to the x-oscillation. This makes sense if the higher frequency is exactly three times higher than the basic frequency. Here "exact" means that the 3x-multiple of the basic frequency must be within the FWHM-bandwidth of the higher frequency. For the crystal simulated in Fig. 5 it was possible to tune the first harmonic from initially 90.8 kHz to 91.4 kHz by grinding the x-length from initially 28.8mm

to 28.6mm, such that the frequency was one third of the eigen-frequency 274.2 kHz (Petkovsek et. al. 2011).
Fig. 6 shows as a further example the yz-shear mode of a y-excited BBO-crystal with dimensions 8.6 x 4.05 x 4.5mm at 131.9 kHz (Bammer & Petkovsek, 2011). A similar mode is also possible for $LiTaO_3$ and in (Bammer & Petkovsek, 2008) a dual mode operation of a LiTaO3-crystal (21x7x5mm) with the yz-shear mode (381kHz) and the x-mode (127kHz) was demonstrated.
We remark that dual mode operation was also proposed for conventional PEMs by Canit & Pichon, C., 1984, but without the possibility of frequency tuning, as given for SCPEMs.

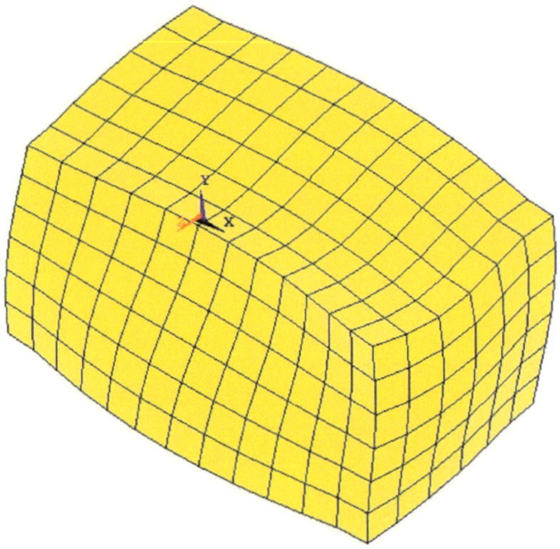

Fig. 6. ANSYS-simulation: yz-shear-eigenmode with 122.85 kHz of a y-excited BBO-crystal with dimensions 8.6 x 4.05 x 4.5mm

What can be now achieved by adding a second frequency-tuned eigenmode? The retardation in Eq. 3 is now modified according to

$$\delta(t) = \delta_1 \sin(\omega t) + \delta_3 \sin(3\,\omega t) \tag{4}$$

For this optimized result the series in Eq. 4 must approximate a square wave function oscillating between $-\pi$ and $+\pi$ yielding the values $\delta_1 = 4$, $\delta_3 = 4/3$. With this inserted in Eq. 2 the transmission course as depicted in Fig. 7 (in comparison with two curves $\delta_1 = \pi$, $\delta_3 = 0$; $\delta_1 = 1.2\pi$, $\delta_3 = 0$ from Fig. 3) can be produced.
Table 1 shows a comparison of the different transmission widths and rise times for the three transmission curves shown in Fig. 7. The values are given first in % of the period time of the base oscillation and second in ns for the 91.4kHz-crystal shown in Fig. 5.

δ_1	δ_3	FWHM-width [%]	FWHM-width [ns]@91.4kHz	Rise-time [%]	Rise-time [ns]@91.4kHz
π	0	16.7	1823	11.3	1241
1.2π	0	13.7	1497	8.8	962
4	4/3	6.5	708	4.2	460

Table 1. FWHM-transmission widths and rise times for a dual mode SCPEM

The proper addition of a second mode decrease transmission width and rise time by a factor of more than 2.

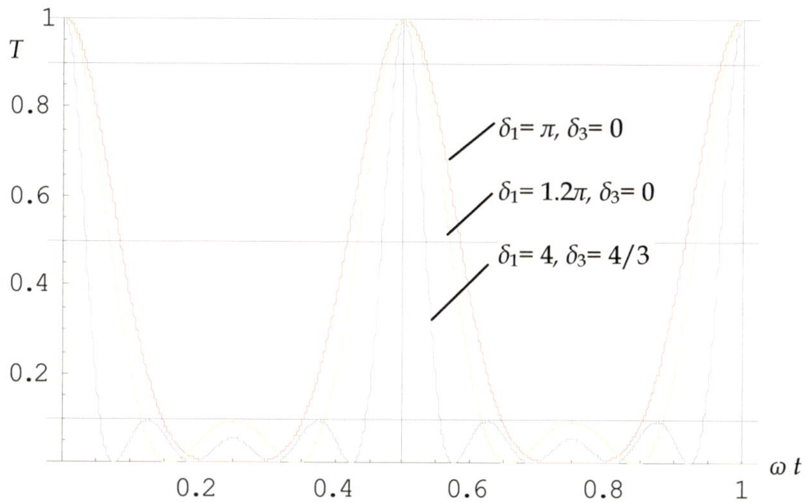

Fig. 7. Transmission curves of a dual mode SCPEM with $\delta_1 = \pi$, $\delta_3 = 0$; $\delta_1 = 1.2\pi$, $\delta_3 = 0$; and $\delta_1 = 4$, $\delta_3 = 4/3$.

3. Simulating SCPEM-Q-switching

In this chapter a simple numerical model for the description of the laser dynamics will be introduced and applied to a Nd:YAG-rod laser.

3.1 The laser photon life time t_p

The life-time t_p of a photon in a laser cavity with intrinsic transmission T_c, outcoupler reflection R_{out}, modulated internal transmission $T(t)$ and roundtrip frequency f_c is given by

$$t_p(t) = -1 \big/ \ln\left(T_c\, R_{out}\, T(t)\right)^{f_c} \tag{5}$$

If the internal transmission is due to a SCPEM with transmission course as in Fig. 7, then a strong modulation of the photon life time t_p is produced. This parameter is of utmost importance in the laser rate equations describing the laser dynamics.

3.2 The laser rate equations

The simplest model to describe laser-activity is based on two coupled rate equations for the average population density n (of the upper laser level) and the average photon density P in the laser gain medium (see for an introduction e.g. Siegmann, 1986).

$$\begin{aligned}\dot{P}(t) &= \Gamma c \sigma P(t)(n(t) - n_{th}) - P(t) / t_p(t) + M n(t) / t_n \\ \dot{n}(t) &= p(1 - n(t) / n_{\max}) - n(t) / t_n - c \sigma P(t)(n(t) - n_{th})\end{aligned} \tag{6}$$

with c...velocity of light in the gain medium, σ...cross-section of induced emission, M...fraction of light emitted by spontaneous emission which travels in a direction that contributes to the laser mode by ASE (amplified spontaneous emission) and is also responsible for the start of laser operation, t_n...life time of the upper laser level, $p = P_{abs}/E_{pp}/V_g$ (P_{abs}...absorbed pumping power, E_{pp}...pump photon energy, V_g...volume of gain medium) ... pumping rate when the upper laser level is empty (number of excited states per m³ and s), $p\,(1 - n(t)/n_{max})$... reduced pumping rate when the upper laser level becomes filled using n_{max} the number of laser active atoms per m³, n_{th}...thermal excitation of the lower laser level (important for quasi-three level systems like Yb:YAG, where the lower laser level has little energetic distance to the ground level and is therefore filled in thermal equilibrium according to the Boltzmann-statistics), Γ ... laser mode overlap factor (He et al., 2006).

Neglecting all losses between laser gain medium and out-coupler the laser power P_l of the laser is given by

$$P_l = \tfrac{1}{2}\, A\, c\, T_{out}\, E_{pl}\, P\,, \tag{7}$$

where A is the effective emitting area, T_{out} is the out-coupler transmission, and E_{pl} is the laser photon energy.

3.3 Simulating a Nd:YAG-rod-laser

Fig. 8 shows a very simple setup based on a side-pumped Nd:YAG-rod with diameter D = 3mm, length L_g = 75mm and dotation 1.1 at. %. Between the laser back mirror and the rod a SCPEM together with a PBSC (polarizing cube beam splitter) are placed. Pumping is done at 808nm, the laser emits at 1064nm.

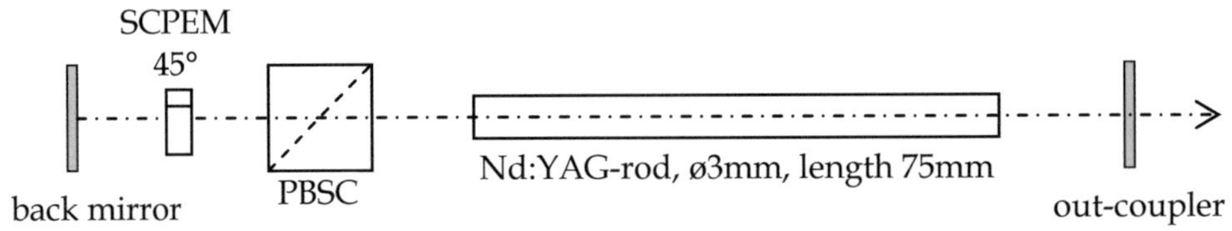

Fig. 8. Setup of a side-pumped DPSSL Q-switched with a SCPEM

The following parameters are used for the simulation:

resonator length L = 0.3m, maximum number of places for Nd-atomes in the YAG-host: 1.36 10^{28} m^{-3} → with dotation 1.1% the number of active atoms becomes n_{max} = 1.496 10^{26} m^{-3}, thermal population of lower laser level n_{th} = 0, out coupling surface $A = D^2\pi/4$, mode factor $\Gamma = L_g\,/(\,n_g\,L_g + L - L_g) = 0.2075$ [2], intrinsic cavity transmission T_c = 0.9 (taking into account the depolarizing effect of the laser rod, leading to high loss with the PBSC, further the PBSC-transmission is rated only >95%), out coupler transmission T_{out} = 0.1, life time of upper laser level t_n = 230μs, cross-section for stimulated emission σ = 28 10^{-24} m^{-3} (Koechner, 1999). ASE-factor M: to cause some amplification one spontaneously emitted photon must go after on laser-mirror-reflection through the whole laser rod. Hence M must be smaller than two

[2] This formula for the mode factor is obtained by assuming a constant laser mode cross-section A along the resonator, equal to the cross-section A of the gain. Further it must be considered that all generated photons are generated on the laser mode volume. A more detailed calculation needs a numerical integration over the photon density in the laser mode.

times the projection of a quarter of the effective out-coupling surface on the unit sphere in the gain medium $2(A/4)/(L/2)^2$ divided by the surface of the unit sphere 4π:

$$M < n_g^2 A/\pi/L^2/2 \sim 0.75\ 10^{-4} \tag{8}$$

This holds for loss free "perfect fitting" spontaneous emission from the centre of the laser cavity, which is never the case. We choose therefore a much smaller $M = 10^{-5}$ and remark that changing this figure has little influence on the results presented latter.
For the SCPEM the chosen parameters are $T_{min} = 0.01$, $T_{max} = 0.99$, first resonance frequency $f_R = 91$kHz, $\delta_1 = 1.2\,\pi$, $\delta_3 = 0$.
The absorbed pumping power is first assumed to be $P_{abs} = 100$W. By setting zero the left sides of the rate equations in Eq. 6 and using $T_{\max}$ for the SCPEM-transmission the stationary values of the laser power P_{l0} and the inversion n_0 can be calculated. The left graph in Fig. 9 shows this stationary laser output power P_{l0} versus the absorbed pumping power P_{abs}, the right graph in Fig. 9 shows the laser output power P_l versus the out-coupler transmission T_{out}.

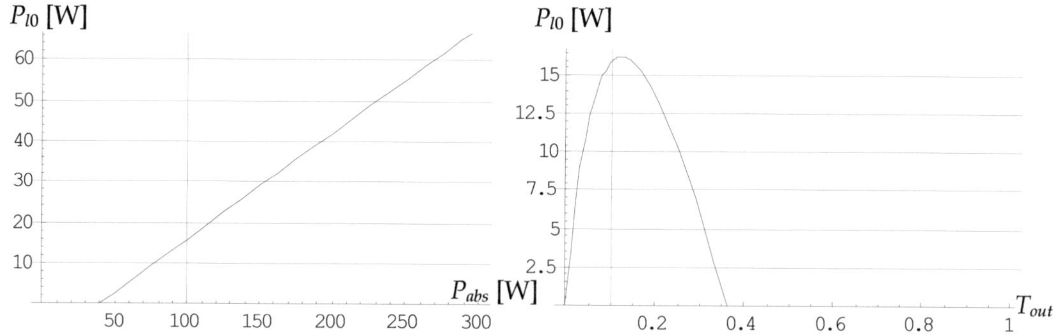

Fig. 9. cw-laser power versus P_{l0} pumping power P_{l0} (left); cw-laser power P_{l0} versus out-coupler transmission (right)

The vertical lines in Fig. 9 indicate the actual values of P_{abs} and T_{out} chosen in the simulation. Obviously the performance of this laser is poor due to the high loss in the cavity ($T_c = 0.9$). Fig. 10 shows now the simulation result of Eq. 6 with the parameters given above.

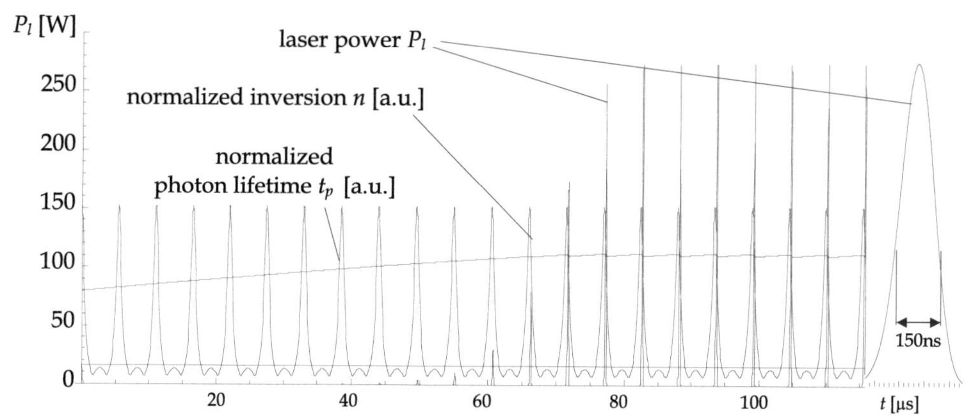

Fig. 10. Simulation results of Nd:YAG-rod laser based on Fig. 8 and Eq. 6.

The simulation starts with the steady state values for SCPEM-off. First the laser power drops to zero and remains zero during several SCPEM-cycles while the inversion n increases. Then pulsing starts and rapidly a quasi-stationary situation sets with constant pulsing at every transmission window with the following pulse parameters: peak power P_{peak} = 275W, FWHM-pulse duration t_{FWHM} = 150ns, pulse energy 46µJ, average power P_{av} = 8.44W. The poor average power is due to the fact that the pulses occur too late, shortly after the transmission peak, such that the laser-light is not polarized linearly and experience strong loss in the PBSC.

3.4 Chaotic behavior

Not every modulator frequency can be imposed to the laser. In many case chaotic behaviour is found, i.e. randomly varying pulse emission. Chaotic behaviour in connection with a SCPEM has interesting attractors in Poincare-maps (indicating deterministic chaos) and bifurcation diagrams showing that low modulator frequencies do not allow stable operation (Fig. 11).

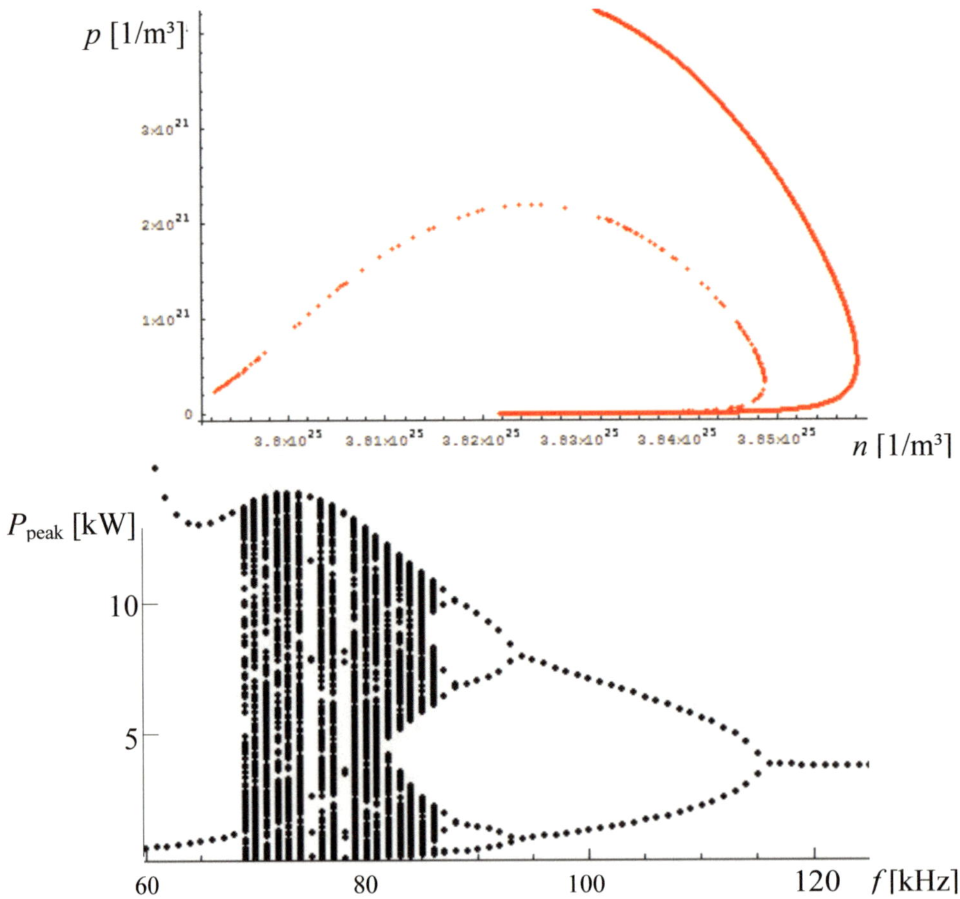

Fig. 11. Numerical analysis of a SCPEM-Q-switched Yb:YAG-slab laser
Poincare-map of photon density p vs. inversion density n for chaotic operation (top)
Bifurcation diagram: Peak powers P_{peak} vs. SCPEM-frequency f (bottom)

In Fig. 11 a side-pumped Yb:YAG-slab laser (slab size 10 x 5 x 0.5mm) with 50mm resonator length and 400W pumping power was simulated based on the rate equations 6.

3.5 Experimental results

Fig. 12 shows one setup that was used for a realization of SCPEM-Q-switching. The gain unit uses a Nd:YAG rod with diameter: 5 mm, length: 110 mm, doping concentration: 1.1% and polished ends (plano-plano ended) with an AR-coating for 1064 nm with reflection < 0.25%. It is side pumped from two sides with linear diode stacks, each one using 8 diode laser bars emitting at 808nm.

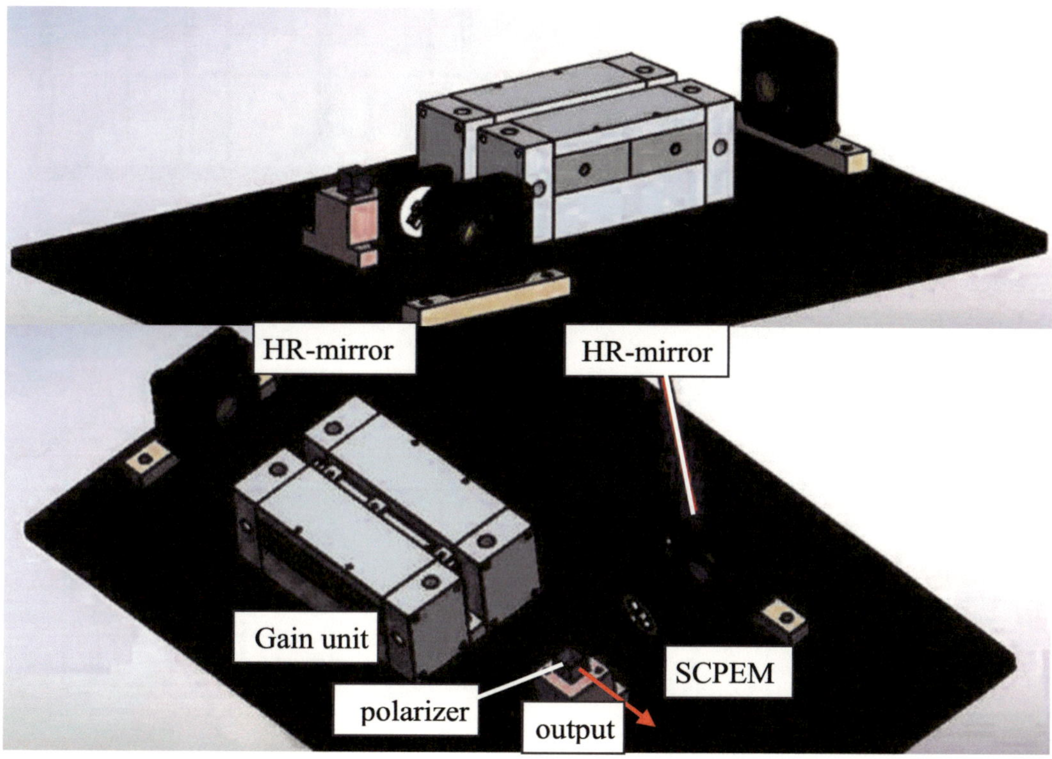

Fig. 12. Nd:YAG-laser for SCPEM-Q-switching

Fig. 13 shows stable pulsed operation with pulse frequency 190.2 kHz, average power 2.1 W, peak power 70 W, and pulse width 333ns. When the modulator is switched off the laser emits continuous wave with 2.8 W. This configuration, however do not allow stable operation at higher power. Fig. 14 shows the laser performance at higher power. The pulses are emitted irregularly with strongly varying pulse parameters.

Higher stable pulsed power can be achieved with a higher modulator frequency since low SCPEM-frequencies tend to produce chaotic output as indicated by Fig. 11.

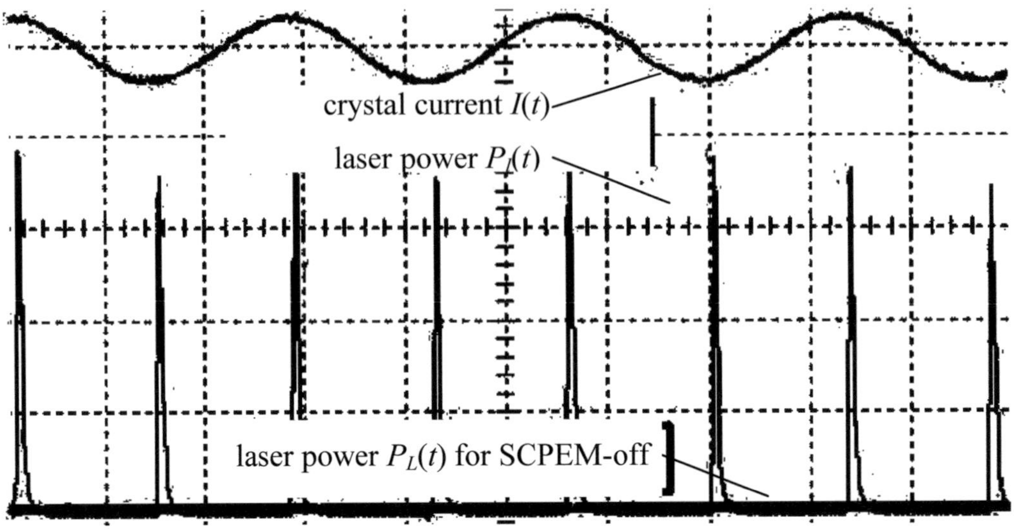

Fig. 13. Pulse sequence of a SCPEM-Q-switched Nd:YAG-laser: pulse frequency 190.2kHz, average power 2.1W, peak power 70W, pulse width 333ns. Crystal current $I(t)$ (upper graph) and laser powers $P_l(t)$ for SCPEM-on and SCPEM-off (with 2.8W cw-power)

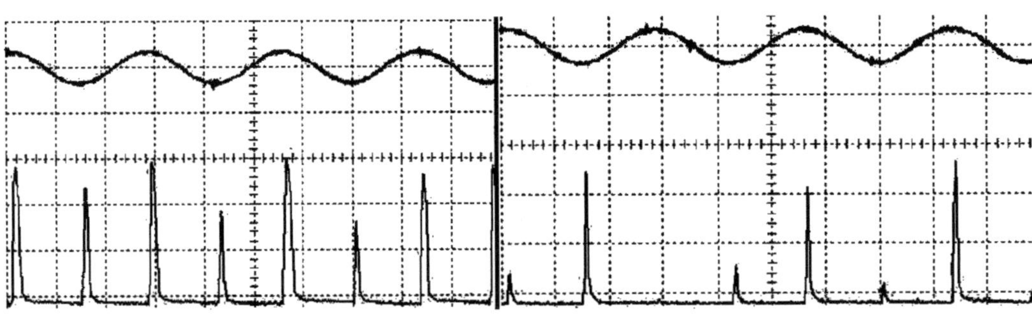

Fig. 14. Chaotic pulse sequences. Left: Average power 4.3 W (6 W with SCPEM-off). Right: Average power 6.5 W (9 W with SCPEM-off).

4. Conclusion

The possible use of SCPEMs for Q-switching was discussed. With a proper choice of parameters stable pulsed operation is possible with high efficiencies, little loss in average power, peak power 100 times higher then the average power, pulse repetition rates 100-300 kHz, and pulse durations down to 25ns. The important advantage is the simplicity of the solution making it interesting for quasi-cw applications, where no change of pulse repetition rate is needed. Especially frequency doubling and tripling should be possible with high efficiency with this type of laser operation.

5. Acknowledgment

This work was supported by the project "SCPEM-Laser" with the project number SPA/03-128 sponsored by the "Bundesministerium für Wissenschaft und Forschung" (Ministry for Science and Research).
Here we would like to thank Mr. Rok Petkovsek from the University of Ljubljana, who was involved in many realizations of SCPEM-Q-switching.

6. References

Badoz, J. & Canit, J.C. (1983). *New Design for a Photoelastic Modulator*, Appl. Opt. 22, 592

Bammer, F. & Holzinger, B. (2006). *A compact high brilliance diode laser*, SPIE-Proceeding of Photonics West 2006, High-Power Diode Laser Technology and Applications IV, Vol. 6104

Bammer, F.; Holzinger, B. & Schumi, T. (2006). *Time multiplexing of high power laser diodes with single crystal photo-elastic modulators*, Optics Express, Vol. 14, No. 8

Bammer, F. (2007). *Photoelastischer Modulator und Anwendungen*, Patent, AT 502 413

Bammer, F. & Holzinger, B. (2007). *Time-multiplexing generates a diode laser beam with high beam quality*, Optics and Laser Technology, 39, p. 1002 - 1007.

Bammer, F.; Holzinger, B. & Schumi, T. (2007). *A single crystal photo-elastic modulator*; SPIE-Proceeding of Photonics West 2007, Optical Components and Materials IV, Vol. 6469

Bammer, F. & Petkovsek, R. (2007). *Q-switching of a fiber laser with a single crystal photo-elastic modulator, Optics Express*, Vol. 15, No. 10

Bammer, F.; Petkovsek, R.; Frede, M. & Schulz, B. (2008). *Q-switching with a Dual Mode Single Crystal Photo-Elastic Modulator*, SPIE-Proceeding 7131, GCL/HPL 2008

Bammer, F.; Petelin, J.; Petkovšek, R. (2011). *Measurements on a Single Crystal Photo-Elastic Modulator*, OSA-Proceeding of CLEO2011

Bammer, F.; Schumi, T.; Petkovsek, R. (2011). *A new material for Single Crystal Modulators: BBO*, SPIE-Proceeding, 8080A-17, to be published

Bass, M. (1995). *Handbook of Optics*, McGraw Hill

Canit, J.C. & Pichon, C. (1984). *Low frequency photoelastic modulator*, App.Opt. 23/13, p.2198

Canit, C.; Gaignebet, E. & Yang, D. (1995). Photoelastic Modulator: *Polarization modulation and phase modulation*, J. Optics (Paris), vol. 26, n° 4, pp. 151-159

Dieulesaint, E. & Royer, D. (2000). *Elastic Waves in Solids*, Springer Publishing Company

Eichler, H.J. & Eichler, J. (2006). *Laser*, Springer Publishing Company

Halido, D. & Weil, R. (1974). *Resonant-piezoelectro-optic light modulation*, Journal of Applied Physics, Vol. 45, No. 5

Kemp, J.C. (1969). *Piezo-Optical Birefringence Modulators*, J. Opt. Soc. Am. 59, 950

Koechner, W. (1999). *Solid State Laser Engineering*, Springer Publishing Company

Krausz, F.; Kuti, Cs. & Turi, L. (1990). *Piezoelectrically Induced Diffraction Modulation of Light*, IEEE Journal of Quantum Electronics, Vol. 26, No. 7

Maldonado, T. A. (1995). Electro-optic modulators, Handbook of optics, McGraw-Hill

Nye, J. F. (1985). *Physical Properties of Crystals*, Oxford University Press

Petkovsek, R.; Saby, J.; Salin, F.; Schumi, T.; Bammer, F. (2011). *SCPEM-Q-switching of a fiber-rod-laser*, to be published

Schumi, T. (2006). *Analyse von photoelastischen Modulatoren aus Lithiumniobat*, Diploma Thesis, Vienna University of Technology

He, F., Price, J. H., Vu, K. T. et al. (2006). *Optimization of cascaded Yb fiber amplifier chains using numerical-modeling*, Optics Express 14, 12846-12858

Siegmann, A.E. (1986). *Lasers,* University Science Books, Mill Valley, California

Yariv, A. (1984). *Optical Waves in Crystals*, New York, Wiley

www.hindsinstruments.com

2

Acousto-Optically Q-Switched CO_2 Laser

Jijiang Xie and Qikun Pan
Changchun Institute of Optics, Fine Mechanics and Physics, Chinese Academy of Sciences
State Key Laboratory of Laser Interaction with Matter
China

1. Introduction

The compacted CO_2 laser with narrow pulse width and high repetition frequency has great potential applications, such as echo-splitting radar (Kariminezhad et al., 2010; Carr et al. 1994), laser processing (Hong et al., 2002), environmental monitor (Zelinger et al., 2009), laser medical instrument (Hedayatollahnajafi et al., 2009), laser-matter interaction (Chang & Jiang, 2009), etc. In fact, currently compacted CO_2 laser is usually implemented with pulsed output by electro-optically Q-switch (Tian et al., 2001) and mechanical Q-switch (Kovacs et al., 1966), and the related theories and technologies are relative maturity. But the method of electro-optically Q-switch often needs high voltage, more between different molecules in the laser gain medium; third, according to various factors (optical loss in cavity, transmittance of output mirror and so on) influencing the output performances of laser, the theoretical and experimental researches are implemented to optimize the design of compacted CO_2 laser; forth, the tunable design with grating has been finished and the spectrum lines are measured in the experiment; fifth, the experiments that laser irradiated HgCdTe detectors are implemented with the acousto-optically Q-switched CO_2 laser.

Both the theory and experiments show that acousto-optically Q-switch is an effective technical method to realize the laser output with narrow pulse width and high repetition frequency. The compacted size acousto-optically Q-switched CO_2 laser which has been optimally-designed realizes 100 kHz repetitive frequency. The full band of wavelength tuning between 9.2μm and 10.8μm is obtained by grating selection one by one. With the further development of acousto-optically modulator, the performance will be greatly improved which will provide an effective method to realize high repetition frequency pulse for CO_2 laser.

2. Acousto-optically Q-switched CO_2 laser

2.1 Acousto-optically Q-switch

Acousto-optically Q-switch is a key component of CO_2 laser to realize pulse output. The acousto-optically Q-switch is composed by driving unit and acousto-optically modulator which is inserted in the cavity. The principle of acousto-optically modulator is the refraction index of crystal changed by using ultrasonic wave as it is transmitting through crystal. The crystal with periodic variation of the refraction index is as the same as a phase grating. When optical beam propagates this crystal, it will create a diffractive wave to realize the optical beam deflection i.e. Bragg diffraction, as Fig. 1(a). The working principle of acousto-optically Q-switch is shown as Fig. 1(b). The ultrasonic wave is generated by radio frequency (RF) signal with several tens MHz through acousto-optically transducer. Therefore, whether the optical beam is in the condition of deflection or not is totally determined by RF signal controlled by transistor-transistor logic level. When the transistor-transistor logic level is located in high level, the ultrasonic equivalent phase grating will make the optical beam deflection. The deflective angle can completely make the optical beam escape the resonator, which the resonator is in the state of high loss and low Q value. The resonator cannot form the oscillation which means the Q-switch "close" laser. When the transistor-transistor logic level is laid on low level, the RF signal suddenly stops and the ultrasonic field in the Q-switched crystal disappears. This means the switch "open" and the resonator resumes the high Q value with oscillated optical beam output. Accordingly, Q value alternates one time that will generate a Q-switched pulse output from laser. At the same time, if the transistor-transistor logic level is carried out an encoded control, it will realize the encoded pulses output of laser.

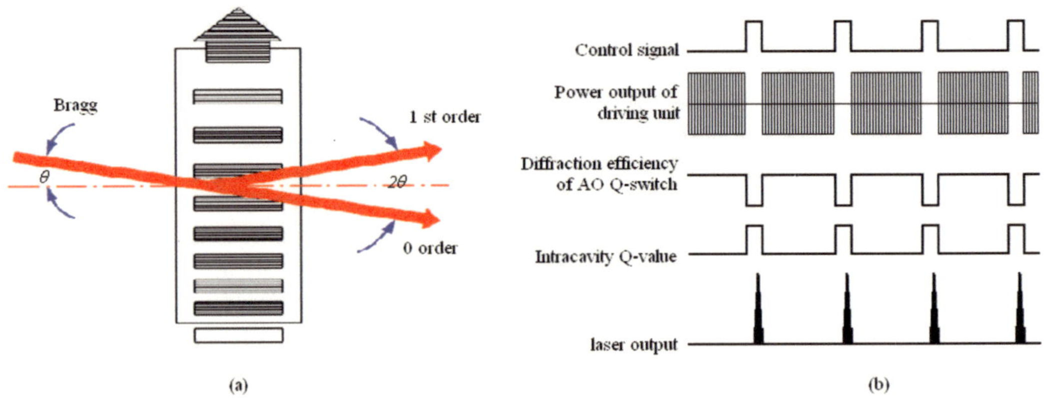

Fig. 1. The scheme of Q-switch

2.2 Basic structure of CO_2 laser

The principle scheme of CO_2 laser is shown in Fig. 2.The laser resonator adopts half external cavity structure with direct current discharge gain area. The diffraction efficiency of acousto-optically modulator is higher when the incident laser is linearly polarized light. Under this condition, the performance of acousto-optically Q-switch is better. This structure is propitious to realize linearly polarized laser output and reduce optical loss, which is helpful for laser output with Q-switch.

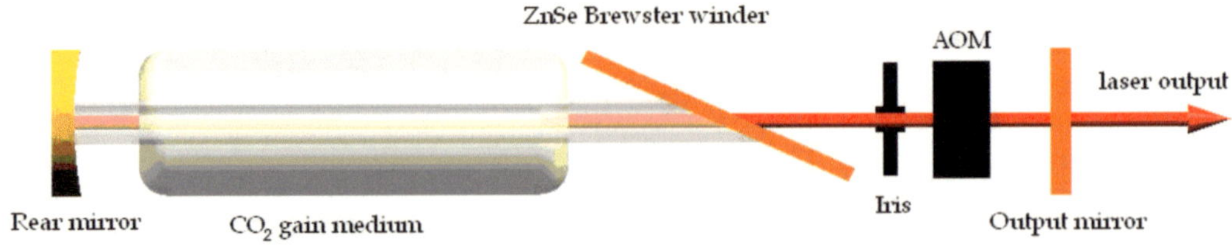

Fig. 2. Schematic diagram of acousto-optically Q-switch CO_2 laser

The discharge tube is made by glass with water cooled pipe. The rear mirror is a spherical mirror coated with gold, and the Brewster window is made by ZnSe material. The working gas is made up of Xe, CO_2, N_2, He. The acousto-optically Q-switch is inserted between the output mirror and the Brewster window. An iris with adjustable aperture size is inserted in the resonator.

3. Dynamical analysis of acousto-optically Q-switched CO_2 laser

3.1 The six-temperature model for acousto-optically Q-switched CO_2 laser

Based on the theory of five-temperature model of the dynamics for CO_2 laser (Manes & Seguin, 1972), the dissociating influence of CO_2 to CO molecules on laser output is concerned. Thus the equivalent vibrational temperature of the CO molecules is taken as a variable quantity of the differential equations for this model. Then the differential equations include six variable quantity of temperature.

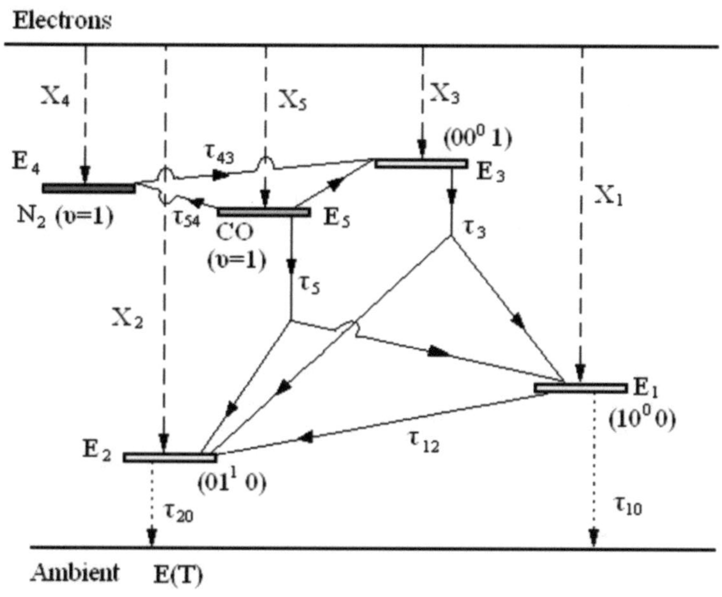

Fig. 3. Schematic energy-level diagram for N_2-CO_2-CO system

In figure 3, we present a schematic diagram of the set of processes of electron collision excitation for CO_2, N_2, and CO molecules, all kinds of energy transfer among molecules, and excited emission and spontaneous emission. The distribution of vibrational energy in different vibrational modes can be described by the following equations (Smith & Thomson, 1978):

$$\frac{dE_1}{dt} = n_e(t) f n_{co_2} X_1 h\nu_1 - \frac{E_1 - E_1(T)}{\tau_{10}(T)} - \frac{E_1 - E_1(T_2)}{\tau_{12}(T_2)} + \frac{h\nu_1}{h\nu_3}\frac{E_3 - E_3(T,T_1,T_2)}{\tau_3(T,T_1,T_2)} + \frac{h\nu_1}{h\nu_5}\frac{E_5 - E_5(T,T_1,T_2)}{\tau_5(T,T_1,T_2)} + h\nu_1 \Delta NWI_{\nu_0} \quad (1)$$

$$\frac{dE_2}{dt} = n_e(t) f n_{co_2} X_2 h\nu_2 + \frac{h\nu_2}{h\nu_3}\frac{E_3 - E_3(T,T_1,T_2)}{\tau_3(T,T_1,T_2)} + \\ + \frac{E_1 - E_1(T_2)}{\tau_{12}(T_2)} - \frac{E_2 - E_2(T)}{\tau_{20}(T)} + \frac{h\nu_2}{h\nu_5}\frac{E_5 - E_5(T,T_1,T_2)}{\tau_5(T,T_1,T_2)} \tag{2}$$

$$\frac{dE_3}{dt} = n_e(t) f n_{co_2} X_3 h\nu_3 + \frac{E_4 - E_4(T_3)}{\tau_{43}(T)} - \frac{E_3 - E_3(T,T_1,T_2)}{\tau_3(T,T_1,T_2)} + \frac{h\nu_3}{h\nu_5}\frac{E_5 - E_5(T,T_3)}{\tau_{53}(T,T_3)} - h\nu_3 \Delta NWI_{\nu_0} \tag{3}$$

$$\frac{dE_4}{dt} = n_e(t) f n_{N_2} X_4 h\nu_4 - \frac{E_4 - E_4(T_3)}{\tau_{43}(T)} + \frac{h\nu_4}{h\nu_5}\frac{E_5 - E_5(T,T_4)}{\tau_{54}(T,T_4)} \tag{4}$$

$$\frac{dE_5}{dt} = n_e(t)(1-f) n_{co_2} X_5 h\nu_5 - \frac{E_5 - E_5(T,T_3)}{\tau_{53}(T,T_3)} - \frac{E_5 - E_5(T,T_1,T_2)}{\tau_5(T,T_1,T_2)} - \frac{E_5 - E_5(T,T_4)}{\tau_{54}(T,T_4)} \tag{5}$$

Where n_{CO_2} and n_{N_2} are, respectively, the number density of CO_2 and N_2 molecules per unit volume and $n_e(t)$ is the number density of electrons per unit volume. T_1 is the equivalent vibrational temperature of the CO_2 symmetrical stretching mode. T_2 is the equivalent vibrational temperature of the CO_2 bending mode. T_3 is the equivalent vibrational temperature of the CO_2 asymmetric mode. T_4 is the equivalent vibrational temperature of the N_2 molecules. T_5 is the equivalent vibrational temperature of the CO molecules. Eqs. (1)-(3) describe the variation in the stored energy in unit volume (erg/cm^3) as a function of time in CO_2 symmetrical, bending, and asymmetrical modes respectively. Eq. (4) expresses the time evolution of the stored energy density in unit volume for N_2 molecules. Eq. (5) expresses the time evolution of the stored energy density in unit volume for CO molecules, which are dissociated by CO_2 molecules. f is the non-dissociated fraction of CO_2 molecules. For simplicity f is considered as a constant in this paper (in general f is a function of time, electrical field intensity and electron number density).

By taking the sum of Eqs. (1)-(5) in the steady state, the following equation which describes the time evolution of the stored energy density in gas mixture CO_2-N_2-He-CO will be obtained:

$$\frac{dE_K}{dt} = \frac{E_1 - E_1(T)}{\tau_{10}(T)} + \frac{E_2 - E_2(T)}{\tau_{20}} + (1 - \frac{h\nu_1}{h\nu_3} - \frac{h\nu_2}{h\nu_3})\frac{E_3 - E_3(T,T_1,T_2)}{\tau_3(T,T_1,T_2)} + \\ + (1 - \frac{h\nu_3}{h\nu_5})\frac{E_5 - E_5(T,T_3)}{\tau_{53}(T,T_3)} + (1 - \frac{h\nu_1}{h\nu_5} - \frac{h\nu_2}{h\nu_5})\frac{E_5 - E_5(T,T_1,T_2)}{\tau_5(T,T_1,T_2)} + \\ + (1 - \frac{h\nu_4}{h\nu_5})\frac{E_5 - E_5(T,T_4)}{\tau_{54}(T,T_4)} \tag{6}$$

Where the total gas kinetic energy E_k per unit volume is:

$$E_k = \left(\frac{5}{2}n_{N_2} + \frac{5}{2}n_{CO} + \frac{3}{2}n_{He} + \frac{5}{2}fn_{CO_2}\right)kT \tag{7}$$

Where $n_{CO} = (1-f)\, n_{CO_2}$ is the number density of CO molecules per unit volume.
Taking into consideration stimulated emission, spontaneous emission, and losses in the cavity yields the equation that describes the time evolution of the cavity light intensity:

$$\frac{dI_{v_0}}{dt} = \frac{-I_{v_0}}{\tau_c} + chv_0\left(\frac{\Delta NWI_{v_0}}{h} + n_{001}P(J)S\right) \tag{8}$$

Where c is the light velocity, h is Planck constant, τ_c is photon life time in the cavity:

$$\tau_c = -\frac{2L}{c\left(\ln R_1 + 2\ln R_2\right)} \tag{9}$$

Where R_1 is the reflection coefficient of rear mirror, R_2 is the reflection coefficient of the output mirror, L is the resonator length. The expressions of W, S in the equation of (8) are:

$$W = \frac{\lambda^2 F}{4\pi^2 v_0 \Delta v_L \tau_{sp}} \tag{10}$$

$$S = \frac{2\lambda^2 \Delta v_N}{\pi A \tau_{sp} \Delta v_L} \tag{11}$$

Where λ is the laser wavelength, υ_0 is the laser frequency, $\Delta\upsilon_L$ is the laser transition line width, $\Delta\upsilon_N$ is the laser natural line width, τ_{sp} is the spontaneous emission rate, A is the cross section of the laser beam, $F=l/L$ is the filling factor, l is the length of gain media.
When the Q-switch is closed, the loss in the cavity is so high that there is no laser output and the laser intensity in the cavity is nearly zero. So the Eqs. (1)-(5) equal zero. In the five equations mentioned above, E_1 (the energy per unit volume stored in the CO_2 symmetrical stretching mode), E_2 (the energy per unit volume stored in the CO_2 bending mode), E_3 (the energy per unit volume stored in the CO_2 asymmetrical modes), E_4 (the energy per unit volume stored in the N_2 molecules), E_5 (the energy per unit volume stored in the CO molecules) are defined respectively by T_1, T_2, T_3, T_4, T_5. A computer program processed in MATLAB is used to solve the five nonlinear equations. The five temperatures and an estimated ambient temperature are the initial values of the laser after the Q-switch is opened.
After the Q-switch is opened, the transmittance of output mirror is t. Therefore the population number of upper levels falls sharp; the laser oscillates rapidly in a short time, and then engenders a giant pulse output. The mathematical model of Q-switched CO_2 laser consists of seven differential expressions, which are dE_1/dt, dE_2/dt, dE_3/dt, dE_4/dt, dE_5/dt, dE_k/dt, $dI\upsilon_0/dt$ respectively. The variables in these expressions are T_1, T_2, T_3, T_4, T_5, T, $I\upsilon_0$.There is six temperature variables in the differential expressions, therefore it is called six-temperature model.

The specific expressions of the seven coupled differential equations are represented by Eqs. (1)- (6) and (8). Based on the Runge–Kutta theory, the seven variables (T_1, T_2, T_3, T_4, T_5, T, and $I\upsilon_0$) can be obtained through solving the differential equations. The laser output power can be obtained from $I\upsilon_0$:

$$P_{out} = -\frac{A}{2}\ln(R_2)\times\frac{1-R_2-\alpha}{1-R_2}\times I(t)\times 10^{-7} \quad (12)$$

Where α is the loss coefficient of output mirror.

3.2 The numerical calculation

3.2.1 The parameters of CO_2 laser

The experimental schematic diagram is shown in Fig. 2. The key parameters of laser just as follows: the inner diameter is 8 mm, the gain area length of discharge tube is 800mm and the mixed gas pressure is 3.3kPa. The gas ratio is Xe: CO_2: N_2: He=1:2.5:2.5:17.5. The radius of curvature of rear mirror is 3m and its reflectivity is 98.5%. The length of optical cavity resonator is 1200mm. The medium of acousto-optically modulator used in our experiment is Ge single crystal whose single pass transmittance is 90% for the wavelength 10.6μm. The first order diffraction efficiency of acousto-optically medium for linearly polarized light is about 80% in horizontal direction and optical aperture is 6×10 mm^2. The transmittance of output mirror is 39%.

3.2.2 The numerical calculation of six temperature model

The initial values of Q-switched CO_2 laser have been considered before the Q-switch is opened. Based on the measured results of the laser and related values of the equation parameters which are shown in table 1, the six-temperature model is used for simulating the process of dynamical emission in the acousto-optically Q-switched CO_2 laser. A computer program processed in MATLAB is used to solve the differential equations based on the Runge–Kutta theory. Considering the open time of acousto-optically Q-switch, the establishing time of laser pulse should be added the corrected value 0.85μs (Xie et al., 2010). Under those conditions, the variation of output power with time is shown in Fig. 4(a). At the same conditions, the calculation of rate equations is shown in Fig. 4(b).

Parameter	Numerical value	Parameter	Numerical value
υ_1/c [1]	1337 cm^{-1}	h	6.626×10^{-27}erg·s
υ_2/c [1]	667 cm^{-1}	B_{CO_2} [2]	0.4cm^{-1}
υ_3/c [1]	2349 cm^{-1}	τ_{sp} [2]	0.2s
υ_4/c [3]	2330 cm^{-1}	λ	10.6μm
υ_5/c [3]	2150 cm^{-1}	c	2.998×10^{10}cm·s^{-1}
X_1 [2]	$5\times10^{-9}cm^3\cdot s^{-1}$	M_{CO_2} [3]	7.3×10^{-23}g
X_2 [2]	$3\times10^{-9}cm^3\cdot s^{-1}$	M_{CO} [3]	4.6×10^{-23}g
X_3 [2]	$8\times10^{-9}cm^3\cdot s^{-1}$	M_{N_2} [3]	4.6×10^{-23}g
X_4 [2]	$2.3\times10^{-8}cm^3\cdot s^{-1}$	M_{He} [3]	6.7×10^{-24}g
X_5 [2]	$3\times10^{-8}cm^3\cdot s^{-1}$	k	1.38×10^{-16}erg·K^{-1}

Table 1. Parameters used in the calculations 1 See reference (Rossmann et al., 1956), 2 See reference (Soukieh et al., 1998), 3 See reference (Smith & Thomson, 1978)

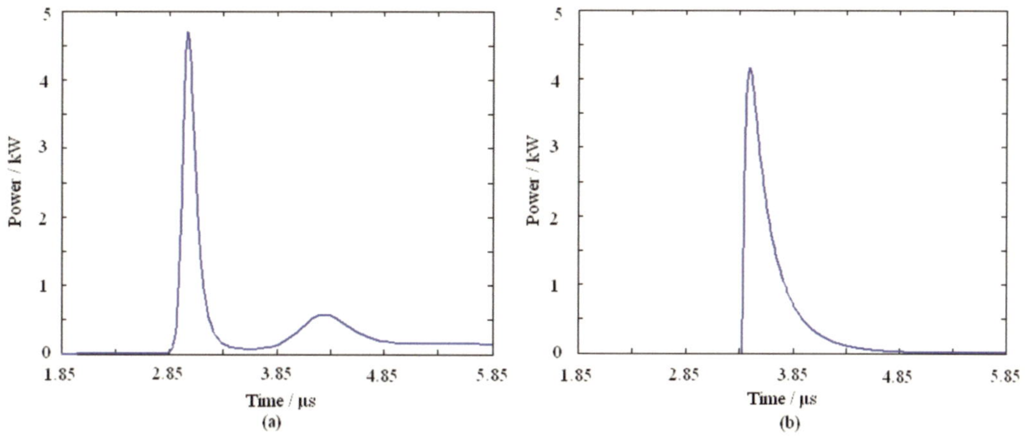

Fig. 4. Theoretical shapes of the pulsed laser (a) Calculated by six temperature model (b) Calculated by rate equations

3.2.3 Experimental device

The diagram of experimental device is shown in Fig.4. The path of laser output will be changed by the mirror first, and then the laser is divided into two paths by a beam splitter (a coated ZnSe mirror whose transmittance is 30%). One path of laser is accepted by the detector, and then the laser pulse waveform will be displayed by the oscilloscope after enlarged by the amplifier. The other path of laser is monitored by the power meter at the same time.

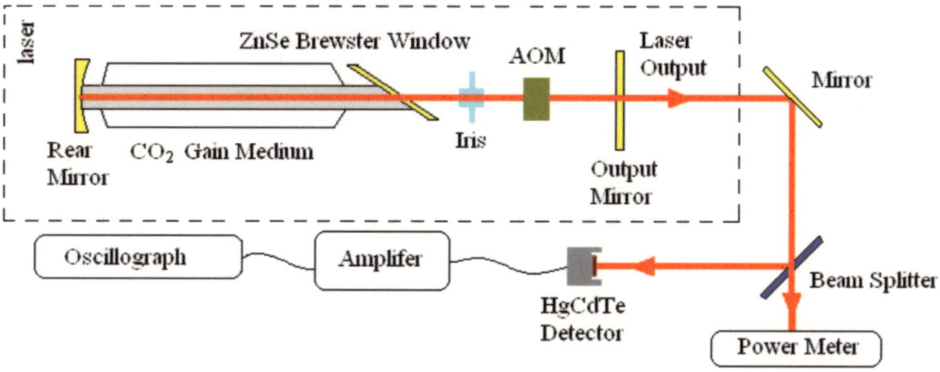

Fig. 5. Diagram of experimental device

When the transmittance of output mirror is 39% and the repetitive frequency is 5 kHz, the Q-switched pulsed laser is detected by a photovoltaic HgCdTe detector with the model of PV-10.6 made by VIGO Company, and the laser waveform is monitored on a TDS3052B digital storage oscilloscope with 500-MHz bandwidth. The shapes of the pulsed laser are shown in Fig.6, with a 160 ns pulse width and a 3μs delay time. The average power is 1.14 W which is measured by the power meter with the model of LP-3C made by Bejing WuKe Photo Electricity Company. Thus the average output power of this device is 3.8 W (light splitting ratio is 3:7), and the peak power of this laser is 4750 W.

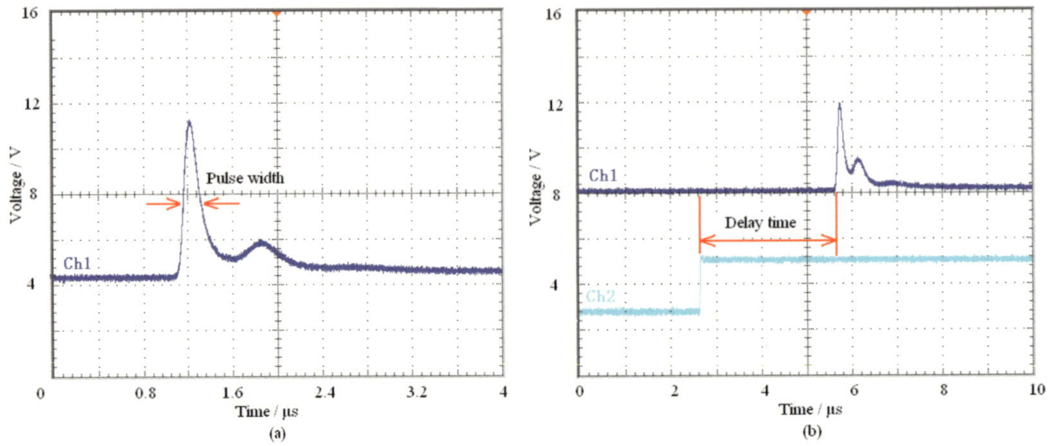

Fig. 6. Measurement shapes of the pulsed laser: (a) pulse width, (b) delay time. (Ch1 is Laser pulse waveform; Ch2 is transistor-transistor logic trigger signal)

A conclusion can be drawn by comparing figure 4 with figure 6. The measured results of peak power, pulse width and establishing time of laser pulse are in agreement with the theoretical calculations by rate equations as well as six-temperature model. The tail phenomenon is obvious in the calculation by six-temperature model, which is more consistent with the experimental result. The six-temperature model more perfectly explains the behavior of energy transfer between different molecules in the laser gain medium, and analyzes the influence of gas temperature on laser output. However the rate equations can only explain the particle population transfer between up and down energy levels. Therefore the six-temperature model gives a more correct analysis than the rate equations. The comparison of theoretical calculations and experimental results are shown in table 2.

Comparing results	rate equations	six temperature model	experimental result
Pulse width/ns	200	166	160
Peak power/kW	4.15	4.7	4.75
Pulse delay time/μs	3.35	2.9	3
Tail phenomenon	Not obvious	Obvious	Obvious

Table 2. The comparison of theoretical calculations and experimental results

4. The optimum design of acousto-optically Q-switched CO_2 laser

Some parameters (The optical loss in cavity, the transmittance of output mirror, etc) of laser have great influence on laser output. When an acousto-optically Q-switched CO_2 laser is to be designed, those factors must be considered. In the following section, the further theoretical calculations and experimental methods will be introduced to guide the optimum design of laser.

4.1 The influence of acousto-optically Q-switch on laser output

The medium of acousto-optically modulator used in our previous experiments is Ge single crystal whose single pass transmittance is 90% for the wavelength 10.6μm. The influence of absorption coefficient of acousto-optically crystal on laser output has been described in detail (Xie et al., 2010). And the high absorption coefficient is one of the most important factors to hinder the development of acousto-optically Q-switched CO_2 laser. In order to further clarify the influence degrees of the absorption coefficient on laser output, the variation of output power with single pass transmittance(δ) has been simulated by six-temperature model when other parameters of laser are invariable just as Fig.7. The calculations show that when the acousto-optically modulator has higher single pass transmittance, the peak power of pulse laser will be higher, and the establishing time of laser pulse will be shorter. When the single pass transmittance is increased by 2 percent, the increment of peak power of pulse laser is about 6 percent. With the development of crystal process technology, the crystal with higher single pass transmittance would be manufactured (The maximum value of best acousto-optically crystal is higher than 94% presently), and then the performance of acousto-optically Q-switched CO_2 laser output would be further enhanced.

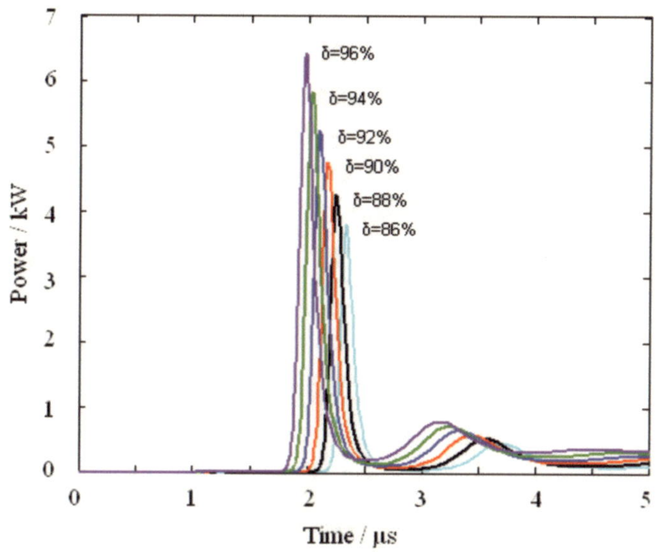

Fig. 7. Output power of the laser versus time at different single pass transmittance

4.2 The influence of output mirror transmittance on laser output

When the transmittance (t) of output mirror is 18%, 25%, 32%, 39%, 46%, and 53% separately, a computer program processed in MATLAB is used to solve the differential equations based on the Runge-Kutta theory. Figure 8 shows the variation of light intensity in the cavity with time. The light intensity is reduced with increasing of the transmittance gradually. And when the increment of the transmittance is 7 percent, the decrease in light intensity is almost 28 percent. Then the laser pulse waveform is calculated by Eq. (12) at different transmittances as in figure 9, which shows that the influence of transmittance on output power is obvious. The peak power is a function of output mirror transmittance when other conditions remain unchanged (just as shown in imaginable line in figure 9). The theoretical results show that the peak power has a maximum when it changes with transmittance, which lays a theoretical foundation for the optimization of laser parameter.

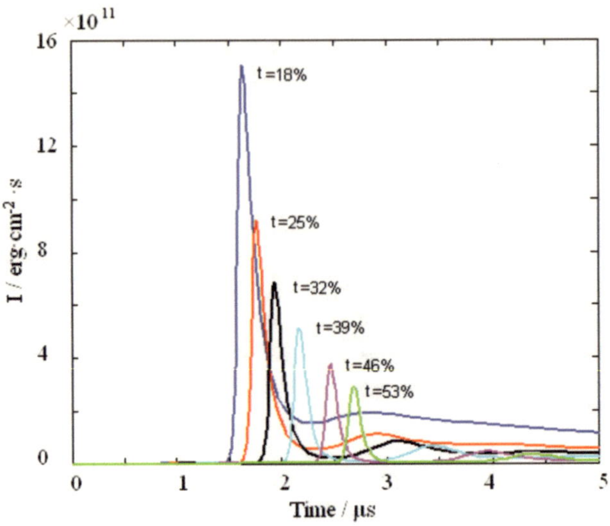

Fig. 8. Light intensity in the laser cavity versus time at different transmittance

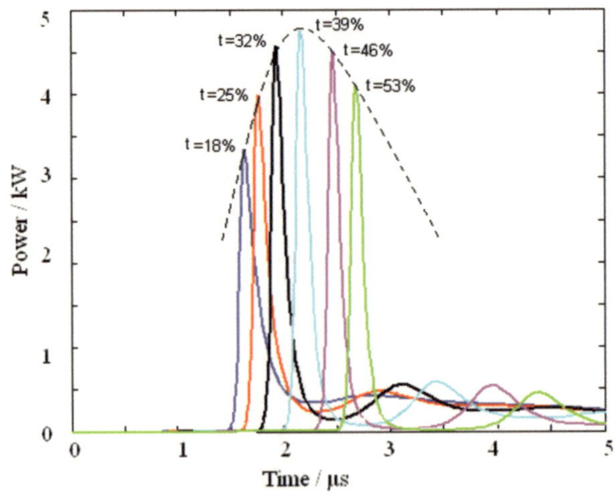

Fig. 9. Output power of the laser versus time at different transmittance

According to the simulation results above, our laboratory has manufactured a group of ZnSe output mirrors (the transmittances are 18%, 25%, 32%, 39%, 46%, 53% separately, and the aperture size all are 30mm). Performance characteristics have been investigated as a function of output mirror transmittance with an acousto-optically Q-switch. At 5 kHz pulse repetition frequency, the average power and pulse width are measured for each mirror independently, and then the peak power of laser is calculated. The quadratic fitting curves of peak power and pulse width versus the transmittance of output mirror are shown in Fig.10. Compared with imaginable line in Fig.9, the experimental results agree well with that of the six-temperature model theory. The transmittance of output mirror has a significant influence on the peak power and pulse width of acousto-optically Q-switched CO_2 laser, and the optimal value of transmittance is 39% in the experiment.

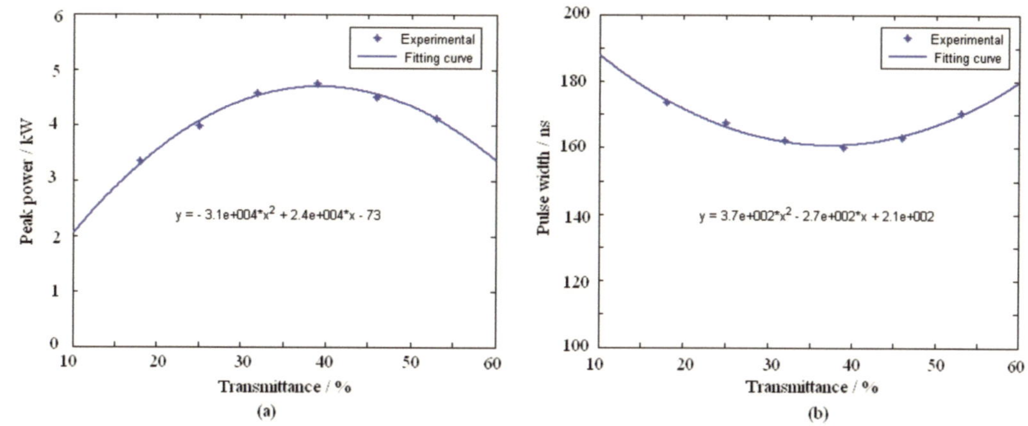

Fig. 10. The fitting curves of peak power (a) and the pulse width (b) versus transmittance of the output mirror

4.3 The influence of iris on output laser

Higher requirement of laser quality has been tendered in the fields of scientific research and engineering applications. In order to meet this demand, an iris is inserted in the resonator to obtain laser with basic mode. The influence of aperture size of iris on laser output is investigated when the transmission of output mirror is 39%. The fitting curves of peak power and pulse width versus the pulse repetition frequency are shown in Fig.11 individually at different aperture size. Just as Fig.11 shows, at the same pulse repetition frequency, when the size of aperture is decreased, the peak power is reducing and the pulse width is increasing. That is because the marginal laser oscillation is limited by the iris, the mode volume of laser cavity is reduced. Therefore, the better quality of laser, the lower peak power will be obtained.

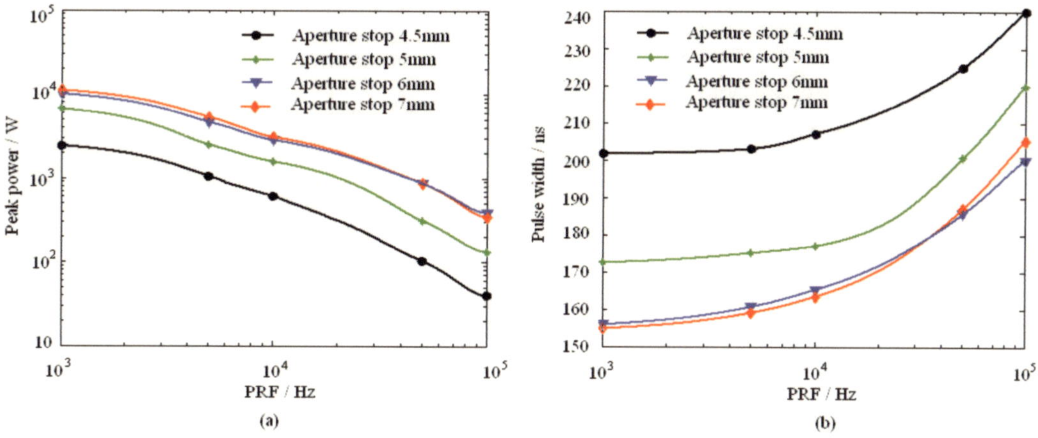

Fig. 11. Peak power (a), pulse width (b) of the CO_2 laser as a function of apertures stop

There are many vibrational-rotational levels in CO_2 molecules. The calculated results show that the total energy available for extraction is the energy stored in all the rotational levels for the case of long weak pulse, whereas for the case of very short pulses only the energy stored in the active rotational sublevel is available (Smith & Thomson, 1978). When the aperture size of iris is small, the iris will limit oscillation of marginal laser. Under this condition, more rotational levels are needed to participate in excited emission so as to realize laser output, therefore the pulse width of laser is wide and the peak power is low. The pulse width is reduced gradually with the increasing of aperture size. When the aperture size is over 6 mm, the pulse width will increase with the increasing of aperture size at high repetition frequency (pulse repetition frequency ≥ 40 kHz). That is to say, when the aperture size is over 6 mm, the iris just limits laser oscillations between some weaker levels. Under this condition, a bigger gain volume is obtained, at the same time, the pulse width is effectively compressed and the quality of laser is improved. The optimal aperture size of iris is 6mm in our experiment. The laser mode is shown in Fig.12.

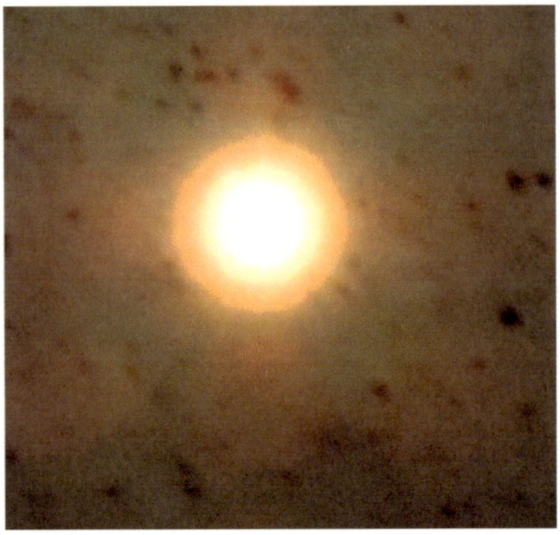

Fig. 12. Laser mode

4.4 Output performance of pulse repetition frequency

According to theoretical and experimental research mentioned above, the performance of laser output is better when the transmittance of output mirror is 39% and the aperture size of iris is 6mm. The discharge current is from 8mA to 16mA, and maximum output power is 22 W. When inserting the acousto-optically modulator, the maximum output power is decreasing to 9.5 W. The laser output mode is TEM_{00}. The pulsed frequency range is from 1 Hz to 100 kHz. The pulsed waveforms of 10.6μm with repetition rate 1 kHz measured by HgCdTe detector (VIGO PC-10.6) are shown as Fig.13. The output pulsed width is about 156 ns as shown in Fig. 13(a) (channel 1) and channel 2 is transistor-transistor logic trigger signal. The pulsed establishing time is 2.9μs, at that time the measured average power is 1.56 W. The output stability of laser is quite well in high repetition frequency, which the pulsed amplitude difference is smaller than ±5% as shown in Fig. 13(b).

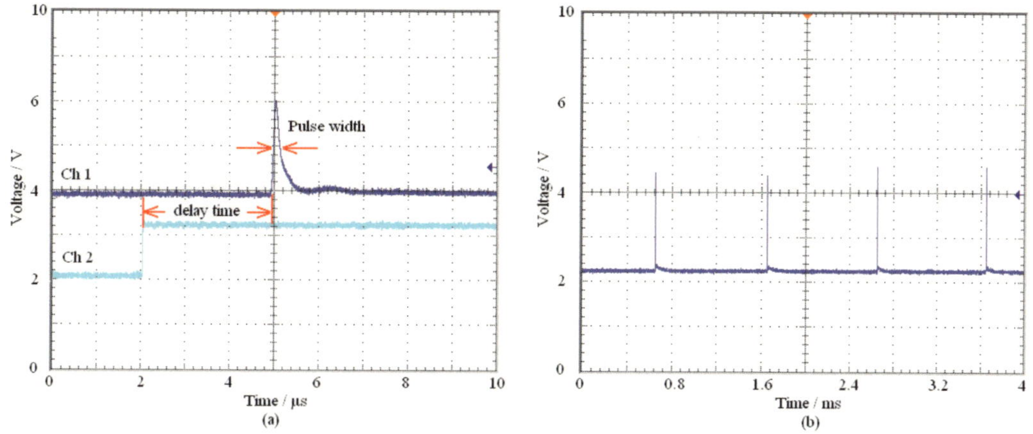

Fig. 13. Output waveforms of CO_2 laser at 1 kHz

5. Tunable acousto-optically Q-switched CO_2 laser

The CO_2 laser has more than 100 spectral lines between 9-11μm, among which several spectral lines next to 10.6μm have the maximum gain (Qu et al., 2005; Ma et al., 2002). Therefore the output spectral line for non-tunable CO_2 laser is 10.6μm. Through the technology of wavelength tuning, the spectral lines between 9-11μm would be obtained one by one. The technologies of wavelength tuning for pulse CO_2 laser include injection locking (Menzise et al., 1984), Fabry-Perot (F-P) etalon (Wu et al., 2003), and grating tuning (Izatt et al, 1991). The injection locking often can not obtain single line output, and it needs complicated techniques, so this method is not suitable for compacted CO_2 laser in engineering applications. Although tuning with F-P etalon has the advantage of simple structure and easy operation, but it only applies to the case when the gain length of laser is shorter, so it can not obtain high power laser output. The diffracting character of grating determines that it has a good capability of wavelength tuning. By changing the incident angle on the grating, the laser output wavelength can be selected. Therefore we designed tunable acousto-optically Q-switched CO_2 laser using the grating as a tuning device.

5.1 The principle of grating tuning

The relationship between diffraction angle, grating constant and wavelength can be obtained by the grating equation just as follows.

$$d(\sin\alpha + \sin\beta)=m\lambda \tag{13}$$

Where d is the grating constant, α is the incident angle of laser, β is the diffraction angle of laser, m is the order of diffraction, λ is the laser wavelength. The resonator is composed by rear mirror (or lens) and grating, and it usually works under the condition of Littrow autocollimator (Wang, 2007). That is to say, the direction of first order diffraction of grating is in consistent with optical axis. At that time, the incident angle α equals diffraction angle β, and the grating equation (13) will be replaced by equation (14).

$$2d\sin\alpha=m\lambda \tag{14}$$

Equation (14) shows that under the Littrow autocollimator condition, the spectral lines which satisfy the grating equation could sustain oscillation in the resonator, and the other spectral lines could not sustain oscillation because of the bigger diffracting loss. Incident angle (α) is determined by the grating constant (d). And the diffraction efficiency or the light distribution of each order is depends on the shape of grating grooves, the character of grating incidence surface, the polarized character of incident light, etc. (Bayin et al, 2004). An effective method to improve diffraction efficiency of single-order is to reduce the diffraction series. Through limiting the grating constant (d), the diffraction angles of other diffraction orders except the first order would go beyond the grating reflective surface, so its corresponding diffracted light do not exist in fact. To meet that condition, the grating constant that is corresponding with the selected laser wavelength should satisfy the relationship just as follows.

$$0.5\lambda < d < 1.5\lambda \tag{15}$$

Thus, the grating constant (d) corresponding with wavelength tuning range of CO_2 laser (9μm-11μm) would be 5.5μm- 13.5μm and the value of grating groove is 74 - 181 lines /mm.

5.2 The device of grating tuning

The output mirror is replaced by grating to realize the wavelength tuning of acousto-optically Q-switched CO_2 laser. Based on the principles of grating tuning, laser wavelength tuning device is designed just as Fig.14. The grating and mirror1 which are placed on the horizontal rotary table must be vertical to the surface of the rotary table and ensure that their intersection coincides with the axis of rotary table. The direction of diffraction grating is horizontal, and the angle between grating and mirror1 is 60°. At this time, the angle between the laser output reflected by mirror1 and the optical axis of laser is 60° too. Then the output laser will be changed direction by the mirror2 to obtain a reasonable direction. The working way of grating is first order oscillation and zero order output. According to the results of theoretical and experimental analysis, the metal engraved grating (120lines/mm) is manufactured with blaze wavelength of 10.6μm and its first order diffraction efficiency is about 60%.

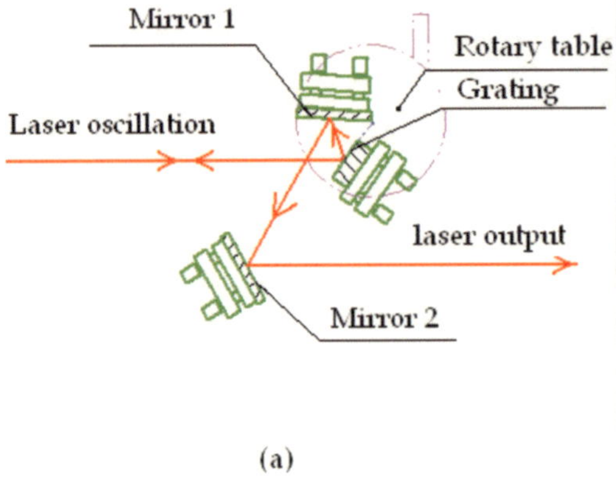

Fig. 14. Device of grating tuning. (a) The schematic, (b) The Photoshop

5.3 Output performance of grating tuning

The spectrum lines are measured by the spectrum analyzer under the working condition of Q-switch, just as shown in figure 15. There are 67 lines which average power of single line is over 2.5W among 9.18μm -10.88μm, and the highest average power is about 8 W. According to the measured spectrum distribution, the spectrum lines of 10R/10P are abundant, and the average power is higher compared with the spectrum lines of 9R/9P. The output power of laser greatly different in different wave bands since the different laser gains in different wavelength and the grating diffraction efficiency variation.

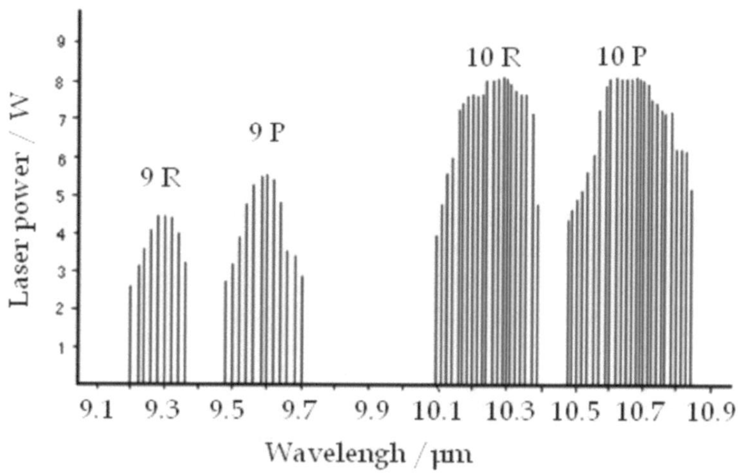

Fig. 15. Laser Spectral Lines

6. The application of acousto-optically Q-switched CO_2 laser

One of the outstanding CO_2 laser features making it suitable for optoelectronic countermeasure applications relates to its distinctive emission spectrum composed of several tens of particular output lines situated in the 9-11μm wavelength range (Atmospheric window)(Yin & Long, 1968). And the laser jamming is a priority research area of optoelectronic countermeasure (LIQIN et al., 2002; Wang et al., 2010). Thresholds for laser jamming in detectors depend on jamming mechanisms, irradiation time, beam diameter, detector dimensions, optical and thermal properties of the materials used in detector construction, quality of thermal coupling to the heat sink, etc. (Bartoli et al., 1976; Evangelos et al., 2004; Sun et al., 2002). The optoelectronic detector (high sensitivity, high signal-to-noise ratio) is irradiated by CO_2 laser, which would induce the saturation effect and lead to signal loss of detector (Bartoli et al., 1975). Since these parameters can vary considerably from one detector to another, it would be very costly and impractical to measure damage thresholds for every case of interest.

A compacted multifunctional acousto-optically Q-switched CO_2 laser is constructed just as shown in Fig.16. The range of repetition frequency could adjust from 1 Hz to 100 kHz, and the tuning range of output wavelength is from 9.2μm to 10.8μm. The spot diameter is 5 mm and the laser mode is basic mode.

Fig. 16. Tunable acousto-optically Q-switch CO_2 laser

6.1 Experimental device of laser irradiate detector

Fig.17 shows the schematic diagram of detector irradiated by pulse laser. HgCdTe detectors with the model of PV-10.6 and PC-10.6 made by VIGO Company are researched in our experiments separately. The spot diameter would be limited at 5 mm by the iris. The power density reaching to the photo-sensitive surface of detector is continuously adjustable through adjusting the attenuators inserted in the light path. The beam splitter is a coated ZnSe plat mirror whose transmittance is 30%. The optical components need to be adjusted to make the laser pass the geometric center of every component.

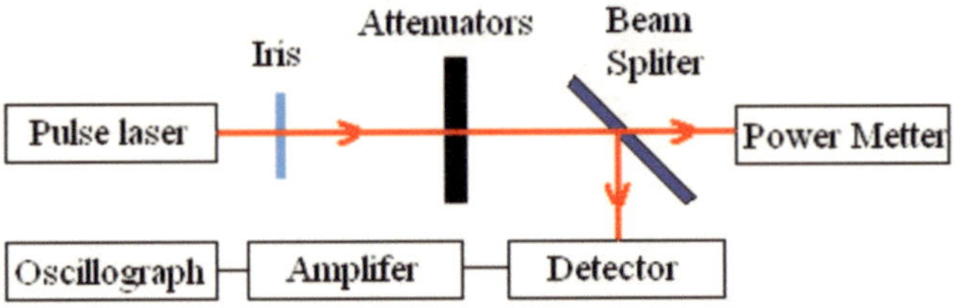

Fig. 17. Experimental schematic of detector irradiated by pulse laser

6.2 The experiment of laser jamming

During laser irradiation, the incident laser power density could increase continuously. The criterion of interference and saturation phenomenon for HgCdTe detector is as follows: when the top of output voltage waveform of detector emerge distorted phenomenon, the detector has been subjected to disturbance; when the peak value of output voltage keeps invariant as the incident laser power grows, the detector is saturated. The average power of the laser is measured by power meter and it is changed into the average power at the photo-sensitive surface of detector, and then the power density could be obtained.

6.2.1 The experiment of laser jamming for PC-10.6 detector

Before Irradiation, the transmittance of attenuators should be adjusted to minimum so as to protect the detector. The detector is irradiated by pulse laser with pulse repetition frequency of 1 kHz. The duration of laser irradiation on detector is 0.5s each time, and then the output voltage waveform of detector would be recorded by the oscilloscope after each irradiation. The detector would be continuously irradiated by pulse laser after it return to nature. The voltage waveform of detector is shown in figure 18(a) when it works at normal condition. The output voltage waveform is monitored as the attenuation is decreased continuously so as to judge the interference phenomenon. The laser power density of photo-sensitive surface is the disturbance threshold as soon as the detector is been disturbed. At this time, the output voltage waveform is shown just as figure 18(b), and the disturbance threshold is 0.452W/cm^2. Then decrease the attenuation continuously and observe the output voltage waveform to judge the saturation phenomenon. The laser power density of photo-sensitive surface is the saturation threshold as soon as the detector is saturated. At this time, the output voltage waveform is shown just as figure 18(c), and the saturation threshold is 1.232W/cm^2. Henceforth, the output voltage amplitude of the detector could not increase obviously with the increasing of laser power density. The output voltage waveform is shown just as figure 18(d), and the detector is saturated completely. At this time, the laser power density that reaches to the photo-sensitive surface of detector is 12.883 W/cm^2. Pause a moment, the detector will restore again, therefore the detector has not been damaged.

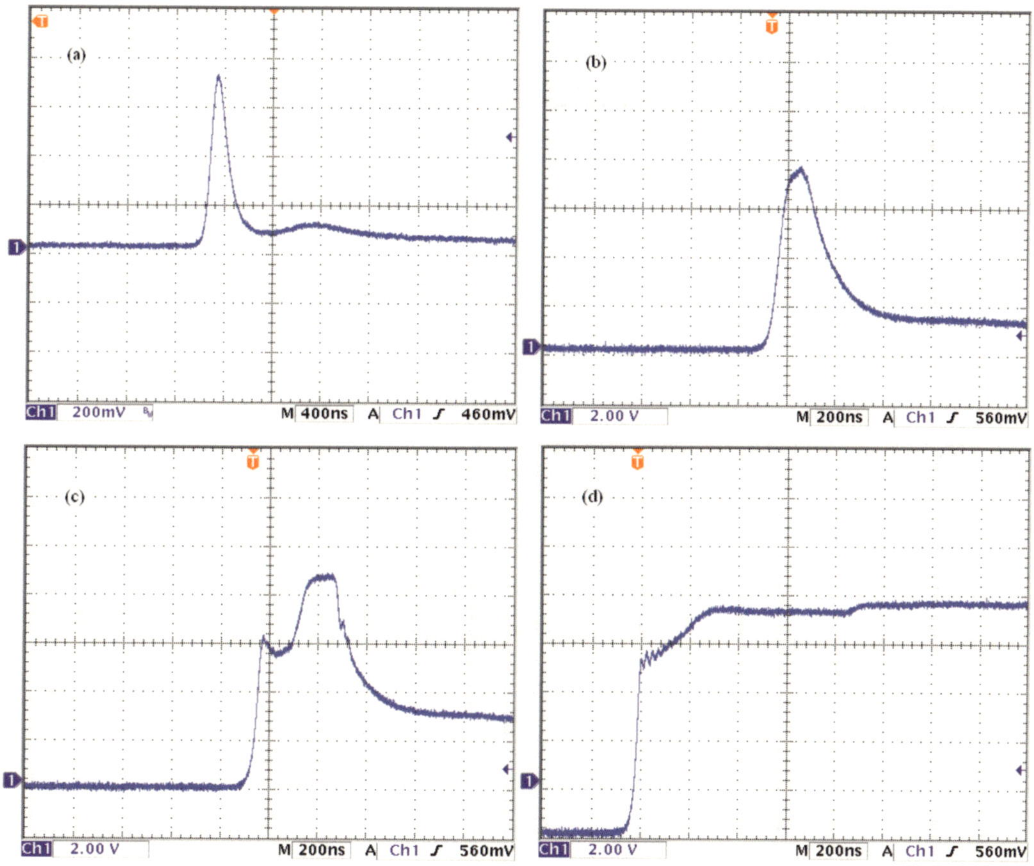

Fig. 18. Experimental results of PC-10.6 detector irradiated by pulse laser: (a) Normal working, (b) Disturbance phenomenon, (c) Saturation phenomenon, (d) Saturation completely

The response curve of PC-10.6 detector irradiated by pulse laser is shown in Fig.19. The output voltage amplitude increases linearly with the increase of power density, and when the power density increases to a certain value (saturation threshold), the output voltage amplitude will drive to plateau. At this time, the detector is saturated.

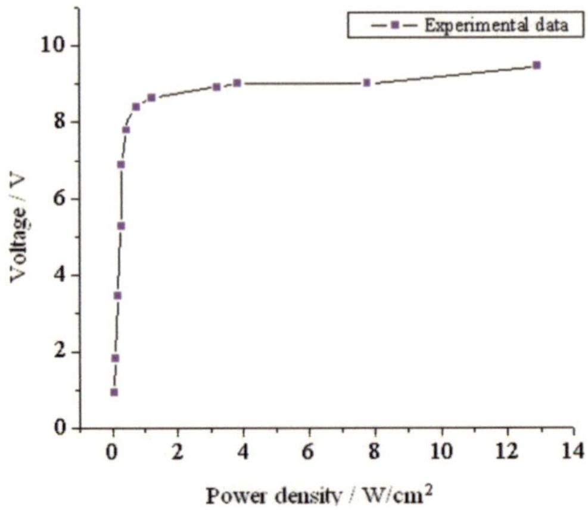

Fig. 19. The response curve of PC-10.6 detector irradiated by pulse laser

6.2.2 The experiment of laser jamming for PV-10.6 detector

The jamming experiments on PV-10.6 detector have been done by using the same method with that of PC-10.6 detector. The measured results are as follows: the disturbance threshold is 0.452W/cm², the saturation threshold is 0.345 W/cm², and the power density of complete saturation is 2.820 W/cm². The output voltage waveforms of detector are shown in figure 20.

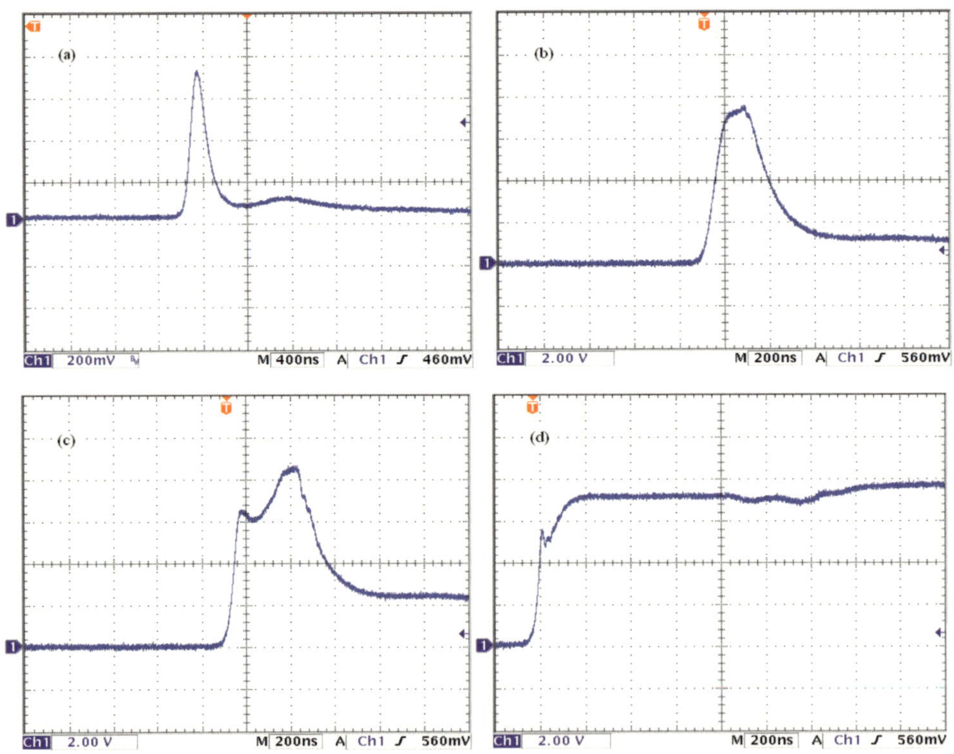

Fig. 20. The experimental results of PV-10.6 detector irradiated by pulse laser: (a) Normal working, (b) Disturbance phenomenon, (c) Saturation phenomenon, (d) Saturation completely

The response curve of PV-10.6 detector irradiated by pulse laser is shown in Fig.21. Its variable tendency is similar to the PC-10.6 detector, but its disturbance threshold and saturation threshold are lower than the PC-10.6 detector. The PV-10.6 detector is prone to be disturbed by the pulse laser.

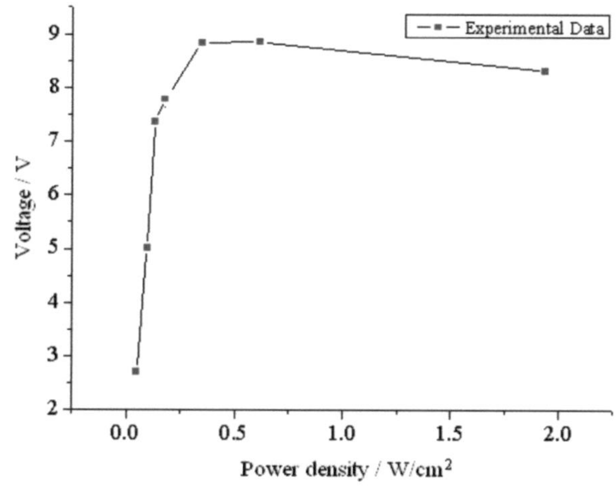

Fig. 21. The response curve of PV-10.6 detector irradiated by pulse laser

6.3 The influence of pulse repetition frequency on the threshold of detector

The pulse repetition frequency of CO_2 laser has great influence on disturbance threshold and saturation threshold of detector. Based on the experimental method mentioned above, the PC-10.6 HgCdTe detector is irradiated by pulse laser with pulse repetition frequency at 1 kHz, 5 kHz, 10 kHz, 20 kHz, 30 kHz, 40 kHz, and 50 kHz respectively. Then the disturbance threshold and saturation threshold of detector are obtained at different pulse repetition frequency. The variations of detector threshold with pulse repetition frequency are shown in Fig.22. Both the disturbance threshold and saturation threshold are reduced with the increase of pulse repetition frequency, and the decline of threshold is even faster at high frequency. When the detector is working at normal condition, the current carriers of heat balance would keep unchanged, and the resistance of detector depends on the photo-generated carriers only. Laser irradiation with multi-pulse is a course of thermal accumulation. When the detector is irradiated by multi-pulse laser, it could not be cooled quickly in one pulse period so that the temperature of detector will arise. At that time, the powerful pulse laser not only engenders the photo-generated carriers, but also engenders current carriers of heat balance. The current carriers of heat balance will lead to the higher dark resistance which would disturb the normal working of detector. The higher the frequency, the more obvious the thermal accumulation, and the detector would be disturbed more easily. Therefore the development of laser with high pulse repetition frequency used in the field of laser irradiation has great scientific significance.

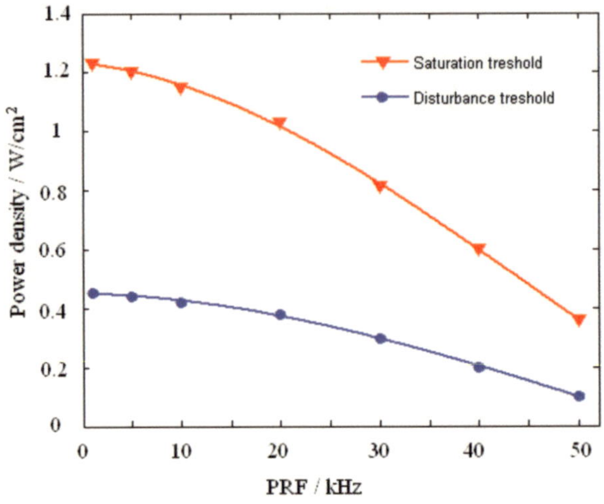

Fig. 22. Variations of detector threshold with pulse repetition frequency

7. References

Bartoli, F. J.; Esterowitz, L.; Kruer, M. R. & Allen, R. E. (1975). Thermal recover processes in laser irradiated HgCdTe (PC) detectors. *Applied Optics*. Vol.14, No. 10, (May 1975), PP. 2499-2507, ISSN 0003-6935

Bartoli, F.; Esterowitz, L.; Allen, R. & Kruer, M. (1976). A generalized thermal modal for laser damage in infrared detectors. *Journal of Applied Physics*. Vol. 47, No.7, (June 1975), PP. 2875-2881, ISSN 0021-8979

Bayanheshig; Gao, J. X.; Qi, X. D. & Li, C. Q. (2004). Manufacture for High Diffraction Efficiency Grating in the First-order Output of 10.6μm Laser. *Journal of Optoelectronics Laser*. Vol. 15, No.10, (January 2004) PP.1137-1140, ISSN 1005-0086, (In Chinese)

Carr, L.; Fletcher, L. ; Crittenden, M.; Carlisle, C. ; Gotoff, S. ; Reyes, F. & Francis, D. (1994). Frequency-agile CO_2 DIAL for environmental monitoring. *SPIE*, Vol.2112, PP.282-294, ISBN 9780819414038, Atlanta, USA

Chang, L. & Jiang, Y. (2009). Effect of Laser Irradiation on $La_{0.67}Ba_{0.33}MnO_3$ Thin Films. *Acta Physica Sinica*.Vol.58, No.3, (July 2008), PP. 1997-2001. ISSN 1000-3290 (In Chinese)

Evangelos, T. Juntaro, I. & Nigel, P. (2004). Absolute linearity measurements on HgCdTe detectors in the infrared region. *Applied Optics*.Vol.43, No.21, (January 2004), PP.4182-4188, ISSN 0003-6935

Hedayatollahnajafi, S.; Staninec, M.; Watanable, L.; Lee, C. & Fried, D. (2009). Dentin bond strength after ablation using a CO_2 laseroperating at high pulse repetition rates. *SPIE*. Vol.7162, PP.71620F0-9, ISBN 9780819474087, San Jose, USA

Hong, L.; Li, L. & Ju, C. (2002). Investigation of cutting of engineering ceramics with Q-switched pulse CO_2 laser. *Opticsand Lasers in Engineering*. Vol.38, (June 2001), PP. 279-289, ISSN 0143-8166

Izatt, J. R.; Rob, M. A. &Zhu. W. S. (1991). Two-and three-grating resonators for high-power pulsed CO_2 lasers. *Applied Optics*. Vol. 30, No. 30, (January 1990), PP.4319-4329, ISSN 0003-6935

Kariminezhad, H.; Parvin, P.; Borna, F. &Bavali, A. (2010). SF_6 leak detection of high-voltage installations using TEA-CO_2 laser-based DIAL. *Optics and Lasers in Engineering*. Vol.48, (May 2009), PP. 491-499, ISSN 0143-8166

Kovacs, M. A.; Flynn,C. & Javan, W. A. (1966) . Q witching of molecular laser transitions. *Applied Physics Letter*. Vol. 8, No. 3, (December 1965), PP. 61-62, ISSN 0003-6951

LIQIN, M.; XIANGAI, X.; XIAOJUN, X.; WENYU, L. & QISHENS, L. (2002). Chaos in photovoltaic HgCdTe detectors under laser irradiation. *Applied physics B*. Vol75, (April 2002), PP.667-670, ISSN 0946-2171

Ma, Y. & liang, D. (2002). Tunable and frequency-stabilized CO_2 waveguide laser. *Optical Engineering*. Vol.41, No.12, (February 2002), PP. 3319-3323, ISSN 0091-3286

Manes, K. R. & Seguin, H. J. (1972). Analysis of the CO_2 TEA laser. *Journal of Apply Physics*. Vol.43, No.12, (March 1972) PP.5073-5078, ISSN 0021-8879

Menzise, R. T.; Flamant, P. H.; Kavaya, M. J. & Kuiper, E.V. (1984). Tunable mode and line selection by injection in a TEA CO_2 laser. *Applied Optics*.Vol.23, No.21, (May 1984), PP.3854-3861, ISSN 0003-6935

Qu, Y.; Ren, D.; Hu, X.; Liu, F.; Zhang, L. & Chen, C. (2005). Wavelength measurement of tunable TEA CO_2 laser. *SPIE*. Vol. 5640, PP. 564-567, ISBN 9780819455956, Beijing, China

Rossmann, K.; France, W. L.; Rao, K. N. & Nielsen, H. H. (1956). Infrared Spectrum and Molecular Constants of Carbon Dioxide. *The Journal of Chemical Physics*. Vol. 24, No.5, (July 1955), PP.1007-1008, ISSN 0021-0606

Smith, K. & Thomson, R. M. (1978). *Computer Modeling of Gas Laser*, Plenum Press, ISBN 0-306-31099-6, New York

Soukieh, M.; Ghani, B. A. & Hammadi, M. (1998). Mathematical modeling of CO_2 TEA laser. *Optical & laser Technology*, Vol.30, (August 1998), PP. 451-457, ISSN 0030-3992

Sun, C. W.; Lu, Q. S.; Fan, Z. X.; Chen, Y. Z; Li, C. F.; Guan, J. L. & Guan, C. W. (2002). *Effects of laser irradiation*. National defense industry Press, ISBN 7-118-02451-0 Beijing, (In Chinese)

Tian, Z. S.; Wang, Q.; & Wang, C. H. (2001). Investigation of the pulsed heterodyne of an electro-optically Q-switched radio-frequency-excited CO_2 waveguide laser with two channels. *Applied Optics*.Vol.40, No.18, (June 2000), PP.3033-3037, ISSN 0003-6935

Wang, S. W.; LI, Y.; Guo, L. H. & Guo, R. H. (2010). Analysis on the disturbance of CO_2 laser to long-wave infrared HgCdTe detector. *J. infrared Millim. Waves*. Vol.29, No.2, (March 2009), PP.102-104, ISSN1001-9014 (In Chinese)

Wang, Z. J. (2007). *Practical Optics Technical Manual*. Machine Press, ISBN 9787111198178 Beijing, (In Chinese)

Wu, J.; Wan, C.; Tan, R.; Wang, D. & Tang Y. (2003). High repetition rate TEA CO_2 laser with randomly coded wavelength selection. *Chinese optics letters*. Vol. 1, No.10, (June 2003), PP.601-603, ISSN1671-7694

Xie, J. J.; Guo, R. H.; Li, D. J.; Zhang, C. S.; Yang, G. L. & Geng, Y. M. (2010). Theoretical calculation and experimental study of acousto-optically Q-switched CO_2 laser. *Optics Express*. Vol.18, No.12, (March 2010), PP.12371-12380, ISSN 0925-3467

Yin, P. K. L. & Long. P. K. (1968). Atmospheric Absorption at the Line Center of P (20) CO_2 Laser Radiation. *Applied optics*. Vol.7, No. 8, (February 1968), PP. 1551-1553, ISSN 0003-6935

Zelinger, Z.; Strrizik,; Kubat, M.; Civis, P.; Grigorova, E.; Janeckova, R.; Zavila, O.; Nevrly, V.; Herecova, L.; Bailleux, S.; HorKa, V.; Ferua, M.; Skrinsky, J.; kozubkova, M.; Drabkova, S. &Janour, Z. (2009). Dispersion of Light and Heavy Pollutants in Urban Scale Models: CO_2 Laser Photoacoustic Studies. *Applied Spectroscopy*. Vol.63, No.4, (August 2008), PP.430-436, ISSN 0003-7028

3

Mode-Locked Fibre Lasers with High-Energy Pulses

S.V.Smirnov[1], S.M. Kobtsev[1], S.V.Kukarin[1] and S.K.Turitsyn[2]
[1]*Novosibirsk State University*
[2]*Aston University*
[1]*Russia*
[2]*UK*

1. Introduction

The recent explosive development of the physics and technology of fibre lasers as well as a vast expansion of their application areas greatly stimulate the quest for and the study of more advanced operational modes of such lasers, including generation of ultra-short and high-energy pulses. Since the invention of the laser researchers have continuously strived to generate shorter laser pulses. Simultaneously achieving a short duration and a high pulse energy is certainly more challenging than improving one of these parameters independently. However, it is this combination that becomes increasingly important in a wide range of scientific, technological, medical, and other applications. High energies and ultra-short pulse durations are both associated with high field intensity that often makes physical system non-linear. In fibre lasers, there are specific properties relevant to both shortening of the pulse duration and increasing their energy. The main obstacles on the road to shorter pulses are a relatively high dispersion and non-linearity of fibre resonators. In addition, the path to high-energy ultra-short pulses is, typically, further complicated by relatively low energy damage thresholds of standard fibre components, such as splitters, isolators, and so forth. While the effects of dispersion can be compensated by different rather advanced means, the nonlinear effects in fibres are much more difficult to manage. Thus, nonlinearity plays a critical role in the design of advanced fibre laser systems, but paradoxically, it is somewhat undesirable to many engineers because of its very limited controllability. Substantial efforts have been made to reduce the resonator nonlinearity, *e.g.* by using large-mode-area fibres, and this direction presents an important modern trend in laser technology. On the other hand, understanding and mastering nonlinear physical fibre systems offer the potential to enable a new generation of laser concepts. Therefore, it is of great importance to study physics and engineering design of laser systems based on nonlinear photonic technologies. In particular, new nonlinear approaches and solutions pave a way for development of advanced mode-locked fibre lasers with ultra-short high-energy pulses.

Presently, passive mode locking is one of the key methods of ultra-short pulse (USP) generation. As recently as a few years ago, femto- and pico-second pulses extracted directly from the master oscillator operating in a passive mode-locking regime had relatively low energies, typically, not exceeding at few dozens of nJ and, in some special cases, hundreds of nJ. In order to radically boost the pulse energy, additional optical amplifiers were used or, otherwise, a completely different method of short pulse generation, Q-switch was employed, which allowed considerably higher per-pulse energy, albeit at the expense of longer duration, typically, over several nanoseconds and more.

The combined Q-switching and mode locking in one cavity has also been successfully employed for generation of high-energy pulses of laser radiation (Lin *et al* 2008, Jabczyński *et al*, 2006). Another way to increase per-pulse energy of output radiation is the cavity dumping technique (Johnson *et al*, 1976), which can be used in all the mentioned above types of lasers. In order to increase the intra-cavity pulse energy, the cavity dumping method uses a closed multi-path resonator, into which a so-called cavity dumper is inserted that allows picking single high-energy pulses out of the cavity at a frequency lower than the original pulse repetition rate. More powerful pump sources may also increase per-pulse energy in passively mode-locked lasers, but this may be achieved with certain combinations of pulse and cavity parameters only (Akhmediev *et al*, 2008; Chang *et al*, 2008). Another method traditionally utilised in most high-energy laser systems of different types relies on extra-cavity optical amplification.

A completely different physical approach to achieving higher energy of USP generated in lasers with passive mode locking is based on elongation of the laser cavity. The pulse repetition rate of a mode-locked laser is inversely proportional to its resonator length, this is why using longer cavities leads to lower pulse repetition rate and, consequently, to higher pulse energy at the same average output power. The following simple relation describes this design principle:

$$E_{\mathrm{p}} \propto P_{\mathrm{ave}}\, T_{\mathrm{R}} \propto P_{\mathrm{ave}}\, nL/c.$$

Here E_{p} is the pulse energy of a mode-locked laser, and P_{ave} is the average radiation power, T_{R} is the resonator round trip time, n – the refraction index of the cavity medium, c – the speed of light, L – the resonant cavity length. This method is well suited for fibre lasers whose resonator length may reach dozens (Ania-Castañón *et al*, 2006; Ivanenko *et al*, 2010) and even hundreds of km (Ania-Castañón, 2009). This allows generation in mode-locked lasers of the highest-energy pulses possible for this type of lasers. Of course, this high pulse energy coming out of the master oscillator can be further increased in an optical amplifier.

The first studies of long mode-locked fibre lasers performed by many research groups around the world showed that the simultaneous solution of the problems of short duration of pulses and of their relatively high energy is a non-trivial task. In this chapter, we will discuss the physical conditions required for high-energy pulse generation as well as the problems and limitations related to cavity lengthening up to several kilometeres or even dozens of kilometeres. We will consider the physical mechanisms of mode locking including new types of mode locking that takes place in such lasers. We will overview the recent publications in this rapidly growing area and will analyse the prospects of long-resonator mode-locked fibre lasers.

2. Review of recent progress in mode-locked fibre lasers with high-energy pulses

First experiments on considerable cavity elongation up to 100 and 400 m in solid-state (Kolev *et al*, 2003) and fibre (Kang *et al*, 1998; Fong *et al*, 2006; Fong *et al*, 2007) mode-locked lasers have shown that in such relatively long resonators, it is possible to achieve stable passive mode locking. As a result, the pulse repetition frequency can be reduced by more than an order of magnitude (down to ~1.7 MHz) and the per-pulse energy can be raised by the same factor at the same average power of output radiation. Recently, a further increase of mode-locked laser cavity length by approximately an order of magnitude was demonstrated (Kobtsev *et al*, 2008c). As a result, a stable mode-lock regime was achieved in a fibre laser with optical length of the cavity 3.8 km. In these experiments, a laboratory sample of a ring Yb fibre laser was used, its diagram being shown in Fig. 1. Pumping of a 7-m active Yb-doped fibre with a 7-μm core was performed (Grudinin *et al*, 2004) with a 980-nm laser operating at up to 1.5 W of CW output power. The Yb-doped GTWave fibre used in the laser is a convenient choice because it allows using inexpensive diode lasers with multi-mode 100-μm output fibres for pumping. Microscope objectives were employed in order to guide radiation in and out of the free-space laser resonator with discrete elements. The ring configuration of the free-space portion of the resonator was formed by three broad-band ($\Delta\lambda$ ~ 100 nm) highly reflective mirrors. Coupling of radiation out of the resonator was done with a polarisation beam splitter that also ensured linear polarisation of the output radiation. Laser generated unidirectional radiation pulses despite the fact that no optical diode was used in the resonant cavity. The propagation direction of the generated pulses coincided with that of the pumping radiation inside the active fibre.

Mode locking of the laser was achieved by using the effect of non-linear rotation of radiation polarisation (C.J. Chen *et al*, 1992; Matsas *et al*, 1992; Chong *et al*, 2008). Control over the polarisation was carried out with the help of three phase plates inserted into the laser cavity. Upon initial alignment of these plates and start of mode locking, this operation was henceforth self-activated as soon as pumping was switched on.

In this experiment the laser cavity comprises no elements with negative (anomalous) dispersion in the spectral range of generation. Mode-locked generation was achieved in an all-positive-dispersion laser configuration. At the minimal length of the resonator 9.4 m, the laser generated single chirped pulses with 3.5-ps duration (Fig. 2(a)) and 300-mW average output power at 1075 nm. The pulse repetition rate was 22 MHz, the spectrum width being 2 nm. Extra-cavity pulse compression by two diffraction gratings (1,200 lines/mm, spaced by ~15 mm from each other) lead to reduction of the output pulse duration to 550 fs (Fig. 2(b)). The product $\Delta\nu\Delta\tau$~0.33 demonstrates that the compressed pulses were transform-limited. The highest per-pulse energy reached 14 nJ. It is relevant to note that a stable mode-locked operation of this laser was achieved without using any additional elements for limitation of laser radiation spectrum, unlike it was reported by (Chong *et al*, 2006) and (Chong *et al*, 2007). In order to elongate the laser cavity a 2.6-km stretch of All-Wave (Lucent) fibre was used. This physical length of the fibre corresponds to the 3.8 km optical length of the laser cavity. The laser diagram with elongated cavity is given in Fig. 3. Pumped with the same amount of power (1.5 W), the laser generated unchanged average output radiation power, 300 mW. However, mode-locked operation of this laser with ultra-long cavity behaved differently depending on the adjustment of the phase plates (Fig. 4).

It was possible to run the laser stably both in the mode of single nanosecond pulses (Fig. 4(a)) or in that of nanosecond pulse trains (Fig. 4(b)), as well as in the mode of nanosecond pulses with noticeable microsecond pedestal (Fig. 4(c)). When generating trains of nanosecond pulses, the train duration was up to 300 ns and the train contained up to 20 pulses, each about 1 ns long. When generating single nanosecond pulses, their duration was 3 ns, repetition rate 77 kHz, and per-pulse energy 3.9 μJ. To the best of our knowledge, the energy of pulses we have generated (3.9 μJ) is the highest to-date achieved directly from a mode-locked fibre laser without application of Q-switching or/and cavity dumping techniques.

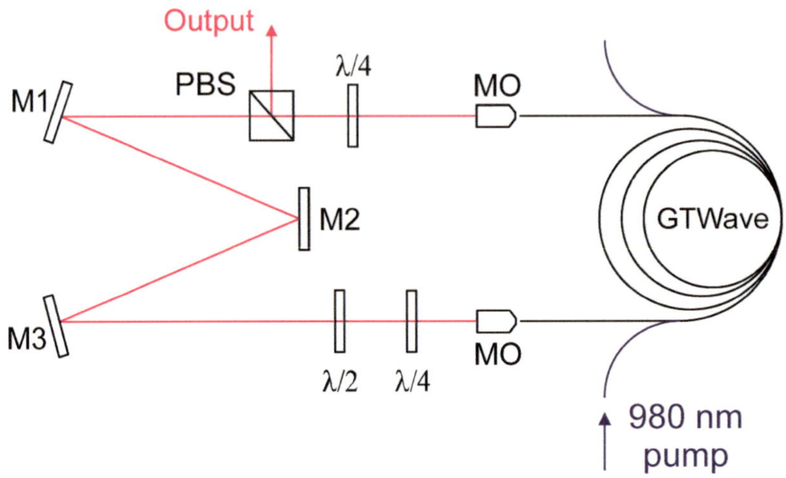

Fig. 1. Schematic of the fibre laser: MO — microscope objective, PBS — polarizing beam splitter, M1–M3 — high-reflectivity mirrors, λ/4 - quarter-wave plate, λ/2 - half-wave plate.

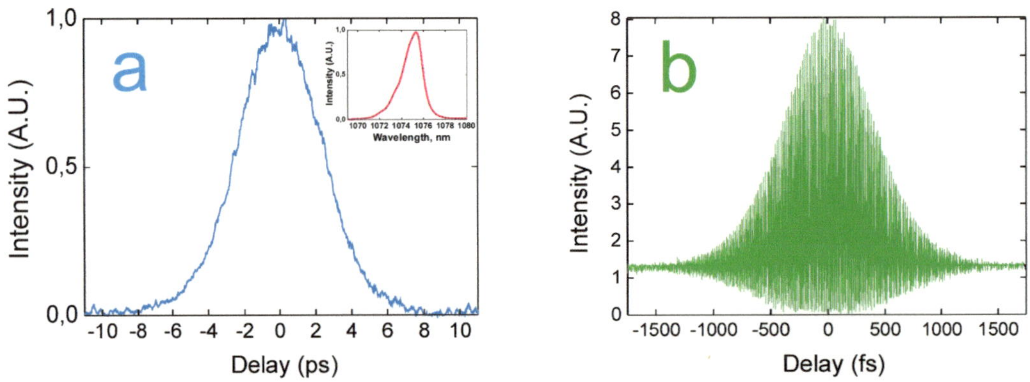

Fig. 2. (a) Background-free autocorrelation trace of chirped pulses from laser output, inset: optical spectrum of the laser; (b) Interferometric autocorrelation trace of de-chirped laser pulses.

In Fig. 5, the radiation spectra of ultra-long mode-locked laser generating single pulses and multiple pulse trains are shown. It can be seen that the width of the spectra differs significantly. The spectrum width of laser radiation in the case of single pulses (Fig. 5(left)) amounted to 0.35 nm, suggesting that the obtained pulses can be subsequently compressed into picosecond range, which possibility, however, was not experimentally verified in this study. Spectrum width of the multiple pulse train radiation (Fig. 5(right)) was almost 20 nm and this indicates that these pulses have different radiation wavelengths. Radiation spectra of these pulses overlap and form relatively wide resulting spectrum. Note an approximately 11-nm shift of laser spectrum into the long-wavelength range in the case of multiple pulse train generation. Laser spectra of single pulse radiation both in short and in ultra-long cavity are quite smooth-shaped and they exhibit no steep wings with peaks at the edges, which are typical for all-normal dispersion fibre lasers with strong spectral filtering.

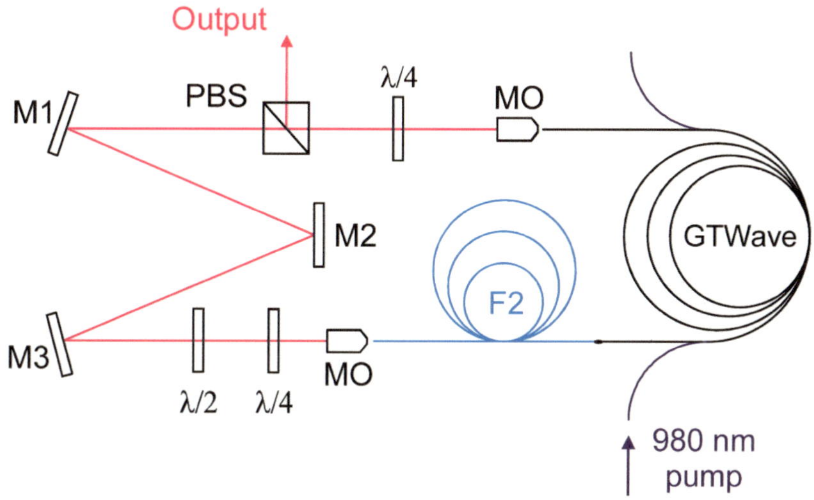

Fig. 3. Schematic of extra-long mode-locked fibre laser: F2 — All-wave fibre, length of 2,6 km.

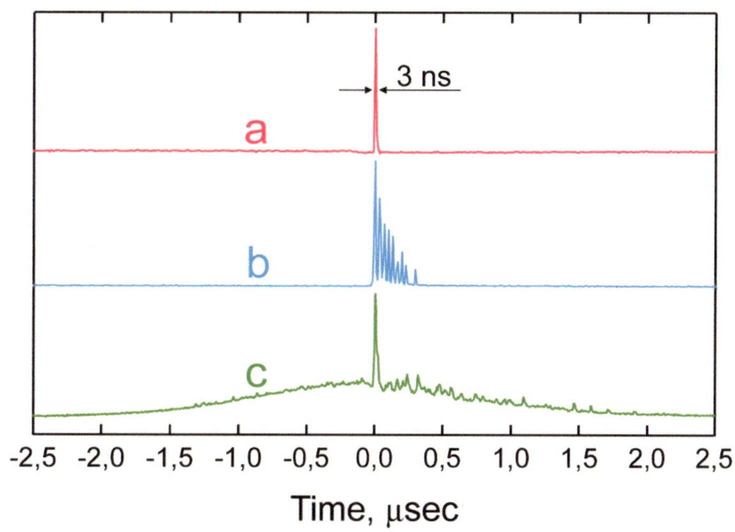

Fig. 4. Temporal distribution of laser radiation intensity in different types of mode-locked operation: a — generation of single 3-ns pulses, b — generation of multiple nanosecond pulse trains, c - generation of single nanosecond pulses with noticeable microsecond-long pedestal. For all types of operation the pulse period (or period of pulse trains in case b) was ~ 13 μs.

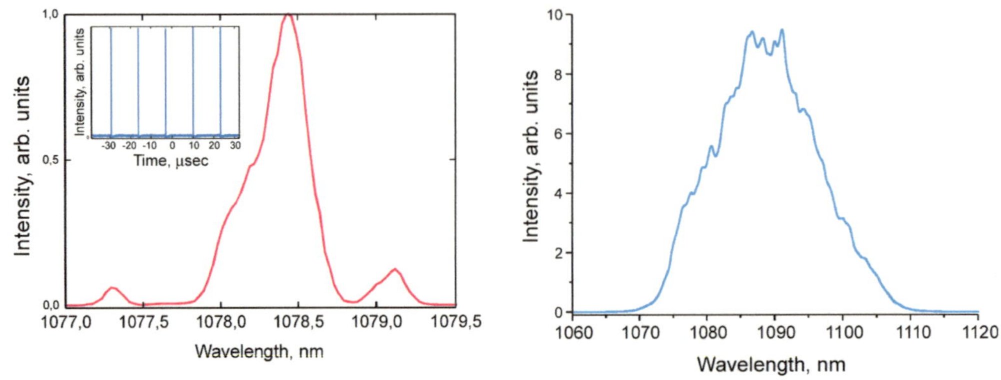

Fig. 5. Optical spectra of the ultra-long mode-locked Yb-doped fibre laser. On the left: spectrum in the case of single 3-ns pulses generation, inset: real-time oscilloscope trace of single pulse train; on the right: spectrum in the case of multiple nanosecond pulse train generation.

New results obtained in the experiments (Kobtsev *et al*, 2008c) with relatively high energy of pulses directly in a passively mode-locked laser stimulated further investigations in this area (Tian *et al*, 2009; Zhang *et al*, 2009; Lin *et al*, 2011; Song *et al*, 2011; Ai *et al*, 2011; Tian *et al*, 2009; Kobtsev *et al*, 2009; Senoo *et al*, 2010; Kelleher *et al*, 2009; L. Chen *et al*, 2009; Wang *et al*, 2011; Nyushkov *et al*, 2010; Kobtsev *et al*, 2010a; Ivanenko *et al*, 2010). In particular, demonstrated experimentally in (Kelleher *et al*, 2009) were 1.7-ns-long pulses with a giant chirp generated in 1,2-km-long fibre laser mode-locked due to nanotube-based saturable absorber. The time-bandwidth product was about 236, which was ~750 times the transform limit, assuming a sech^2 profile. In (Senoo *et al*, 2010), passive mode-lock regime was also used in a full-PM fibre cavity of novel θ-configuration with the use of saturable absorber (single-wall carbon nanotubes). Generation in an ultra-long cavity with anomalous dispersion was reported in (Li *et al* 2010). In (L. Chen *et al*, 2009; Tian *et al*, 2009), a mode-lock regime was obtained in an ultra-long laser with semiconductor saturable absorber mirrors. In (Kobtsev *et al*, 2009), a numerical and analytical study of different generation regimes and mechanisms of switching between them was reported for lasers passively mode-locked due to nonlinear polarisation evolution (NPE).

3. Applications of high-energy pulses

Despite the fact that studies of long-cavity mode-locked fibre lasers with high-energy pulses have started only recently, these lasers have already found interesting practical applications. The first applications of such lasers were demonstrated in high-energy super-continuum (SC) generation (Kobtsev *et al*, 2010c).

One of the most obvious approaches to the generation of high-energy super-continuum pulses is based on an increase in the peak power of the pumping pulses (e.g., using several amplification stages). Note that the spectral broadening grows as the pumping power increases, since the super-continuum generation results from the simultaneous action of several non-linear optical effects. However, the corresponding energy losses due to the stimulated Raman scattering and the linear loss related to the propagation in optical fibre also become larger. Thus, the output super-continuum power is saturated at higher pumping powers, so that a further increase in the pumping power does not lead to a corresponding increase in the super-continuum power (J.H. Lee, 2006). Therefore, alternative methods are required for a further increase in the SC pulse energy. It is highly desirable to supply longer pump pulse duration rather than higher peak power and this can be very naturally implemented with the novel long cavity all-fibre all-positive-dispersion lasers.

Figure 6 demonstrates the diagram of the discussed long-cavity all-fibre ring laser. The Yb fibre that is free of linear birefringence and that is cladding-pumped by means of a multimode coreless fibre serves as the active medium (Grudinin *et al*, 2004). The length of the Yb-doped active fibre is 10 m, and the core diameter is 7 μm. The active fibre is pumped by a multimode diode laser with an output power of up to 1.5 W at the wavelength of 980 nm. The mode locking results from nonlinear polarisation evolution (NPE). An increase in the cavity length using an SMF-28 fibre leads to a decrease in the pulse repetition rate and hence, to a respective increase in the pulse energy at the same mean power.

The demonstrated laser makes it possible to generate pulses with duration of 10 ns and energy of 4 μJ at the repetition rate of 37 kHz. The mean output power (150 mW) is limited by the working range of the fibre polarisation splitter that provides the out-coupling of radiation. The FWHM of the pulse spectrum is 0.5 nm, and the corresponding duration of the bandwidth-limited pulse is 2 ps. This indicates the gigantic chirp of the generated 10-ns pulses. Note that single-pass dispersion broadening of 2-ps bandwidth-limited pulses resulting in 10-ns pulses is possible in the SMF-28 fibre with a length of about 500 km. In the laser, the significant pulse broadening is reached at a substantially smaller cavity length (8 km).

For a further increase in the pump-pulse energy, we employ an additional amplification stage based on a cladding-pumped Yb-doped fibre (Grudinin *et al*, 2004). The energy of the amplified pulses is 80 μJ, and the mean power of the amplified radiation is 3 W, the 10-ns pulse duration remaining unchanged. The amplified pulses are fed to a 30-m segment of SC-5.0-1040 micro-structured fibre, where the SC radiation is generated in the spectral range 500–1750 nm. (Note that the measurements are limited by the optical spectrum analyser at the long-wavelength boundary.) The spectrum of the SC radiation at the exit of the fibre exhibits a high-intensity peak in the vicinity of the pumping wavelength, which is also typical of the SC in the case of the CW pumping (Kobtsev & Smirnov, 2005). The presence of such a peak is related to an extremely high probability of low-energy soliton generation upon the decay of the continuous wave or the long pumping pulse owing to modulation instability (note the L-shaped distribution of solitons with respect to energy (Dudley *et al*, 2008). No self-frequency shift is observed for low-energy solitons with a relatively low peak power. Therefore, a significant fraction of the input energy remains unconverted and the spectral peak emerges in the vicinity of the pumping wavelength. Another feature of the SC spectrum (Fig. 7) that is important for several practical applications is the presence of a wide plateau in the spectral interval of 1125–1550 nm. Within this interval, the SC power spectral density is varied by no more than 1 dB.

In the above mode of SC generation using nanosecond pumping pulses, the temporal structure of the output pulses is retained on the pulse time scale in contrast to the generation mode using femtosecond and picosecond pumping. Indeed, the SC generation under femtosecond pumping in the range of the anomalous dispersion is initiated by the decay of input pulses into a series of solitons whose duration also falls into the femtosecond range(Kobtsev & Smirnov, 2008b). During the subsequent spectral broadening, the solitons

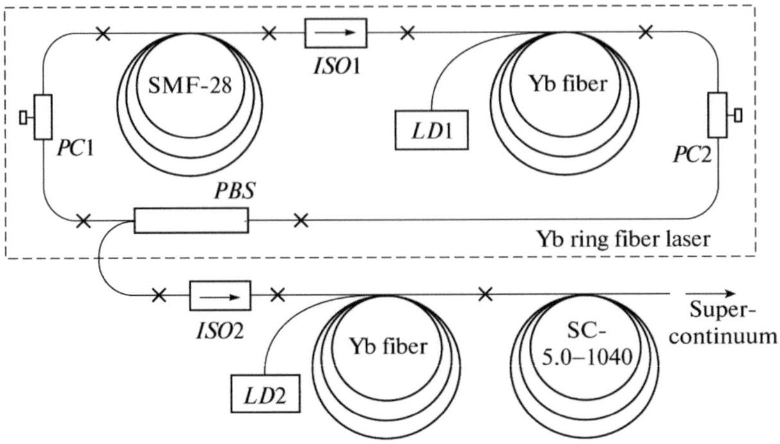

Fig. 6. Schematic layout of the all-fibre SC generator: LD1, LD2 —pumping laser diodes, PC1, PC2 — fibre polarisation controllers, ISO1, ISO2 — polarisation-insensitive optical isolators, and PBS — fibre polarisation beam splitter.

are spread in time and give rise to a wave packet with a complicated temporal structure, such that the duration of the wave packet can be greater than the duration of the femtosecond pumping pulses by a factor of tens. In the case of picosecond pumping pulses, the initial stage of SC generation is characterised by the development of modulation instability, so that the picosecond pumping pulses decay into a stochastic series of sub-pulses (solitons) (Kobtsev & Smirnov, 2008b), whose duration (10–100 fs) depends on the pumping power and the dispersion of the fibre. As in the case of femtosecond pumping, the propagation of soliton sub-pulses along the fibre is accompanied by a temporal spread owing to the group-velocity dispersion. The duration of the resulting wave packet of SC radiation can be several times greater than the duration of the picosecond pumping pulses.

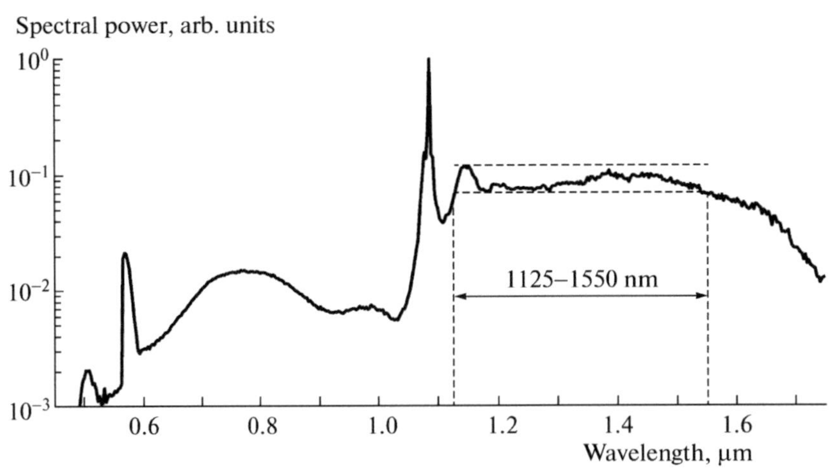

Fig. 7. Spectrum of high-energy SC.

Similar processes induced by the modulation instability and the decay of the excitation pulses into soliton sub-pulses correspond to SC generation in the presence of nanosecond pumping.

In contrast to the femtosecond and picosecond scenarios, the pulse duration at the entrance of the fibre is significantly greater than the characteristic spread of the SC soliton components, so that the duration of SC wave packets at the exit of the fibre is almost the same as the pumping pulse duration. Figure 8 illustrates this effect by presenting the results of numerical modelling of the solution to the generalized nonlinear Schrödinger equation for the pumping pulse with duration of 1 ns and energy of 50 μJ in a 30-cm long SC-5.0-1040 micro-structured fibre. To reduce the computation time, we set the pump-pulse duration and the fibre length to values that are smaller than the corresponding experimental parameters. It can be seen that the SC radiation at the exit of the fibre represents a complicated wave packet containing a large number of sub-pulses (optical solitons) (see inset to Fig. 8). The duration of such solitons is 10 fs, their peak power being higher than the peak power of the pumping pulses by no less than an order of magnitude. In the course of propagation, the positions of solitons inside the wave packet evolve owing to the fibre dispersion, but the characteristic scale of such variations does not exceed few dozens of picoseconds and the duration of the nanosecond wave packet is generally maintained. In the experimental measurements of the intensity time distribution, single solitons are not resolved and we observe the output power averaged over a large number of them. Note that averaging soliton spectra (Kobtsev & Smirnov, 2005) yields a wide smooth SC spectrum in experiments. Numerical simulations show that the time-averaged power is only weakly varied in the process of spectral broadening, so that the shape of the nanosecond pulse remains almost unchanged during the SC generation.

It is pertinent to note that various types of generation in high-energy mode-locked lasers may substantially differ from each other in their efficiency of spectral broadening and in the shape of their super-continuum spectra (Wang *et al*, 2011). This circumstance makes the lasers in question particularly interesting from the viewpoint of development of controllable super-continuum generation whose parameters could be changed dynamically in the process of operation.

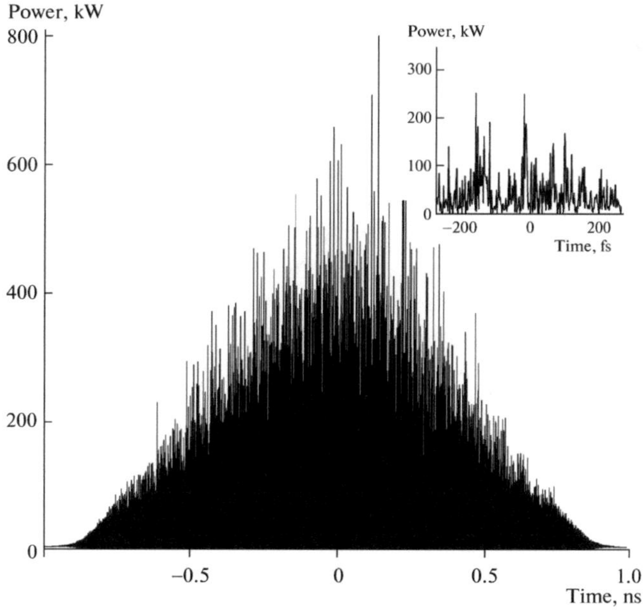

Fig. 8. Numerically simulated plot of the SC radiation intensity *vs* time. The inset shows a fragment of the time distribution on a smaller scale.

4. Underlying physics of high-energy pulse generation

In addition to having record parameters for mode-locked lasers (ultra-low pulse repetition rate, high pulse energy), fibre lasers with greatly extended cavity represent very interesting objects for fundamental physical science. Notwithstanding their relative simplicity of design and of the mathematical models describing them, the lasers with mode-locking achieved due to the effect of non-linear polarisation rotation exhibit an exceptional variety of generation modes and leave unanswered a plethora of questions, which will be the subject of future investigations.

The general principle of mode locking in the considered lasers is rather simple (Hofer *et al*, 1991; Matsas *et al*, 1992; Fermann *et al*, 1993). When travelling along a span of optical fibre, laser pulses with sufficiently high power level will rotate their polarisation ellipse due to the Kerr effect (dependence of the refraction index on the radiation intensity). The rotation angle of the polarisation ellipse determines the radiation losses in the polarisation splitter, the component that couples one of the linearly polarised radiation components out of the cavity. Thus, the optical losses related to the rotation angle of the polarisation ellipse are tied to the radiation intensity. In other words, the optical fibre together with the polarisation splitter form a system that works like a saturable absorber: the losses of the generated pulse inside the polarisation splitter are minimal, but the existing fluctuations are suppressed because of their higher level of losses. In addition to the polarisation splitter and the optical fibre, the laser optical train includes polarisation controllers or phase plates serving for adjustment of the laser (independently of the intensity transformations of the polarisation ellipse).

Two approaches are usually employed for mathematical description of the lasers that make the subject of this study. One of them is based on the solution of a system of coupled non-linear Schrödinger equations for the two orthogonal radiation polarisations. The other one is more heuristic and is based on the solution of the Ginsburg-Landau equation (Haus, 2000).

Both experiment and the numerical modelling demonstrate a great variety of generation modes differing in the number of pulses inside the cavity, in their duration, energy level, in the character of their auto-correlation functions (ACF), as well as in the shape and width of their pulse spectra. The transition of the laser from one generation mode to another may occur both when the pumping power is changed and when the tuning of the intra-cavity polarisation components, – phase plates or polarisation controllers, – is adjusted. The large diversity of possible laser operation modes has not yet been properly catalogued in the literature. A number of attempts to describe different generation regimes have been undertaken (see, for instance (Wang *et al*, 2011)), however the proposed classifications are incomplete. One of the possible systematic descriptions of generation regimes can be done on the basis of their classification according to the pulse-to-pulse stability.

Taking this as a starting point, a vast variety of results produced in experiments and in numerical modelling may be divided into two main types of generation regimes. The first type presents a well-known generation of isolated single pulses (Fig. 9 b, d) with bell-shaped auto-correlation function and steep spectral edges (see Fig. 9 a, c). As the numerical modelling has shown, the pulse parameters obtained in this generation regime are stable and do not change with round trips after approaching those asymptotic values after the initial evolution stage. Since the cavity has all-positive dispersion, the generated laser pulses exhibit a large amount of chirp (Matsas *et al*, 1992; Chong *et al*, 2008; Wise *et al*, 2008) and may be efficiently compressed with an external diffraction-grating compressor, which is confirmed by both the numerical results and the experiment.

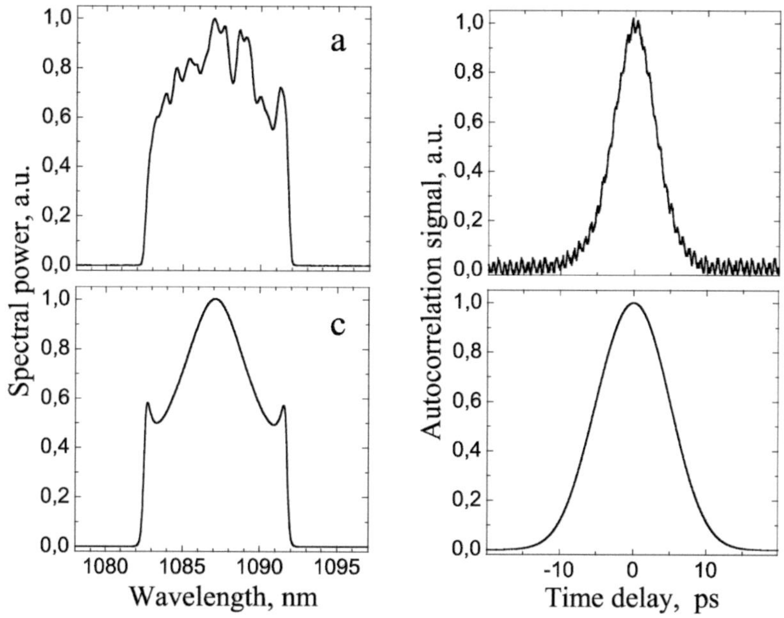

Fig. 9. Stable single-pulse generation: a, b - experimental results; c, d - simulations; a, c - spectra; b, d - ACFs.

It is more interesting, however, that in addition to this standard generation regime, the considered laser scheme also exhibits a different type of operation. In this second mode, an unusual double-structured ACF (femto- and pico-second) can be observed, see Fig. 10 b, e and (Horowitz & Silberberg, 1998). Experimentally measured laser generation spectra of this type usually have a rather smooth bell-shaped appearance (see Fig. 10 a and (Zhao *et al*, 2007)). However, as the numerical simulation demonstrates, such smooth spectra are a result of averaging over a very large number of shorter pulses, whereas the spectrum of an individual pulse contains an irregular set of noise-like peaks (see the un-averaged spectrum in Fig. 10 d shown with a grey line). In the temporal representation this type of generation corresponds to pico-second wave packets consisting of an irregular train of femtosecond sub-pulses (see Fig. 10 f). The peak power and width of such sub-pulses stochastically change from one round trip along the cavity to another, also leading to fluctuations in the wave packet parameters easily noticeable when observing the output pulse train in real time on the oscilloscope screen during experiments (see Fig. 10 c). In other words, such irregular short-scale structures are "breathing" being embedded into a more stable longer-scale pulse envelope. No systematic change in generation parameters such as power or wave-packet duration is observed even after several hours of operation. The absence of systematic drift of pulse parameters is typical for numerical simulations as well (in the latter case, however, much shorter pulse train of about only 5×10^4 is examined.) The discussed regime presents an interesting symbiotic co-existence of stable solitary wave dynamics and stochastic oscillations. Note that similar structures have been studied in different context numerically in the complex cubic-quintic Ginzburg-Landau model in (Akhmediev *et al*, 2001; Komarov *et al*, 2005). Generation of short-scale irregular structures could be attributed to wave collapse regimes triggered within certain parameter ranges (Kramer *et al*, 1995; Chernykh & Turitsyn, 1995). The co-existence of stable steady-state pulses and pulsing periodic, quasi-periodic, or stochastic localised structures is a general feature of multi-parametric dissipative non-linear system (see *e.g.* (Akhmediev *et al*, 2001; Kramer *et al*, 1995) and references therein). Each particular non-linear dynamic regime exists in a specific region of parameter space. Therefore, in systems that possess a capability to switch operation from one region of parameters to another, one can observe very different lasing regimes. The parameter regions where pulsing (periodic, quasi-periodic or stochastic) localized structures do exist might be comparable to or even larger than the regions of existence of conventional steady-state solitons. Note that in the multi-dimensional parameter space of the considered laser scheme, it is almost impossible to explore all the possible operational regimes via direct modelling.

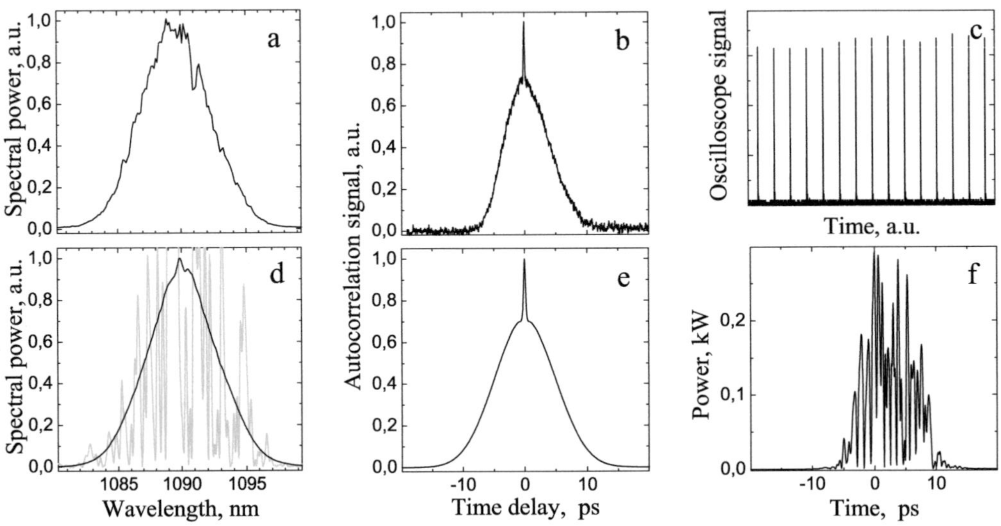

Fig. 10. Quasi-stochastic generation regime: a–c — experimental results; d–f — simulations; a, d — spectra; b, e — ACFs; c — pulse train from oscilloscope; f — non-averaged intensity distribution.

Both experiments and simulations show that wave-packets generated in the double-scale femto-pico-second regime can be compressed only slightly and pulses after compression remain far from spectrally limited. As our experiments have demonstrated, after extra-cavity compression of these complex wave packets with the help of two diffraction gratings, ACF of the resulting pulses has qualitatively the same double-feature shape. We would like to stress, however, that a possibility of compression is not an ultimate condition that should be imposed on any generated pulses with chirp or having more complex structure. Some applications might even be more appropriate for such double scale pulses with an efficient short scale modulation (irregular) of the longer scale pulse envelope.

Switching between different modes of laser generation can be triggered by changing the parameters of polarisation controllers. In order to clarify the physical mechanisms leading to quasi-stochastic oscillations and mode switching we have carried out massive numerical simulations of laser operation in the vicinity of the boundary of the stable single-pulse generation. While performing numerical modelling, we introduced into the cavity fixed-duration pulses with varying energies and analysed the resulting gain coefficient for the pulse over one complete round trip over the resonator (see black line in Fig 11). At the input power $P = 1$ (in arbitrary units) the one-trip gain coefficient equals unity, which on the negative curve slope corresponds to stable generation. Starting with $P \sim 1.75$ a.u., single-pulse generation becomes unstable (corresponding to positive slope of the black curve in Fig. 11). In this region, an exponential growth of small intensity fluctuations can be observed and, over only several round trips of the cavity, an isolated picosecond pulse is decomposed into a stochastic sequence of femtosecond pulses. Because in this process, a substantial change in the pulse form takes place the gain curves in Fig. 11 no longer correspond to the real situation. Exponential power growth is quickly quenched and over a number of trips along the cavity (as a rule, from dozens to hundreds) an isolated picosecond pulse with stable parameters is formed in the cavity again. So, in our numerical results there was no bi-stability, which could be expected on the basis of the curve shapes in Fig. 11: at any fixed set of cavity parameters only one of the two generation types could be stable irrespective of the initial field distribution within the resonator.

For reliable switching of the generation type it is necessary to adjust polarisation controllers or to change other cavity parameters. For instance, curves 1–3 in Fig. 11 correspond to different resonator length (12.0, 12.3, and 12.6 m accordingly). It can be seen that as the resonator parameters are shifted towards the boundary of single-pulse generation, the width of the stable domain is reduced. Indeed, for curve 3 in Fig. 11 the unity-gain point almost coincides with the extremum, so that even small intensity fluctuations may bring the system into the unstable region and lead to decay of the pulse. If the resonator length is further increased the stable single-pulse generation becomes impossible and the laser starts generating quasi-stochastic wave packets.

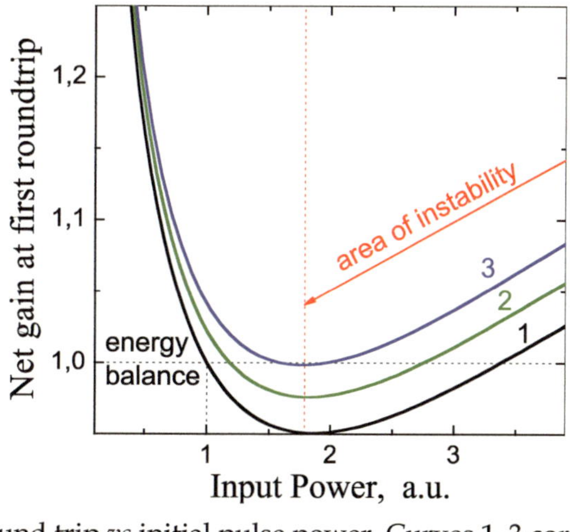

Fig. 11. Net gain per round trip *vs* initial pulse power. Curves 1–3 correspond to slightly different cavity parameters close to boundary of the single-pulse generation regime area.

Conclusions that we drew on the basis of the above analysis are also valid for the effect of generation type switching observed in experiments when changing parameters of polarisation controllers. It should also be noted that the set of curves presented in Fig. 11 for dependence of gain per round-trip on the pulse power is qualitatively very well reproduced in analytical treatment of laser generation, in which every optical element of the laser corresponds to a 2×2 unitary matrix.

It is also worth noting that, besides the two above-mentioned basic regimes, intermediate (transient) modes possessing intermediate parameters of their pulse-to-pulse stability can be observed both in experiment and in numerical modelling. Such modes distinguish themselves in experiment by the height of the peak of their auto-correlation function (the better the stability, the smaller the ratio of the heights of the peak and the pedestal) and by the attainable pulse compression coefficients.

5. Prospects and limitations

The mode-locked fibre lasers with high-energy pulses considered in this Chapter are potentially very interesting for a range of practical applications such as *e.g.* super-continuum generation, material processing, and others. Notwithstanding, the practical development of these lasers may be complicated by a number of obstacles that we are going to consider in the following discussion.

As it can be seen from the results of the experimental work and numeric modelling that we have conducted, the advantages and demerits of mode-locked fibre lasers are noticeably sensitive to the cavity length. Therefore, we will consider separately short lasers (with cavity length of several meters), long-cavity fibre lasers (one kilometre and more), and lasers with 'intermediate' resonator lengths.

One of the most important problems of mode-locked fibre lasers with high-energy pulses is related to the stability of generation regimes. The stability of mode locking must be considered at different time scales. First, we consider the time interval between pulses (the round-trip time of the cavity, which is about 1 ns for short lasers) and the laboratory time scale, which corresponds to the period of the laser's continuous operation (one to several hours). The stability of laser pulses over the time scale of the inter-pulse interval is used to identify the above regimes (stable and stochastic). In each case, we will consider the stability over the longer time scale and analyse the time drift of parameters and the spontaneous breakdown of the regime.

The experiments show that relatively short lasers (with cavity lengths of several meters) exhibit stable mode locking on both time scales and provide good pulse-to-pulse stability over several hours. The tuning of the polarisation elements leads to the stochastic regime of lasing, which surprisingly is even more stable than the single-pulse mode locking over long time intervals in spite of the pulse-to-pulse fluctuations. This feature means that the parameters of laser pulses exhibit fluctuations at the pulse repetition rate whereas there is no drift of the mean parameters of such fluctuations in the laser. Evidently, the stability is an important advantage of short lasers with regard to practical applications. The disadvantage of short cavities is relatively low pulse energy, which is insufficient for some applications.

An obvious method to increase the pulse energy involves extra-cavity amplification. However, such an approach necessitates significant complications of the laser setup related to the installation of an additional amplification stage with its optical pumping unit. Note also that a pulse stretcher must be used to avoid the decay of the femtosecond and picosecond pulses during amplification. Such comparative complication of the optical trains for the generation of high-energy pulses justifies the interest in the intra-cavity methods for increasing the pulse energy without any additional amplification stages. A method based on an increase in the cavity length was successfully employed in (Kobtsev *et al*, 2008a,c; Kong *et al*, 2010; Nyushkov *et al*, 2010; Kobtsev *et al*, 2010a, b, c). However, the experiments show that the stability of ultra-long lasers with nonlinear polarisation evolution (NPE) mode locking is lower compared to the stability of short lasers on the time scale of the round trip of the cavity and on longer time scales (about one hour). As a matter of fact, we obtain bell-shaped laser spectra using long cavities (see, for example, (Kobtsev *et al*, 2008c)), which indicate the stochastic regime of wave packet generation. In contrast to the operation of short lasers, tuning polarisation elements does not allow the transition of ultra-long lasers into the regime of stable lasing, which is characterised by spectra with steep edges. The experimental data are in agreement with the results of the numerical simulation performed using the method from (Kobtsev *et al*, 2010a, Kobtsev *et al*, 2009). In the experiments with cavity lengths greater than 100 m, we failed to observe stable single-pulse lasing, which is characterised by a bell-shaped ACF and a spectrum with steep edges.

Another aspect of the ultra-long laser instability is the spontaneous lasing suppression (stability over long time scales). In comparison to short lasers, ultra-long lasers exhibit more complicated tuning of the polarisation elements required for the realisation of the regime. A typical scenario is a subsequent spontaneous suppression of lasing over a relatively short time interval (no greater than a few hours). The evident physical effects that lead to the above instability are temperature and the polarisation property drifts typical of long fibres along with inelastic deformations of amorphous optical fibre in polarisation controllers.

Thus, both short- (with lengths of several meters) and long-cavity (with lengths greater than 1 km) mode-locked NPE lasers suffer from significant performance disadvantages from the viewpoint of practical applications: short lasers generate only low-energy pulses and the too long-cavity mode-locking lasers fail to provide stable enough single-pulse lasing. This prompted us to look for a possibility of some optimum that could exist in mode-locked lasers featuring intermediate cavity lengths ranging from several dozens to several hundreds of meters.

The experiments demonstrate gradual rather than steep variations in the parameters of lasers at intermediate cavity lengths. In particular, the pulse mean energy increases almost linearly with the cavity length for both the stable single-pulse lasing and the stochastic regime (Kobtsev & Smirnov, 2011). For almost all of the fixed cavity lengths, the mean pulse energy in the stable regime is less than the mean pulse energy in the stochastic regime in qualitative agreement between the experimental data and modelling.

Another parameter that determines the working characteristics of a passively mode-locked NPE laser is the probability of mode-lock regime triggering at random parameters of the polarisation controllers. Figure 12 shows the results of simulation and demonstrates the number of sets of the polarisation parameters that correspond to the stable single-pulse lasing versus the cavity length. The simulation shows that the probability of activation of stable single-pulse mode-locked lasing monotonically decreases with increasing cavity length. For the cavity length of 100 m, only three points that correspond to the stable pulsed lasing are found in the space of the polarisation element parameters. Numerical simulation shows no stable single-pulse regimes for cavity lengths of greater than 100 m. Note that a decrease in the probability of the mode-locking in numerical simulation agrees with a decrease in the stability of the corresponding regime on long time scales in experiment. In fact, a relatively low probability of the stable regime corresponds to a relatively small region in the parameter space associated with a stable lasing. Even minor temperature fluctuations or fluctuations of the parameters of polarisation elements due to fibre deformation are sufficient to break down the corresponding lasing mode in agreement with the experimental data.

Restrictions on the laser pulse length, apart from the generation (in)stability, can limit specific applications of mode-locked fibre lasers with high-energy pulses. Indeed, for generation of pulses with relatively high energy, lengthening of the resonator is necessary, which, however, leads to increase in the width of generated pulses. In ultra-long lasers with the resonator length about 10 km and more, the duration of the output pulses may be as long as several or even dozens of nanoseconds, thus, being comparable with the pulse duration of Q-switched lasers. This may substantially limit the range of possible applications.

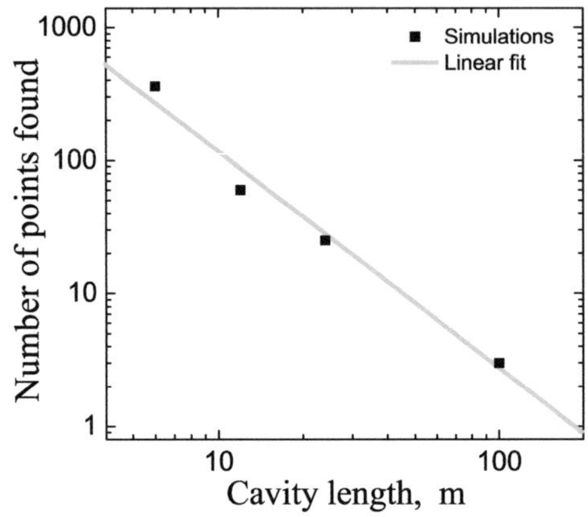

Fig. 12. Plot of the number of points in the parameter space of the polarisation elements that correspond to stable single-pulse lasing in the numerical simulation *vs* the cavity length.

An obvious approach to this problem could be extra-cavity compression of long pulses. In fact, since the resonators of the discussed lasers feature, as a rule, fully normal dispersion, their output pulses may exhibit a considerable amount of chirp, which can be compensated with the help of a compressor. In reality however, such a solution runs into two fundamental difficulties. The first of them comes from the fact that in ultra-long lasers, the pulse chirp may be so high that its compensation becomes technically very difficult or even impossible. The first reason is the absence of standard low-cost fibre-based compressors with anomalous dispersion near 1080 nm, and the second is a more fundamental problem of pulse decomposition in such a fibre due to modulation instability.

Another problem related to pulse compression is the absence of linear chirp in a number of generation modes. For instance, although (Kelleher *et al*, 2009) reports experimentally observed 1.7-ns-long pulses with giant chirp generated in 1.2-km-long fibre laser mode-locked due to nanotube-based saturable absorber, compression of pulses generated in NPE lasers is far from being possible in all the generation regimes. Our experiments and numerical modelling indicate that efficient pulse compression is only possible in a stable generation mode, whereas pulses observed in quasi-stochastic modes of wave packet generation are not easily compressible. In the intermediate (transient) regimes, it may be possible to compress the output pulses by a factor of several units. The shapes of the pulse spectra and of their auto-correlation functions can be used to assess the compressibility of pulses in one or another mode of generation. Efficient compression is only possible, as a rule, in generation regimes characterised by sharp spectrum edges and a bell-shaped auto-correlation function without a narrow peak in the centre.

Thus, the results discussed above indicate a gradual decrease in the stability of lasing associated with a corresponding increase in the cavity length. Stable single-pulse mode-locked lasing can hardly be achieved in an NPE laser with a cavity length exceeding 100 m. However, the experiments reported by (Kelleher *et al*, 2009) show single-pulse lasing in a cavity longer than 1 km when a saturable absorber is used for mode locking. But even in this case, the question of compressibility of pulses characterised by a giant chirp remains open.

6. Conclusions

We have overviewed recent results in a fast developing research area of long-cavity pulsed fibre lasers. Recent research has clearly demonstrated that cavity elongation of mode-locked fibre lasers leads to higher pulse energy directly at the output of the oscillator – up to the level of several μJ. However, the duration of such pulses lies in the nanosecond range and the temporal compressibility of such pulses is yet to be proven. The crux of the problem lies in the fact that, typically, output radiation generated in long resonators has a double-scale form of pulse trains containing sets of sub-pulses rather than that of isolated single pulses. It is this particular feature that distinguishes the output of long mode-locked fibre lasers from that of Q-switched lasers at comparable energies and pulse repetition rates. High-energy nanosecond optical pulse trains filled with stochastic sequences of femtosecond pulses are a hallmark of mode-locked fibre-lasers with ultra-long cavities. We would like to stress that such a very specific temporal characteristic of laser radiation can be used in a variety of applications, including those of Raman amplifier pumping, generation of high-energy super-continuum, material processing, and that this type of output even has advantages over conventional single-pulse radiation in some applications that require temporal modulation of radiation.

7. Acknowledgments

This work was partially supported by the Council of the President of the Russian Federation for the Leading Research Groups of Russia, project No. NSh-4339.2010.2; by the Grant "Scientific and educational staff of innovational Russia in 2009-2013" (Grant No П2490); by the Leverhulme Trust and the Marie Curie FP7 Program IRSES.

8. References

Ai, F. *et al* (2011). Passively mode-locked fibre laser with kilohertz magnitude repetition rate and tunable pulse width. *Optics & Laser Technology*, Vol. 43, No. 3 (April 2011), pp. 501-505, ISSN 0030-3992.

Akhmediev, N., Soto-Crespo, J.M. & Town, G. (2001). Pulsating solitons, chaotic solitons, period doubling, and pulse coexistence in mode-locked lasers: Complex Ginzburg-Landau equation approach. *Physical Review E*, Vol. 63, No. 5 (April 2001), pp. 056602, ISSN 1539-3755 (print), 1550-2376 (online).

Akhmediev, N. *et al* (2008). Roadmap to ultra-short record high-energy pulses out of laser oscillators. *Physics Letters A*, Vol. 372, No. 17 (April 2008), pp. 3124–3128 (2008), ISSN 0375-9601.

Ania-Castañón, J. D. *et al* (2006), . Ultralong Raman fibre lasers as virtually lossless optical media. Physical Review Letters, Vol. 96, No. 2 (January 2006), pp. 023902-023905, ISSN 0031-9007 (print), 1079-7114 (online).

Ania-Castañón, J.D. (2009). Design and simulation of ultralong Raman laser links for optical signal transmission. Proc. SPIE 7386, 73862J (2009).

Dudley, J.M., Genty, G. & Eggleton, B.J. (2008). Harnessing and control of optical rogue waves in supercontinuum generation. *Optics Express*, Vol. 16, No. 6 (March 2008), pp. 3644-3651, ISSN 1094-4087.

Chang, W. *et al* (2008). Dissipative soliton resonances in laser models with parameter management. *Journal of Optical Society of America B*, Vol. 25, No. 12 (November 2008), pp. 1972-1977, ISSN 0740-3224, eISSN 1520-8540.

Chen, C.J., Wai, P.K.A. & Menyuk, C.R. (1992). Soliton fibre ring laser. *Optics Letters*, Vol. 17, No. 6 (March 1992), pp. 417-419, ISSN 0146-9592 (print), 1539-4794 (online).

Chen, L. *et al* (2009). Ultra-low repetition rate linear-cavity erbium-doped fibre laser modelocked with semiconductor saturable absorber mirror. *Electronics letters*, Vol. 45, No. 14 (July 2009), pp. 731–733, ISSN 0013-5194.

Chernykh, A.I. & Turitsyn, S.K. (1995). Soliton and collapse regimes of pulse generation in passively mode-locking laser systems. *Optics Letters*, Vol. 20, No. 4 (February 1995), pp. 398-400, ISSN 0146-9592 (print), 1539-4794 (online).

Chong, A. *et al* (2006). All-normal-dispersion femtosecond fibre laser. *Optics Express*, Vol. 14, No. 21 (October 2006), pp. 10095-10100, ISSN 1094-4087.

Chong, A., Renninger, W.H. & Wise, F.W. (2007). All-normal-dispersion femtosecond fibre laser with pulse energy above 20 nJ. *Optics Letters*, Vol. 32, No. 16 (August 2007), pp. 2408-2410, ISSN 0146-9592 (print), 1539-4794 (online).

Chong, A., Renninger, W.H. & Wise, F.W. (2008). Propeties of normal-dispersion femtosecond fibre lasers. *Journal of Optical Society of America B*, Vol. 25, No. 2 (February 2008), pp. 140-148, ISSN 0740-3224, eISSN 1520-8540.

Fermann, M.E. *et al* (1993). Passive mode locking by using nonlinear polarisation evolution in a polarisation-maintaining erbium-doped fibre. Optics Letters, Vol. 18, No. 11 (June 1993), pp. 894-896, ISSN 0146-9592 (print), 1539-4794 (online).

Fong, K.H. *et al* (2006). Generation of low-repetition rate highenergy picosecond pulses from a single-wall carbon nanotube mode-locked fibre laser. *Optical Amplifiers and their Applications Conference (OAA 2006)*, Whistler, British Columbia, Canada, OMD4, June 2006.

Fong, K.H., Set, S.Y. & Kikuchi, K. (2007). High-energy ultrashort pulse generation from a fundamentally mode-locked fibre laser at 1,7 MHz, *Proceedings of IEEE/OSA Optical Fibre Conference (OFC) 2007*, Anaheim, 2007.

Grudinin, A.B. *et al* (2004). Multi-fibrearrangements for high power fibre lasers and amplifiers. United States Patent 6826335, 30.11.2004.

Haus, H.A. (2000). Mode-locking of lasers. *IEEE Journal of Selected Topics in Quantum Electronics*, Vol. 6, No. 6 (December 2000), pp. 1173-1185, 1077-260X.

Hofer, M. *et al* (1991). Mode-locking with cross-phase and self-phase modulation. *Optics Letters*, Vol. 16, No. 7 (April 1991), pp. 502-504, ISSN 0146-9592 (print), 1539-4794 (online).

Horowitz, M. & Silberberg, Y. (1998). Control of noiselike pulse generation in erbium-doped fibre lasers. *IEEE Photonics Technology Letters*, Vol. 10, No. 10 (October 1998), 1389-1391, ISSN 1041-1135.

Ivanenko, A. *et al* (2010). Mode-locking in 25-km fibre laser, *Proceedings of 36th European Conference on Optical Communication (ECOC)*, ISBN: 978-1-4244-8536-9, Torino, Italy, September 2010.

Jabczyński, J. K. *et al* (2006). Q-switched mode-locking with acousto-optic modulator in a diode pumped Nd:YVO_4 laser. *Optics Express*, Vol. 14, No. 6 (March 2006), pp. 2184-2190, ISSN 1094-4087.

Johnson, R.H. *et al*. Mode-locked cavity-dumped laser. United States Patent 3995231, 30.11.1976.

Kang, J.U. *et al* (1998). Demonstration of supercontinuum generation in a long-cavity fibre ring laser. *Optics Letters*, Vol. 23, No. 17, pp. 1375-1377, ISSN 0146-9592 (print), 1539-4794 (online).

Kelleher, E.J.R. *et al* (2009). Generation and direct measurement of giant chirp in a passively mode-locked laser. *Optics Letters*, Vol. 34, No. 22 (November 2009), pp. 3526-3528, ISSN 0146-9592 (print), 1539-4794 (online).

Kobtsev, S.M. & Smirnov, S.V. (2005). Modelling of high-power supercontinuum generation in highly nonlinear, dispersion shifted fibres at CW pump. *Optics Express*, Vol. 13, No. 18 (September 2005), pp.6912-6918, ISSN 1094-4087.

Kobtsev, S.M. & Smirnov, S.V. (2008b). Temporal structure of a supercontinuum generated under pulsed and CW pumping. *Laser Physics*, Vol. 18, No. 11 (November 2008), pp.1260-1263, ISSN 1054-660X (print) 1555-6611 (online).

Kobtsev, S., Kukarin, S. & Fedotov, Y. (2008a). High-energy Q-switched fibre laser based on the side-pumped active fibre. *Laser Physics*, Vol 18, No. 11 (November 2008), pp. 1230-1233, ISSN 1054-660X (print) 1555-6611 (online).

Kobtsev, S., Kukarin, S. & Fedotov, Y. (2008c) Ultra-low repetition rate mode-locked fibre laser with high-energy pulses. *Optics Express*, Vol. 16, No. 26 (December 2008), pp. 21936-21941, ISSN 1094-4087.

Kobtsev, S. *et al* (2009). Generation of double-scale femto/pico-second optical lumps in mode-locked fibre lasers. *Optics Express*, Vol. 17, No. 23 (October 2009), pp. 20707-20713, ISSN 1094-4087.

Kobtsev, S.M. *et al* (2010a). High-energy mode-locked all-fibre laser with ultralong resonator. *Laser Physics*, Vol. 20, No. 2 (February 2010), pp. 351-356, ISSN 1054-660X (print) 1555-6611 (online).

Kobtsev, S.M. & Kukarin, S.V. (2010b) All-fibre Raman supercontinuum generator. *Laser Physics*, Vol. 20, No.2 (February 2010), pp. 372-374, ISSN 1054-660X (print) 1555-6611 (online).

Kobtsev, S.M., Kukarin, S.V. & Smirnov, S.V. (2010c). All-fibre high-energy supercontinuum pulse generator. *Laser Physics*, Vol. 20, No.2 (February 2010), pp. 375-378, ISSN 1054-660X (print) 1555-6611 (online).

Kobtsev, S.M. & Smirnov, S.V. (2011). Fibre lasers mode-locked due to nonlinear polarisation evolution: golden mean of cavity length. *Laser Physics*, Vol. 21, No. 2 (February 2011), pp. 272–276, ISSN 1054-660X (print) 1555-6611 (online).

Kolev, V. Z. *et al* (2003). Passive mode-locking of a Nd:YVO4 laser with an extra-long optical resonator. *Optics Letters*, Vol. 28, No. 14, pp. 1275-1277, ISSN 0146-9592 (print), 1539-4794 (online).

Komarov, A., Leblond, H. & Sanchez, F. (2005). Quintic complex Ginzburg-Landau model for ring fibre lasers. *Physical Review E*, Vol. 72, No. 2 (August 2005), pp. 025604, ISSN 1539-3755 (print), 1550-2376 (online).

Kong, L.J., Xiao, X.S. & Yang, C.X. (2010). Low-repetition-rate all-fibre all-normal-dispersion Yb-doped mode locked fibre laser. *Laser Physics Letters*, Vol. 7, No. 5 (May 2010), pp. 359-362, ISSN 1612-2011 (print), 1612-202X (online).

Kramer, L. *et al* (1995). Optical pulse collapse in defocusing active medium. *JETP Letters*, Vol. 61, No. 11, pp. 887-892.

Lee, J.H., Han, Y.G. & Lee, S. (2006). Experimental study on seed light source coherence dependence of continuous-wave supercontinuum performance. Optics Express, Vol. 14, No. 8 (April 2006), pp. 3443-3452, ISSN 1094-4087.

Li, X. *et al* (2010). Long-cavity passively mode-locked fibre ring laser with high-energy rectangular-shape pulses in anomalous dispersion regime. *Optics Letters*, Vol. 35, No. 19 (September 2010), pp. 3249-3251, ISSN 0146-9592 (print), 1539-4794 (online).

Lin, J.H. *et al* (2008). Stable Q-switched mode locked Nd^{3+}:LuVO4 laser by Cr^{4+}:YAG crystal. *Optics Express*, Vol. 16, No. 21 (October 2008), pp. 16538-16545, ISSN 1094-4087.

Lin, J.H., Wang, D. & Lin, K.-H. (2011). High energy pulses generation with giant spectrum bandwidth and submegahertz repetition rate from a passively mode-locked Yb-doped fibre laser in all normal dispersion cavity. *Laser Physics Letters*, Vol. 8, No. 1 (January 2011), pp. 66–70, ISSN 1612-2011 (print), 1612-202X (online).

Matsas, V.J. *et al* (1992). Selfstarting passively mode locked fibrering soliton laser exploiting non linear polarisation rotation. *Electronics Letters*, Vol. 28, No. 15 (July 1992), pp. 1391–1393, ISSN 0013-5194.

Nyushkov, B.N. *et al* (2010). Generation of 1.7-uJ pulses at 1.55um by a self-modelocked all-fibre laser with a kilometeres-long linear-ring cavity. *Laser Physics Letters*, Vol. 7, No. 9 (September 2010), pp. 661-665, ISSN 1612-2011 (print), 1612-202X (online).

Saito, L., Romero, M. & Souza, E. (2010). 48.8 km Ultralong Erbium fibre laser in active mode-locking operation. *Optical Review*, Vol. 17, No. 4 (July 2010), pp. 385-387, ISSN 1340-6000 (print) 1349-9432 (online).

Senoo, Y. *et al* (2010). Ultralow-repetition-rate, high-energy, polarisation-maintaining, Er-doped, ultrashort-pulse fibre laser using single-wall-carbon-nanotube saturable absorber. *Optics Express*, Vol. 18, No. 20 (September 2010), pp. 20673-20680, ISSN 1094-4087.

Song, R. *et al* (2011). A SESAM passively mode-locked fibre laser with a long cavity including a band pass filter. Journal of Optics, Vol. 13, No. 3 (March 2011), pp. 035201.

Tian, X. *et al* (2009). High-energy laser pulse with a submegahertz repetition rate from a passively mode-locked fibre laser. *Optics Letters*, Vol. 34, No. 9 (April 2009), pp. 1432-1434, ISSN 0146-9592 (print), 1539-4794 (online).

Tian, X. *et al* (2009). High-energy wave-breaking-free pulse from allfibre mode-locked laser system. *Optics Express*, Vol. 17, No. 9 (April 2009), pp. 7222-7227, ISSN 1094-4087.

Wang, L. *et al* (2011). Observations of four types of pulses in a fibre laser with large net-normal dispersion. *Optics Express*, Vol. 19, No. 8 (April 2011), pp. 7616-7624, ISSN 1094-4087.

Wise, F.W., Chong, A., & Renninger, W.H. (2008). High-energy femtosecond fibre lasers based on pulse propagation at normal dispersion. *Laser & Photonics Reviews*, Vol. 2, No. 1-2 (April 2008), pp. 58-73.

Zhang, M. *et al* (2009). Mode-locked ytterbium-doped linear-cavity fibre laser operated at low repetition rate. *Laser Physics Letters*, Vol. 6, No. 9 (September 2009), pp. 657-660, ISSN 1612-2011 (print), 1612-202X (online).

Zhao, L.M. *et al* (2007). Noise-like pulse in a gain-guided soliton fibre laser. *Optics Express*, Vol. 15, No. 5 (March 2007), pp. 2145-2150, ISSN 1094-4087.

4

All-Poly-Crystalline Ceramics $Nd:YAG/Cr^{4+}:YAG$ Monolithic Micro-Lasers with Multiple-Beam Output

Nicolaie Pavel[1,2], Masaki Tsunekane[1] and Takunori Taira[1]
[1]*Institute for Molecular Science (IMS), Laser Research Center, 38 Nishigonaka, Myodaiji, Okazaki*
[2]*National Institute for Laser, Plasma and Radiation Physics, Solid-State Quantum Electronics Lab., Bucharest*
[1]*Japan*
[2]*Romania*

1. Introduction

Laser-induced ignition of air-fuel mixtures in internal combustion engines is a subject that has been investigated extensively during last years. In the beginning, experiments were performed with sized and robust, commercial available lasers that delivered pulses with energy in the range of tens to a few hundreds of mJ and several ns pulse duration (Ma et al., 1998; Phuoc & White, 1999; Weinrotter et al. 2005a; Weinrotter at al. 2005b). These investigations revealed that laser-induced ignition offers significant advantages over a conventional spark-ignition system, such as higher probability to ignite leaner mixtures, reduction of erosion effects, increase of engine efficiency, or shorter combustion time. Thus, developing of an engine ignited by laser could address, even partially, the increase concern of humanity for protecting global environment and preserving fossil resources.

Subsequent research (Kofler et al., 2007) concluded that a suitable laser configuration for engine ignition is a Nd:YAG laser, passively Q-switched by Cr^{4+}:YAG saturable absorber (SA). Q-switched laser pulses with energy up to 6 mJ and 1.5-ns duration were obtained from an end-pumped, 210-mm long Nd:YAG-Cr^{4+}:YAG laser. Furthermore, side-pumping technique was employed to realize a Nd:YAG laser passively Q-switched by Cr^{4+}:YAG SA with 25 mJ energy per pulse and pulse duration around 3 ns (Kroupa et al. 2009); the laser resonator was around 170 mm. However, the length of these lasers make difficult to accomplish compactness of an electrical spark plug used in the automotive industry.

In recent works our group has realized Nd:YAG-Cr^{4+}:YAG micro-lasers and demonstrated laser ignition of an automobile engine with improved performances in comparison with ignition induced by a conventional spark plug (Tsunekane at al. 2008; Tsunekane et al. 2010). The strategy was to shorten the pulse duration by decreasing the resonator length, and to maximize the laser pulse energy by optimizing the pump conditions, the Nd:YAG doping level and length, as well as Cr^{4+}:YAG initial transmission (T_0) and the output mirror transmission (T) (Sakai et al., 2008). A passively Q-switched Nd:YAG-Cr^{4+}:YAG micro-laser with 2.7-mJ energy per pulse and 600-ps pulse duration was realized. This laser, which included optics for pumping, an 11-mm long resonator, as well as optics that collimated and focused the beam to dimension required for fuel ignition, was assembled in a device that matched the dimensions of an electrical spark plug (Tsunekane at al., 2010).

Various papers have also reported that multi-point ignition increases significantly the combustion pressure and shortens the combustion time compared to single-point ignition (Weinrotter et al., 2005a; Phuoc, 2000; Morsy et al., 2001). The experiments employed combustion chambers in which two laser beams were inserted through different windows, and thus distance between the ignition points was adjusted easily. However, the use of a single laser beam that was focused in three points with a diffractive lens failed to demonstrate improved combustion (Weinrotter et al., 2005a), opposite to the two-point ignition experiments. The result was attributed to the short distance between the ignition points. Therefore, study of the influence of multi-point ignition on the performances of a real car engine would require realization of passively Q-switched Nd:YAG/Cr^{4+}:YAG lasers with multiple-beam output and with size close to that of an electrical spark plug.

In this work we report passively Q-switched Nd:YAG/Cr^{4+}:YAG micro-lasers with multiple (two and three)-beam output, each beam inducing air-breakdown in points at adjustable distance. Opposite to the previous realized lasers that used discrete Nd:YAG and Cr^{4+}:YAG single-crystals components (Koefler et al., 2007; Kroupa et al., 2009; Tsunekane et al., 2008; Tsunekane at al., 2010; Sakai et al., 2008), these lasers consist of composite, all-ceramics Nd:YAG/Cr^{4+}:YAG monolithic media that were pumped by similar, independent lines.

This work is organized as follows. Section 2 presents a continuous-wave (cw) pumped Nd:YAG laser passively Q-switched by Cr^{4+}:YAG SA with emission at 1.06 μm. Although the laser pulse energy (E_p) was low, of 270 μJ at the repetition rate of ~9 KHz, and the pulse peak power was of only 16 kW, this device was the first passively Q-switched laser realized in our laboratory. Furthermore, it was used to demonstrate the first passively Q-switched Nd:YAG-Cr^{4+}YAG laser with generation into green visible spectrum at 532 nm by intracavity frequency doubling with LiB_3O_5 (LBO) nonlinear crystal. In these experiments, both active Nd:YAG gain medium and Cr^{4+}:YAG SA were of single-crystal nature. Section 3 is dedicated to repetitively-pumped, passively Q-switched Nd:YAG-Cr^{4+}:YAG lasers with high pulse energy and few-MW level peak power. Results obtained with single-crystals, Nd:YAG and Cr^{4+}:YAG discrete elements are given in Section 3.1. A detailed investigation of laser emission obtained with ceramics Nd:YAG and Cr^{4+}:YAG was performed (Section 3.2): This was a step toward establishing ceramics materials as solution for a microchip laser used in laser ignition of an engine. Lastly, composite, all-ceramics Nd:YAG/Cr^{4+}:YAG monolithic laser with two- and three-beam output were realized. Various characteristics of these devices are given and discussed in Section 3.3. The paper conclusions are presented in Section 4. The lasers described in this work will enable studies on the performances of internal combustion engines with multi-point ignition.

2. Infrared at 1.06 μm and green at 532 nm passively Q-switched Nd:YAG lasers

Passive Q-switching technique is attractive particularly for scientific, medical, or industrial applications that do not require temporal accuracy better than microseconds range. This method yields lower output compared to electro-optic or acousto-optic Q-switched lasers, but has the advantages of a simple design, with good efficiency, reliability and compactness. The first cw, diode end-pumped Nd:YAG laser passively Q-switched by Cr^{4+}:YAG SA

crystal delivered pulses with 11-μJ energy and 337-ps duration, at 6-kHz repetition rate (Zayhowsky and Dill III, 1994). The laser included a composite structure that was made of a thin piece of Nd^{3+}:YAG single-crystal gain medium bonded to a short Cr^{4+}:YAG SA single crystal. Later, laser pulses with increased energy of 100 μJ and duration of 36 ns at 15-kHz repetition rate were obtained from a passively Q-switched Nd:YAG-Cr^{4+}:YAG laser, with medium average output power (Agnesi et al., 1997). Furthermore, laser pulses with high energy E_p= 3.4 mJ, but long duration of 99 ns, were achieved by employing side-pumping geometry of a Nd:YAG-Cr^{4+}:YAG laser (Song et al., 2000).
A sketch of the cw-pumped Nd:YAG-Cr^{4+}:YAG laser developed in our laboratory is shown in Fig. 1 (Pavel et al., 2001a). The gain medium was a composite Nd:YAG rod that was fabricated by diffusion bonding of a Nd:YAG single crystal (thickness of 5.0 mm, 1.1-at.% Nd doping) to an undoped, 1.0-mm thick YAG. The concept of combining doped and undoped components was used, in the beginning, to modify the configuration of the thermal field induced by pumping in solid-state laser rods, and successfully employed to improve the output performances of Nd:YAG (Hanson, 1995), Nd:YVO_4 (Tsunekane et al., 1997), or Yb:YAG (Bibeau et al., 1998) lasers. This method has also found applications in the passive Q-switching technique. Thus, microchip structures that consisted of undoped YAG caps, Nd:YAG and Cr^{4+}:YAG SA bonded together to form a monolithic resonator were demonstrated to produce linearly polarized, single-longitudinal mode output pulses in quasi-cw or pulsed pumping regimes (Zayhowski et al., 2000; Aniolek et al., 2000).

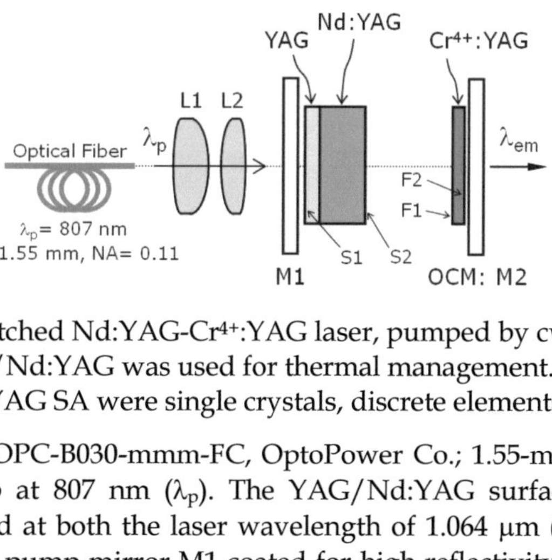

Fig. 1. A passively Q-switched Nd:YAG-Cr^{4+}:YAG laser, pumped by cw diode laser is shown. Composite YAG/Nd:YAG was used for thermal management. The gain Nd:YAG medium as well as Cr^{4+}:YAG SA were single crystals, discrete elements.

A fiber-bundled diode (OPC-B030-mmm-FC, OptoPower Co.; 1.55-mm diameter, 0.11 NA) was used for the pump at 807 nm (λ_p). The YAG/Nd:YAG surfaces (S1 and S2) were antireflection (AR) coated at both the laser wavelength of 1.064 μm (λ_{em}) and λ_p. A plane-plane resonator with the pump-mirror M1 coated for high reflectivity (HR) at λ_{em} and high transmission (HT) at λ_p, and that was placed very close to Nd:YAG, was used. A collimating lens (L1) and a focusing lens (L2) were used to image the fiber bundle end into Nd:YAG, to a diameter of 800 μm. The focusing point was 2.0 mm below surface S1 of Nd:YAG. Cr^{4+}:YAG SA (CASIX Inc., China) with initial transmission T_0 of 0.89, 0.85, and 0.80 and AR coated at λ_{em} on both surfaces (F1 and F2) were used for Q-switching. Each Cr^{4+}:YAG SA crystal was placed close to the out-coupling mirror (OCM) M2.
Figure 2 presents characteristics of Q-switched laser emission obtained from a 40-mm long resonator and an OCM with T= 0.10. A maximum average power of 3.8 W resulted for the Cr^{4+}:YAG with T_0= 0.89 (Fig. 2a) with a beam factor M^2 of 1.4. The laser ran at a frequency as

high as 24.8 kHz. The pulse energy and the pulse duration (t_p, FWHM definition) were E_p= 152 μJ and t_p= 25.4 ns (Fig. 2b), respectively; the pulse peak power was ~6.0 kW. When a SA with a lower transmission T_0 was used, pulses with higher peak power were generated with a reduced average output power. For a Cr^{4+}:YAG SA with T_0= 0.85, pulses with t_p= 18.0 ns at a repetition rate of 16.1 kHz resulted for the maximum absorbed power of 21.0 W. Energy E_p and pulse peak power was 213 μJ and 11.8 kW, respectively. For a Cr^{4+}:YAG with T_0= 0.80 an average power of 2.6 W at the absorbed pump power of 18.6 W resulted. The laser ran at 9.1 kHz repetition rate with pulses of E_p= 272 μJ energy and 16.2 kW peak power.

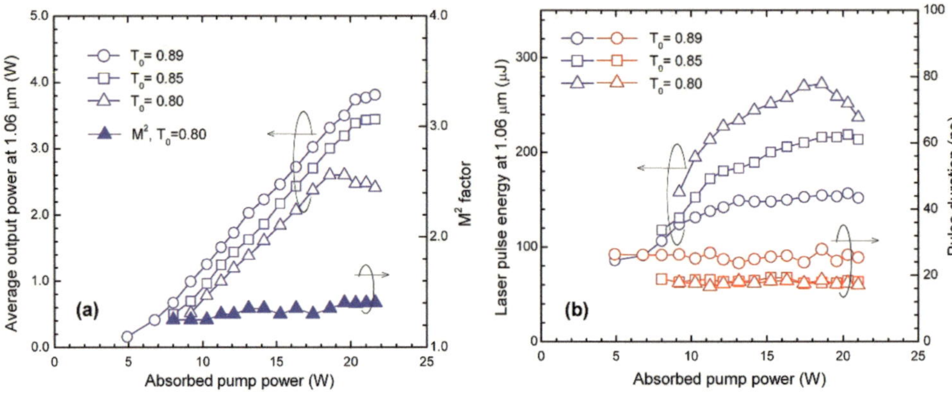

Fig. 2. Cw pumped, passively Q-switched Nd:YAG-Cr^{4+}:YAG laser: (a) Average output power and the laser-beam M^2 factor; (b) Laser pulse energy and laser pulse duration.

The Q-switched laser performances were evaluated with a rate equation model (Pavel et al., 2001a; Degnan, 1995; Zhang et al., 1997). The laser pulse energy is given by general relation:

$$E_p = \frac{h\nu \cdot A_g}{2\gamma_g \sigma_g} \cdot \ln R \cdot \ln\left(\frac{n_{gf}}{n_{gi}}\right) \tag{1}$$

where hν is the photon energy at 1.06 μm, σ_g represents Nd:YAG stimulated emission, γ_g is the inversion reduction factor, A_g is the effective area of the laser beam in Nd:YAG, and the OCM reflectivity is R= (1-T). The initial population inversion density, n_{gi} is:

$$n_{gi} = \frac{-\ln R + L - \ln T_0^2}{2\sigma_g \ell_g} \tag{2}$$

where L represents the resonator round-trip residual loss and ℓ_g is the Nd:YAG length. The final population inversion density, n_{gf} and n_{gi} are related by equation:

$$\left(1-\frac{n_{gf}}{n_{gi}}\right)+\left[1+\frac{(1-\delta)\cdot\ln T_0^2}{(-\ln R+L-\ln T_0^2)}\right]\cdot\ln\left(\frac{n_{gf}}{n_{gi}}\right)+\frac{1}{\alpha}\cdot\frac{(1-\delta)\cdot\ln T_0^2}{(-\ln R+L-\ln T_0^2)}\cdot\left[1-\left(\frac{n_{gf}}{n_{gi}}\right)^{\alpha}\right]=0 \tag{3}$$

$\delta = \sigma_{ESA}/\sigma_{SA}$, with σ_{SA} and σ_{ESA} the absorption cross section and exited-state absorption cross section of Cr^{4+}:YAG, respectively. Parameter α is $\alpha = (\gamma_{SA}\sigma_{SA})/(\gamma_g\sigma_g)\times(A_g/A_{SA})$; γ_{SA} is the inversion reduction factor for Cr^{4+}:YAG and A_{SA} is the laser beam effective area in Cr^{4+}:YAG.

Figure 3 compares experimental (symbols) and calculated (continuous lines) pulse energy for two pump rates. In simulation, spectroscopic parameters of Nd:YAG and Cr^{4+}:YAG were σ_g= 2.3×10^{-19} cm^2 , σ_{SA}= 4.3×10^{-18} cm^2, and σ_{ESA}= 8.2×10^{-19} cm^2 (Shimony, et al., 1995). The laser beam sizes in the gain crystal and in Cr^{4+}:YAG were evaluated by the PARAXIA software package (Sciopt Enterprises, San Jose, California), in which the active medium was described as a thin lens. For the absorbed power of 9.2 W (near to the threshold) the calculated active element focal length was 33.8 cm and the ratio A_g/A_{SA} amounted to 1.1, while for the absorbed pump power of 17.5 W (close to the maximum pump power) the focal length decreases to 13.2 cm and A_g/A_{SA} was 1.3. Good agreement between the experimental results and the calculated values was obtained.

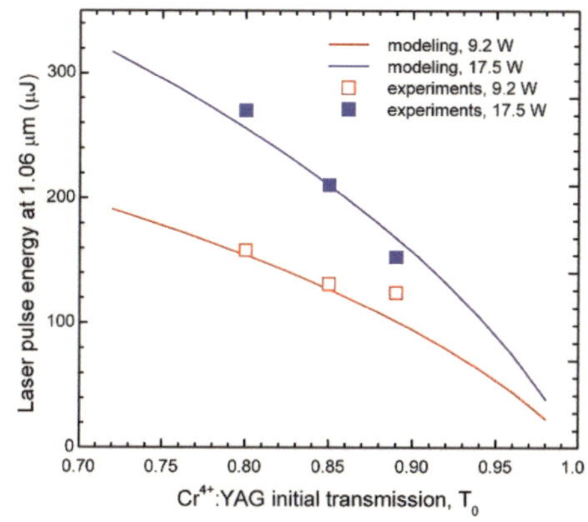

Fig. 3. Laser pulse energy versus transmission T_0 of Cr^{4+}:YAG. Symbols are experimental data (open and filled signs for absorbed pump power of 9.2 W and 17.5 W, respectively), whereas lines represent modeling.

In order to realize a passively Q-switched Nd:YAG-Cr^{4+}:YAG laser with generation into green visible spectrum at 532 nm ($\lambda_{2\omega}$), the set-up of Fig. 1 was modified such to include a nonlinear crystal. We used a V-type resonator, as shown in Fig. 4 (Pavel et al., 2001b). The YAG/Nd:YAG crystal, the Cr^{4+}:YAG SA and a glass plate (BP) positioned at Brewster angle for polarization were placed in the resonator arm (of 80-mm length) made between mirrors M1 and M2. The nonlinear crystal was a 10-mm long LBO (type I, θ= 90^0, ϕ= 11.4^0; operation at 25°C) that was placed between mirrors M2 and M3 (the arm length was 90 mm). The LBO surfaces were AR coated at both λ_{em} and $\lambda_{2\omega}$ wavelengths. The concave mirror M3 has a radius of 50 mm. This arrangement makes use of the high peak power available inside the cavity, and enables high conversion efficiency of the fundamental wavelength λ_{em}.
Figure 5 presents characteristics of the green laser pulses. For a Cr^{4+}:YAG with T_0= 0.90, the maximum average power at 532 nm was 0.95 W (Fig. 5a) at the absorbed pump power of 13.1 W; the laser beam quality was characterized by an M^2 factor of 1.8. The green pulse energy was 226 μJ (Fig. 5b), and the laser runs with a 4.2-kHz rate of repetition and pulse duration of 86 ns. A slightly higher average power of 1.0 W was obtained with the Cr^{4+}:YAG of T_0= 0.85. However, the pulse energy reduced at 131 μJ and the pulse duration increased at 96 ns. Green pulse peak power reached 2.6 kW for the Cr^{4+}:YAG with T_0= 0.90 (Fig. 5b).
The characteristics of Q-switched laser pulses at 532 nm were described with a model of rate equation for photon density inside the resonator (φ), for the inversion of population in

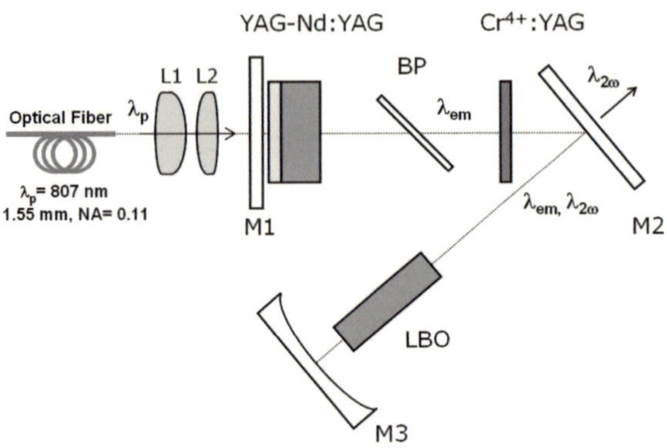

Fig. 4. A passively Q-switched Nd:YAG-Cr^{4+}:YAG laser, intra-cavity frequency doubled by LBO nonlinear crystal.

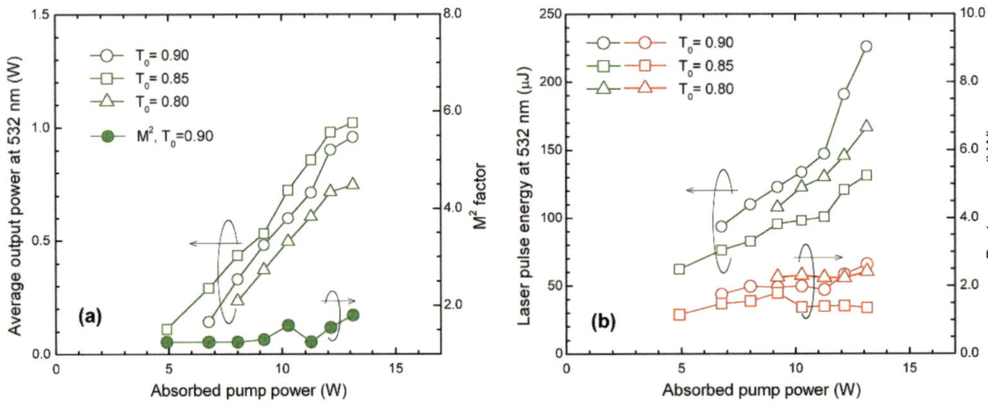

Fig. 5. Characteristics of the laser emission obtained from the Nd:YAG-Cr^{4+}:YAG-LBO laser: (a) Average output power and laser beam M^2 factor; (b) Laser pulse energy and peak power.

Nd:YAG (n_g) and for the population density in Cr^{4+}:YAG (n_{SA}), and in which out-coupling of the cavity field by frequency conversion was considered (Pavel et al., 2001b). The initial population inversion density, n_{gi} is given by relation:

$$n_{gi} = \frac{L - \ln T_0^2}{2\sigma_g \ell_g} \tag{4}$$

The final inversion density n_{gf} and n_{gi} are related by the transcendental equation:

$$\left[1+\frac{(1-\delta)\cdot\ln T_0^2}{L-\ln T_0^2}\right]\cdot\left[1-\left(\frac{n_{gf}}{n_{gi}}\right)^d\right]-\frac{d}{d-1}\cdot\frac{n_{gf}}{n_{gi}}\left[1-\left(\frac{n_{gf}}{n_{gi}}\right)^{d-1}\right]=\frac{(1-\delta)\cdot\ln T_0^2}{L-\ln T_0^2}\cdot\frac{d}{d-\alpha}\cdot\left(\frac{n_{gf}}{n_{gi}}\right)^{\alpha}\cdot\left[1-\left(\frac{n_{gf}}{n_{gi}}\right)^{d-\alpha}\right] \tag{5}$$

where $d= k/(c\gamma_g\sigma_g)$, and the coefficient k is given by the second-harmonic generation theory (Eimerl, 1987; Honea et al., 1998). The green pulse energy is $E_{2\omega} = h\nu_{2\omega} A\ell_c k \int \varphi^2(t)dt$ with $A_{2\omega}$ the green beam effective area, $h\nu_{2\omega}$ the photon energy at $\lambda_{2\omega}$, and ℓ_c the optical length of the resonator. Finally, the analytical expression deduced for the pulse energy $E_{2\omega}$ was:

$$E_{2\omega} = \frac{h\nu_{2\omega}}{2\gamma_g \sigma_g} \cdot A_{2\omega} \cdot (L - \ln T_0^2) \cdot \left\{ \left[1 + \frac{(1-\delta)\ln T_0^2}{(L - \ln T_0^2)} \right] \times \ln \frac{n_{gf}}{n_{gi}} + \frac{d}{d-1}\left(1 - \frac{n_{gf}}{n_{gi}}\right) \right.$$
$$\left. + \frac{d}{\alpha(d-\alpha)} \cdot \frac{(1-\delta)\ln T_0^2}{(L-\ln T_0^2)} \cdot \left[1 - \left(\frac{n_{gf}}{n_{gi}}\right)^{\alpha}\right] - \left[\frac{1}{d(d-1)} + \frac{\alpha}{d(d-\alpha)} \cdot \frac{(1-\delta)\ln T_0^2}{(L-\ln T_0^2)}\right] \times \left[1 - \left(\frac{n_{gf}}{n_{gi}}\right)^{d}\right] \right\} \quad (6)$$

Figure 6 shows modeling of the green pulse energy, for the Cr^{4+}:YAG with T_0= 0.90, versus the absorbed pump power, at various values of the losses L. Agreement with experimental results is good, especially if uncertainties in evaluation of L, or of laser beam variation inside the optical resonator are considered.

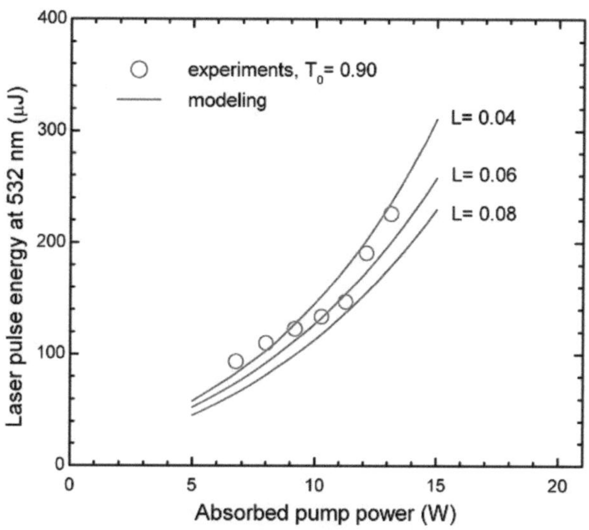

Fig. 6. The green pulse energy versus absorbed pump power for the Cr^{4+}:YAG SA with T_0= 0.90. Signs represent experiments and modeling is given by the continuous lines.

This was the first passively Q-switched Nd:YAG-Cr^{4+}:YAG laser intra-cavity frequency doubled with LBO nonlinear crystal. The laser performances (green pulse of 226 μJ energy and 2.6 kW peak power, with ~1 W average power) were much higher than previously developed systems. For example, green laser pulses with 2.5-μJ energy (190-mW average power) were obtained from a Nd:LSB gain medium passively Q-switched by Cr^{4+}:YAG and intra-cavity frequency doubled by $KTiOPO_4$ (KTP) in a linear resonator (Ostroumov et al., 1997). Furthermore, a Nd:YAG laser that was passively Q-switched by GaAs semiconductor and intra-cavity frequency doubled by KTP, in a V-type laser resonator, yielded green laser pulses with 20.5-μJ energy and ~250-mW average power (Kajava and Gaeta, 1997). Later, a Nd:$GdVO_4$-Cr^{4+}:YAG-KTP laser with 21-μJ energy per pulse (average power of ~400 mW) was realized (Liu et al., 2004). More recently, passively Q-switched Nd:YAG (An et al., 2006) or Nd:$LuVO_4$ lasers (Cheng et al., 2011) intra-cavity frequency doubled with KTP were reported. Novelty of these last two devices is the use of two SA crystals, Cr^{4+}:YAG and GaAs, for the purpose of obtaining shorter and more symmetrical pulses, in comparison with those delivered by a single SA crystal.

3. High-peak power passively Q-switched Nd:YAG/Cr^{4+}:YAG lasers

3.1 Nd:YAG-Cr^{4+}:YAG micro-lasers based on single-crystal components

The cw pumped, passively Q-switched lasers have large pulse-to-pulse energy fluctuations and large timing jitters (Huang et al., 1999; Tang et al., 2003) due to thermal and mechanical instabilities. The purpose of our next research was to realize a Nd:YAG that is passively Q-switched by Cr^{4+}:YAG SA and that can be used for ignition of an automobile engine. The operation frequency of igniters in internal combustion engines is less than 60 Hz, corresponding to an engine speed of 7200 rpm; the duty cycle is less than 5% for automobiles. In such a low frequency range, passively Q-switched lasers that are quasi-cw pumped with a low duty cycle are expected to operate stably due to initialization of the thermal and mechanical conditions during pulses.

Figure 7 is a drawing of a passively Q-switched laser module developed in our laboratory for preliminary experiments (Tsunekane et al., 2008). The active medium was a 1.1-at.% Nd:YAG single crystal (Metal Mining Co., Ltd., Japan) with a length of 4 mm. AR (R<0.2%) and HR (R>99.8%) coatings at λ_p and λ_{em}, respectively, were deposited on the pumped surface S1 of Nd:YAG. HR (R>90%) and AR (R<0.2%) coatings at λ_p and λ_{em}, respectively, were deposited on the intra-cavity surface S2 of Nd:YAG. AR coatings at λ_{em} were deposited on both surfaces of a Cr^{4+}:YAG SA (4-mm thick single crystal; Scientific Materials Corp., USA). The output coupler was flat with transmission T= 0.50 at λ_{em}. Cavity length was 10 mm. The Nd:YAG was end pumped by a fiber coupled, conductive cooled, 120-W peak power laser diode (JOLD-120-QPXF-2P, Jenoptik, Germany) with emission at λ_p= 807 nm; fiber core diameter was 600 μm and numerical aperture NA was 0.22. The fiber end was imaged into Nd:YAG to a spot size of 1.1-mm diameter. Pump energy was controlled by changing the pump pulse duration, whereas the peak pump power was maintained constant at 120 W. The maximum pump duration was 500 μs and the repetition rate was 10 Hz.

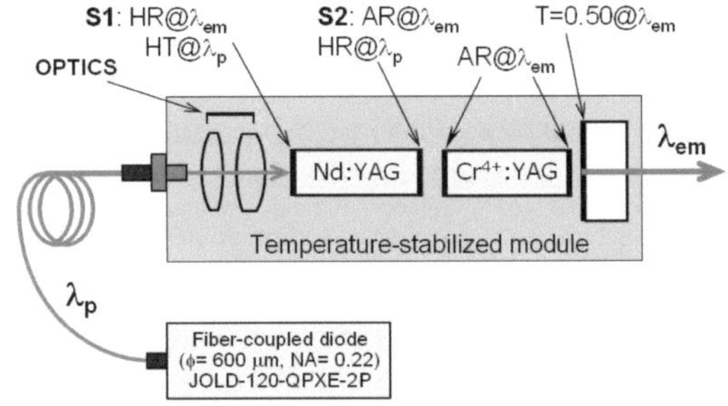

Fig. 7. Schematic drawing of the passively Q-switched Nd:YAG-Cr^{4+}:YAG laser that was build of discrete, Nd:YAG and Cr^{4+}:YAG single crystals.

Figure 8a shows energy of the laser pulse delivered by the Nd:YAG-Cr^{4+}:YAG laser versus initial transmission T_0 of a Cr^{4+}:YAG SA. E_p increases from 0.45 mJ for a Cr^{4+}:YAG with T_0= 0.80 to 4.3 mJ for a Cr^{4+}:YAG with T_0= 0.15. Corresponding pump pulse energy (E_{pump}) was 53 and 5.2 mJ, respectively. The pulse duration was measured with a 10 GHz, InGaAs detector (ET-3500, Electro-Optics Technology, Inc.) and with a 12 GHz oscilloscope (DSO81204B, Agilent Technology). The shortest pulse width of 300 ps was obtained with a Cr^{4+}:YAG of T_0= 0.15 (Fig. 8a). Pulse peak power was 0.16 MW for a Cr^{4+}:YAG with T_0= 0.80 and a record of 14.5 MW for a Cr^{4+}:YAG with T_0= 0.15 (as shown in Fig. 8b).

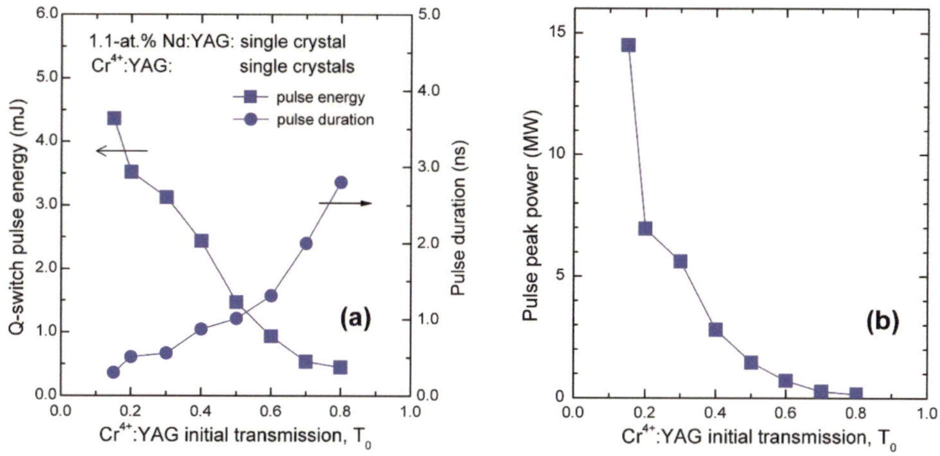

Fig. 8. Characteristics of Q-switched pulses yielded by the Nd:YAG-Cr^{4+}:YAG laser shown in Fig. 7, versus initial transmission T_0 of Cr^{4+}:YAG: a) Energy and duration; b) Peak power.

From the experimental observations, stable breakdown in air was observed for laser pulse energy E_p larger than 1.5 mJ and pulse duration t_p below 1 ns using an aspheric focus lens of 10-mm focal length. Based on these results, a Cr^{4+}:YAG SA single crystal with initial transmission T_0= 0.30 was selected for the laser igniter. The first prototype micro-laser module that was built in our laboratory and that has the same dimensions as a spark plug is shown in Fig. 9. The device includes not only the pumping optics from fiber to the Nd:YAG gain material, but also a beam expanding and focusing optics for ignition. The laser igniter has the same optical design and similar performances as the experimental module shown in Fig. 7, and it is physically possible to ignite a real engine by installing it instead of an electrical spark plug to a plug hole (Tsunekane et al., 2010). For real operation on an engine, however, the mechanical design inside the module has be improved in order to sustain the high temperatures (up to 150°C) and vibrations of a real engine.

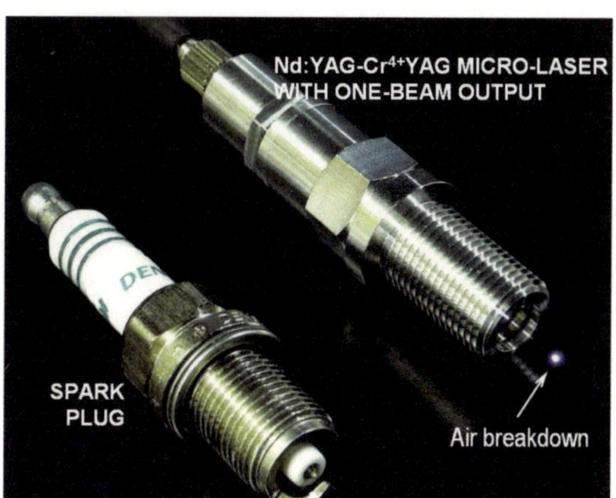

Fig. 9. Nd:YAG-Cr^{4+}:YAG laser with one-beam output. Nd:YAG gain medium and Cr^{4+}:YAG SA were single crystals; all optical components (including the output mirror) were discrete elements. Air breakdown is shown and size comparison is made with a spark plug.

3.2 Ceramics versus single-crystals Nd:YAG-Cr^{4+}:YAG micro-lasers

The Nd:YAG as well as the Cr^{4+}:YAG SA media used in the previous reports were single crystals. The advancement in ceramic techniques has reached a maturity stage, especially in obtaining poly-crystalline cubic laser media of very good optical quality. It is recognized that laser ceramics has become a serious challenge to crystalline optics, especially due to an easier manufacturability and a lower price. The use of poly-crystalline ceramics could decrease the price of the Nd:YAG-Cr^{4+}:YAG laser, which is a critical condition for realizing and using of a laser spark plug for engine ignition. We have therefore conducted an investigation of laser output characteristics obtained from a passively Q-switched Nd:YAG-Cr^{4+}:YAG laser that employs single crystals and poli-crystalline ceramics as Nd:YAG active media as well as Cr^{4+}:YAG SA elements.

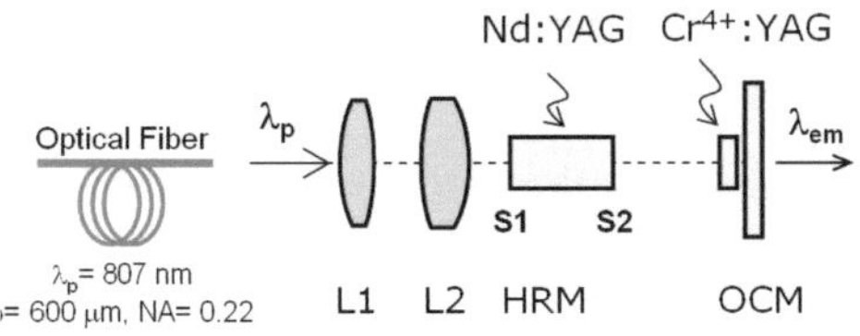

Fig. 10. A sketch of the experimental set-up used for comparative investigation of laser emission with Nd:YAG and Cr^{4+}:YAG single crystals and poly-crystalline ceramics.

A sketch of the experimental set-up is shown in Fig. 10. The laser media were Nd:YAG single crystals with doping level of 1.0-at.% Nd (sample A; Japan) and 2.0-at.% Nd (sample B; Germany), and poly-crystalline Nd:YAG ceramics with 1.1-at.% Nd (sample A^*; Baikowski Japan Co., Ltd.) and 2.0-at.% Nd (sample B^*; Baikowski Japan Co., Ltd.) doping level. The thickness of sample B^* was 3 mm, whereas the other Nd:YAG media had 4 mm in thickness. Side S1 of each Nd:YAG was coated as HR (R> 99.9%) at λ_{em}, and as HT (T> 97%) at λ_p. The other side (S2) was AR coated (T> 99.9%) at λ_{em}, and as HR (R> 95%) at λ_p. The Cr^{4+}:YAG SA had initial transmission, T_0 between 0.80 and 0.20, and were single crystals provided by two different venders (SA1 and SA2, China), as well as poly-crystalline ceramics (SA3; Baikowski Japan Co., Ltd.). Both sides of a Cr^{4+}:YAG SA were AR coated at λ_{em}. Q-switched emission was reported previously in all-ceramics Nd:YAG-Cr^{4+}:YAG (Feng et al., 2004) or Yb:YAG-Cr^{4+}:YAG (Dong et al., 2006; Dong et al, 2007) compact lasers, the pumping being made with diode lasers of low power (few watts) in cw mode. In our experiments, the optical pumping was made with a fiber-coupled diode laser (JOLD-120-QPXF-2P, Jenoptik, Germany) in quasi-cw regime. The pump repetition rate was 10 Hz and the pump pulse duration was fixed at 250 μs; pump energy was controlled by changing the diode current. An optical system made of two L1 and L2 lenses was used to image the fiber end (600-μm in diameter, NA= 0.22) into Nd:YAG to a spot size of 1.1 mm in diameter. A linear resonator made between side S1 of Nd:YAG and a plane OCM was employed.

Figure 11 presents output performances measured in free-generation regime, using a 35-mm long resonator equipped with an OCM of transmission T= 0.20. The slope efficiency, η_s was in the range of 0.65 to 0.61, the highest value being recorded with the 1.0-at.% Nd:YAG single crystal (sample A), as shown in Fig. 11a. The overall optical-to-optical efficiency, η_0 for the maximum pump level of 36.1 mJ per pulse is given in Fig. 11b. Laser pulses with 22.7 mJ energy (optical to optical efficiency η_0 of ~0.63) were measured from the 1.0-at.% Nd:YAG (sample A). Efficiencies η_s and η_0 recorded with the highly-doped Nd:YAG were a little below those measured with Nd:YAG (single crystals or poly-crystalline ceramics) of low concentrations. On the other hand, each Nd:YAG poly-crystalline ceramics showed a slight decrease of these efficiencies, compared with its counterpart Nd:YAG single crystal.

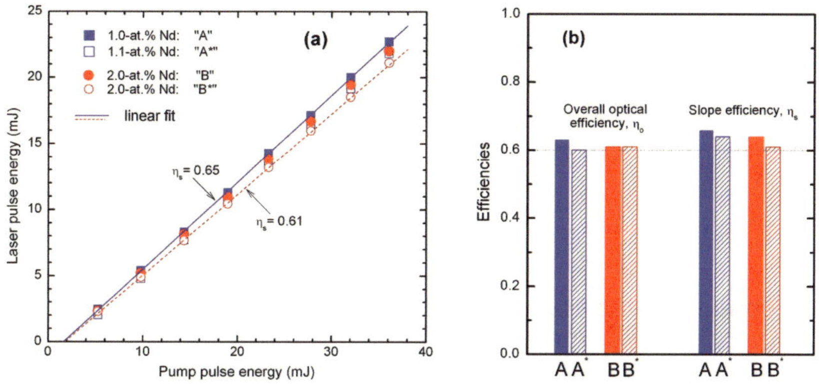

Fig. 11. Characteristics of laser emission in free-generation regime (i.e. without Cr^{4+}:YAG): (a) Laser pulse energy versus pump pulse energy; (b) Overall optical efficiency (η_0) and slope efficiency (η_s) for the available pump pulse energy of 36.1 mJ.

Figure 12 presents characteristics of Q-switched laser pulses obtained with a Cr^{4+}:YAG SA single crystal (SA1) of initial transmission T_0= 0.40, and various OCM transmission T. Laser pulses with energy E_p~1.7 mJ (Fig. 12a) and duration t_p~1.5 ns were yielded by the 1.0-at.% Nd:YAG single crystal (OCM with T= 0.70). The highly-doped 2.0-at.% Nd:YAG single crystal yielded laser pulses with energy E_p= 1.22 mJ and duration t_p= 1.45 ns. The corresponding pulse peak power was 1.1 MW for the 1.0-at.% Nd:YAG and 0.84 MW for the 2.0-at.% Nd:YAG single crystal (Fig. 12c). The Cr^{4+}:YAG single crystal was then replaced with a Cr^{4+}:YAG (SA2) ceramics of the same initial transmission T_0= 0.40. The Q-switched laser pulse energy was E_p= 1.0 mJ for the 1.1-at.% Nd:YAG ceramics (sample A*), and E_p ~1.1 mJ for the 2.0-at.% Nd:YAG ceramics (sample B*) (Fig. 12b). Corresponding pulse peak power was 0.53 and 0.73 MW, respectively. Generally, pulse energy E_p was lower than that measured with the Cr^{4+}:YAG single crystal (SA1), whereas the pulse duration was longer.

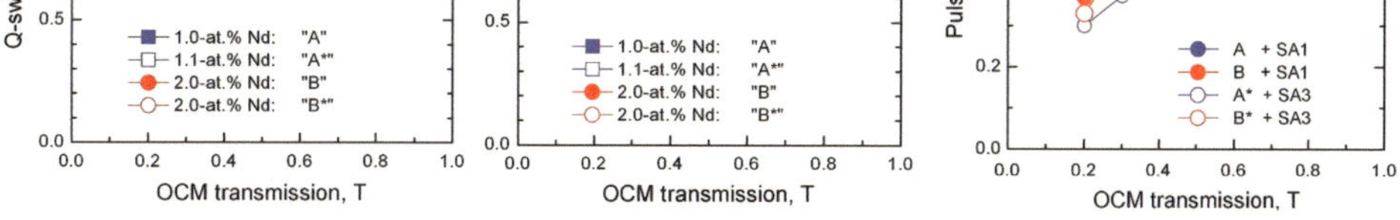

Fig. 12. Q-switched pulse energy versus OCM transmission obtained with Nd:YAG gain media and a Cr^{4+}:YAG SA of T_0= 0.40: (a) The single-crystal Cr^{4+}:YAG (SA1); (b) The poly-crystalline Cr^{4+}:YAG (SA3) ceramics; (c) Laser pulse peak power is shown.

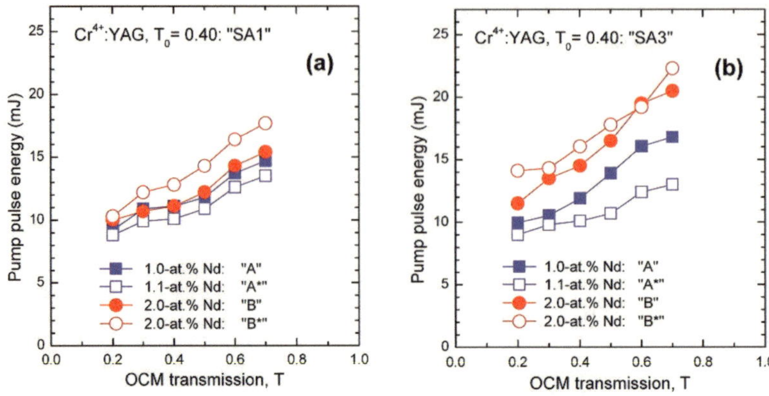

Fig. 13. Pump pulse energy necessary for laser emission with: (a) The single-crystal Cr^{4+}:YAG (SA1); (b) The poly-crystalline Cr^{4+}:YAG (SA3) ceramics.

The pump pulse energy was E_{pump}= 14.7 mJ for 1.0-at.% Nd:YAG (sample A) (Fig. 13a), lower than 15.4 mJ energy of the pump pulse required for Q-switched emission of the 2.0-at.% Nd:YAG sample B (OCM with T= 0.70). On the other hand, E_{pump} necessary for the 1.1-at.% Nd:YAG ceramics (sample A*) was only 13 mJ (Fig. 13b), while E_{pump} required for laser operation of the 2.0-at.% Nd:YAG ceramics was the highest of 22.3 mJ. Generally, higher pump pulse energy was necessary for laser operation of a Nd:YAG gain medium that was Q-switched by Cr^{4+}:YAG ceramics (Fig. 13b), compared with emission of the same laser medium that was Q-switched with a Cr^{4+}:YAG single crystal (Fig. 13a).

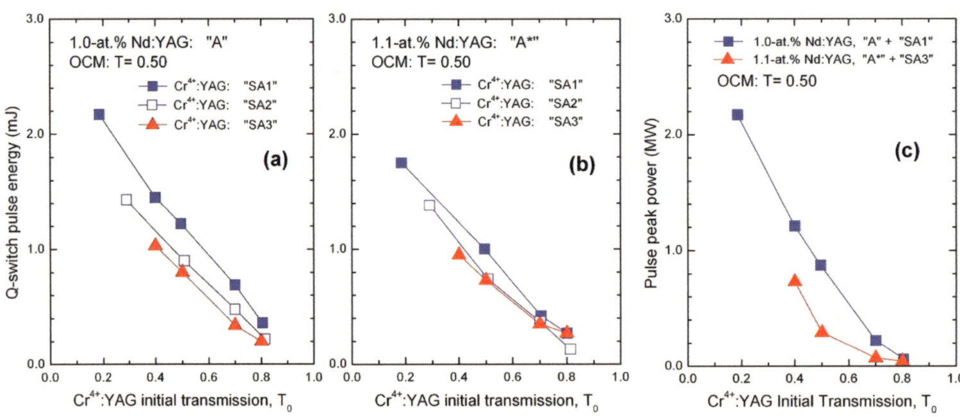

Fig. 14. The influence of Cr^{4+}:YAG initial transmission T_0 on Q-switched laser pulse energy obtained from: (a) The 1.0-at.% Nd:YAG single crystal; (b) The 1.1-at.% Nd:YAG ceramics. (c) Laser pulse peak power is shown.

Figure 14 presents performances of the Q-switched laser pulses obtained with the 1.0-at.% Nd:YAG single crystal (Fig. 14a), the 1.1-at.% Nd:YAG ceramics (Fig. 14b), and using the available Cr^{4+}:YAG SA. The OCM has transmission T= 0.50 and the resonator length was fixed at 35 mm. Differences between pulses obtained with the SA1 and SA2 Cr^{4+}:YAG single crystals were observed at the same initial transmission T_0. Most probably, the final transmissions of the Cr^{4+}:YAG single crystals were a little different, depending of the growth process used by the companies that delivered the SA. Therefore, at this stage of the experiments (Pavel et al., 2010a) and with the available Nd:YAG gain media and Cr^{4+}:YAG SA components, the best laser performances were obtained with single-crystals elements. The laser pulse peak power (shown in Fig. 14c) was also better when we used Nd:YAG single crystal with Cr^{4+}:YAG single crystal (compared with its counterparts ceramics).

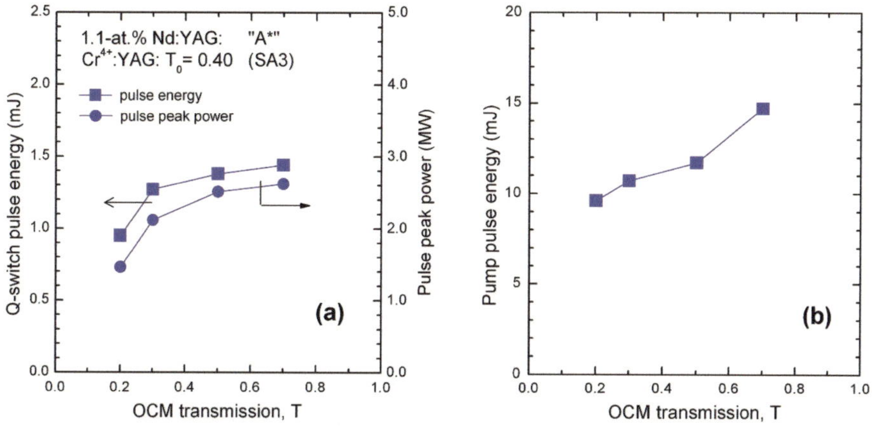

Fig. 15. (a) Q-switched laser pulse energy and peak power versus OCM transmission for the 1.1-at.% Nd:YAG ceramics and the Cr^{4+}:YAG ceramics with T_0= 0.40. The resonator length was 11 mm. (b) Energy of the pump pulse is shown.

In the final experiment the resonator length was reduced to 11 mm. Figure 15 summarizes results obtained with a combination of all-poly-crystalline ceramics, 1.1-at.% Nd:YAG gain medium and Cr^{4+}:YAG SA with T_0= 0.40. Laser pulse energy was around 1.4 mJ when OCM's transmission was higher than T= 0.50, whereas the pulse duration was t_p~ 550 ps. Therefore, corresponding pulse peak power overcomes 2.5 MW (Fig. 15a). Air breakdown was realized with a focusing lens of 11-mm focal length. The pump pulse energy varied between 9.6 mJ when OCM transmission was T= 0.20 (low pulse energy E_p= 0.95 mJ, and t_p= 650 ps) and E_{pump}= 14.7 mJ when OCM transmission was increased at T= 0.70 (Fig. 15b).
It is known that the intensity required for optical breakdown depends on pulse duration. There are not many reports on this subject: According to (Paschotta, 2008), an optical intensity of ~2×10^{13} W/cm^2 is required for air breakdown with laser pulses of 1-ps duration. Therefore, experiments were performed in order to evaluate the optical intensity of ns-duration laser pulses that realizes air breakdown. A Nd:YAG-Cr^{4+}:YAG laser, as shown in Fig. 10, was used. The laser pulse duration was varied by changing the OCM transmission T, and with the help of two Cr^{4+}:YAG that had initial transmission T_0 of 0.39 and 0.29. The resonator length was fixed at 15 mm: The laser beam M^2 factor was ~1.50, as determined by knife-edge method. The air breakdown was observed after a convergent lens of 7.5 mm focal length. A half waveplate and a polarizer were placed after the laser, in order to vary the intensity of the pulse incident on the focusing lens.
Figure 16 presents optical intensity that induced air breakdown (in laboratory conditions). In experiments, the pulse duration t_p could be varied between 0.4 and 1.1 ns; corresponding laser pulse optical intensities that induced air breakdown were ~0.65×10^{13} W/cm^2 and ~0.40×10^{13} W/cm^2, respectively. We therefore concluded that optical intensity of a laser pulse with 1-ns duration that induced air breakdown is ~0.5×10^{13} W/cm^2.

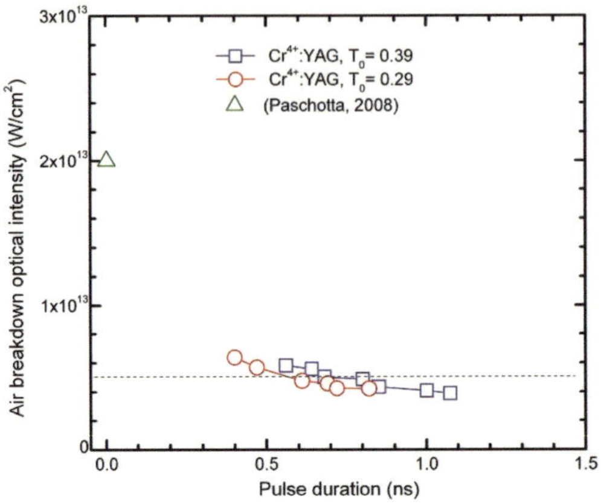

Fig. 16. Optical intensity of a laser pulse with ns duration that induced air breakdown.

3.3 Composite all-ceramics Nd:YAG/Cr^{4+}:YAG monolithic micro-lasers with two and three-beam output for engine ignition

Figure 17 is a sketch of the set-up used to demonstrate, in preliminary experiments, a passively Q-switched Nd:YAG-Cr^{4+}:YAG laser with two-beam output. Generally, one could choose to use one line for pumping, and to divide the high-energy output laser beam into two (ore more) fascicles, which has to be directed at necessary angle and then focused. This solution increases probability of damaging the laser media (due to the high intensity of the laser beam, or due to thermal effects), and could complicate the guiding line. Therefore, our choice was to employ similar, independent, multi-pumping lines, and then to change the optical path of a laser beam before focusing it.

The pump was made at 807 nm (λ_p) with two fiber-coupled (600-µm diameter and numerical aperture NA=0.22) diode lasers (JOLD-120-QPXF-2P, Jenoptik, Germany). Pump repetition rate and pump pulse duration were 5 Hz and 250 µs, respectively. The fiber end was imaged into Nd:YAG to a spot size of 1.1-mm diameter. The two-pump beams were inserted into Nd:YAG with a metal-coated prism. Furthermore, distance between the pumping positions on Nd:YAG input surface was changed by forward and backward translation of this prism.

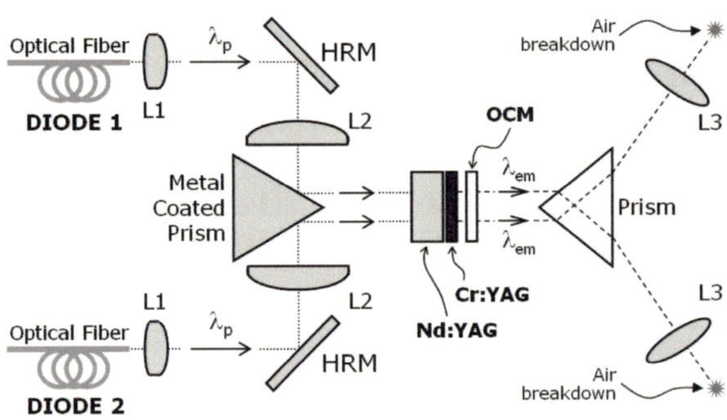

Fig. 17. Schematic of the experimental set-up used for "on table" demonstration of a passively Q-switched Nd:YAG-Cr^{4+}:YAG all-ceramics laser with two-beam output.

The laser medium was a 1.1-at.% Nd:YAG ceramics (Baikowski Japan Co., Ltd.) with thickness of 5 mm and a 10-mm diameter. Surface used for pumping was coated HR at the lasing wavelength of 1.06 μm (λ_{em}) and HT at λ_p. The second surface was coated AR at both λ_{em} and λ_p. Around 90% of the pump radiation was absorbed in Nd:YAG. Cr^{4+}:YAG ceramics with initial transmission T_0 of 0.70, 0.50, and 0.30 were employed for Q-switching, both surfaces of a SA ceramics being coated AR at λ_{em}. The resonator length was 12 mm, and a plane output mirror (OCM) was used for out-coupling. The laser beams path was bent with a prism placed after the resonator, whereas air breakdown was observed behind a lens L3 that has an 11-mm long focal length.

OCM, T	T_0= 0.70			T_0= 0.50			T_0= 0.30		
	E_p (mJ)	t_p (ns)	E_{pump} (mJ)	E_p (mJ)	t_p (ns)	E_{pump} (mJ)	E_p (mJ)	t_p (ns)	E_{pump} (mJ)
0.40	0.8	4.1	9.0	1.0	2.3	12.0	1.7	1.0	18.5
0.50	0.8	2.5	9.5	1.3	1.5	13.0	2.1	0.6	20.0
0.70	0.8	2.4	10.5	1.3	1.4	15.0	2.3	0.7	21.5

Table 1. Characteristics of Q-switched laser pulses obtained from the Nd:YAG-Cr^{4+}:YAG all-poly-crystalline ceramics laser that was build of discrete components.

Characteristics of the Q-switched laser pulses obtained from the Nd:YAG-Cr^{4+}:YAG laser are summarized in Table 1, at various OCM transmission T. Laser pulses of few-ns duration and energy E_p of 0.8 mJ were measured for the Cr^{4+}:YAG with T_0= 0.70. Energy E_p overcame 2 mJ and t_p shortened below 1 ns when combination of Cr^{4+}:YAG with T_0= 0.30 and OCM with T of 0.50 or 0.70 was used. Air breakdown was successfully for the Cr^{4+}:YAG with T_0= 0.30 and all the OCM employed in the experiments.

The next step of our investigations constituted realization of a compact Nd:YAG/Cr^{4+}:YAG laser with two-beam output and dimensions close to an electrical spark plug. Figure 18a shows the experimental set-up. The laser medium was a composite Nd:YAG/Cr^{4+}:YAG ceramics: Research experience of Baikowski Japan Co., as well as available optics on market (for example one could visit: www.thorlabs.com), allowed realization of this medium as a parallelepiped with 10×15 mm^2 surface area, as presented in Fig. 18b. The Nd:YAG doping level was 1.1-at.% Nd, and its length was increased at 8 mm: In this way, the gain medium absorption efficiency at λ_p was better than 0.95, which avoided bleaching effects of Cr^{4+}:YAG by the pump beam (Jaspan et al., 2004). The 3-mm thick Cr^{4+}:YAG SA ceramics had initial transmission T_0= 0.30. Surface S1 of Nd:YAG was coated HR at λ_{em} and HT at λ_p, and the OCM with transmission T= 0.50 was coated on surface S2 of Cr^{4+}:YAG.

The compact pumping line imaged the fiber end to a spot size of 1.0-mm into Nd:YAG. Each laser beam was expanded and then collimated in the "expander" section. Next, the beam was bended with a prism (patent pending), and finally focussed.

The composite, all-ceramics, passively Q-switched Nd:YAG/Cr^{4+}:YAG monolithic laser with two-beam output is shown in Fig. 19a. Figure 19b presents the air breakdown realized with this laser, whereas an electrical spark plug used in industrial gas engine is given for comparison. Each beam delivered Q-switched laser pulses with ~2.5 mJ energy and ~800 ps duration, which corresponds to a peak power of 3.1 MW. Minimal pump pulse energy was ~27 mJ. The laser pulse jitter, which was estimated from 500 consecutive pulses, improved from 3 μs to 1 μs when the pump pulse energy was increased from 27 mJ to 34 mJ respectively, while the pulse standard deviation decreases from 0.53 μs to 0.18 μs. The prism angle was chose such as the distance between the ignition points (φ_c) was 13 mm, whereas the depth of the ignition (b_c) was 9 mm (Pavel et al., 2011a).

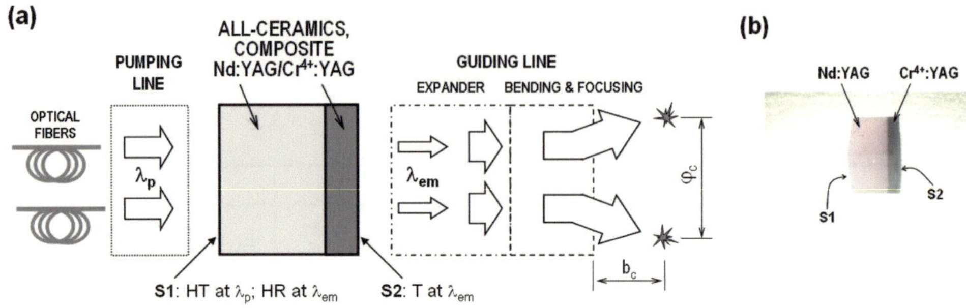

Fig. 18. (a) A sketch of the composite, all-ceramics Nd:YAG/Cr^{4+}:YAG monolithic laser with two-beam output is presented. (b) The rectangular-shaped laser medium is shown.

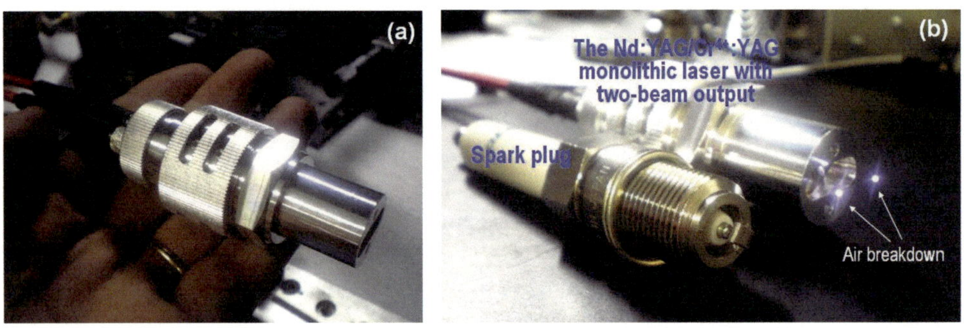

Fig. 19. (a) The Nd:YAG/Cr^{4+}:YAG laser with two-beam output is presented. (b) An electrical spark plug is shown for comparison and air breakdown in two points is illustrated.

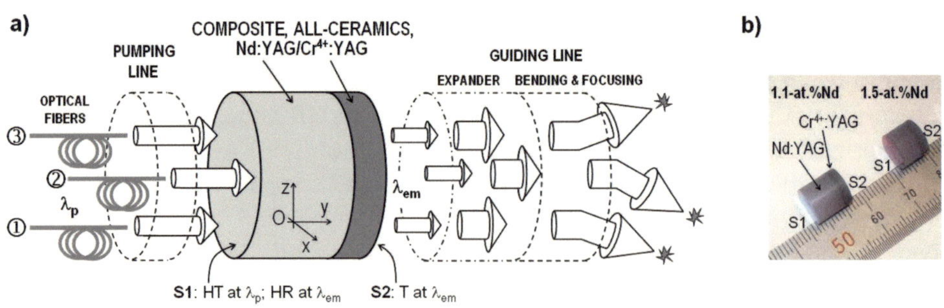

Fig. 20. (a) Schematic of a passively Q-switched, composite, all-ceramics Nd:YAG/Cr^{4+}:YAG monolithic laser with three-beam output. (b) Photo of two composite media is shown.

Once the Nd:YAG/Cr^{4+}:YAG monolithic laser with two-beam output was build, the final goal of our work was realization of a laser device with three-beam output and a size that fits an

electrical spark plug used in automobile industry. The experimental set-up is shown in Fig. 20a. Three composite, all-ceramics Nd:YAG/Cr^{4+}:YAG media (Baikowski Japan Co., Ltd.), each with a 9-mm diameter were prepared for the experiments. While the Cr^{4+}:YAG SA has initial transmission T_0= 0.30 and a thickness of 2.5 mm, the influence of Nd-doping level on Q-switched laser characteristics was investigated by using Nd:YAG with 1.1-at.% Nd (7.5-mm thick), as well as highly-doped 1.5-at.% Nd (thickness of 5 mm) and 2.0-at.% Nd (thickness of 3.5 mm). Again, surface S1 of Nd:YAG was coated HR at λ_{em} and HT at λ_p. The OCM with T= 0.50 at λ_{em} was coated on surface S2 of Cr^{4+}:YAG SA. A photo of two composite Nd:YAG/Cr^{4+}:YAG ceramics is shown in Fig. 20b. The optical pumping was realized through three independent, similar, and compact pumping lines (marked by 1 to 3 in Fig. 20a), each line containing a pair of an aspheric collimating lens and an aspheric focusing lens with short focal length and high NA. We mention that in order to fulfill dimensions of an automobile spark plug, diameter of all lenses was reduced and a new design of the fiber end was made.
The characteristics of the Q-switched laser pulses measured from the Nd:YAG/Cr^{4+}:YAG ceramics are given in Table 2. The energy of the laser pulse yielded by the 1.1-at.% Nd:YAG ceramics was 2.37 mJ, with a pulse peak power of 2.8 MW. The laser beam M^2 factor, which was measured by the knife-edge method, was 3.7.

Nd (at.%)	Pulse energy (mJ)	Pulse duration (ps)	Peak power (MW)	Pump pulse energy (mJ)	M^2 factor
1.1	2.37	850	2.79	27	3.7
1.5	2.03	650	3.12	32.8	4.0
2.0	1.37	660	2.08	32	4.0

Table 2. Characteristics of Q-switched laser pulses obtained from the composite, all-polly-crystalline Nd:YAG/Cr^{4+}:YAG ceramics.

In order to explain the influence of pump-beam spot size on Q-switched laser performances, we used a rate equation model (Zhang et al., 2000; Li et al., 2007) in which the pump beam was assumed to have a top-hat distribution of radius w_p and the laser beam was taken as Gaussian with a spot size of radius w_g. Both w_p and w_g were considered constant along the Nd:YAG/Cr^{4+}:YAG medium. The laser pulse energy is given by Eq. (1), while the initial population inversion density, n_{gi} was written as:

$$n_{gi} = \frac{-\ln R + L - \ln T_0^2}{2\sigma_g \ell_g \cdot \left[1-\exp\left(-2a^2\right)\right]} \quad (7)$$

with $a= w_p/w_g$. The final population inversion density, n_{gf} and n_{gi} are related by relation:

$$\left(1-\frac{n_{gf}}{n_{gi}}\right) + \left[1+\frac{(1-\delta)\cdot\ln T_0^2}{\beta}\right]\cdot\ln\left(\frac{n_{gf}}{n_{gi}}\right)\cdot\frac{(1-\delta)\cdot\ln T_0^2}{\beta}\cdot\left[1-\left(\frac{n_{gf}}{n_{gi}}\right)^{\alpha}\right] = 0 \quad (8)$$

with parameter β: $\beta= (-\ln R+L-\ln T_0^2)/[1-\exp(-2a^2)]$.
Figure 21 presents n_{gf}/n_{gi} versus ratio $a= w_p/w_g$. In simulation losses were L= 0.06 (0.01 for Nd:YAG and 0.05 for Cr^{4+}:YAG final transmission), while spectroscopic parameter of

Nd:YAG was $\sigma_g = 2.63\times10^{-19}$ cm^2 (Taira, 2007). If $w_p<w_g$ the overlap between pump and laser beam is good, but n_{gf}/n_{gi} increases when w_p/w_g decreases. Although the initial inversion of population n_{gi} is high, a small fraction of it used for lasing and therefore Q-switched laser pulse energy is low. If w_p/w_g has a large value, the central part of the inversion of population interacts with laser mode, whereas some outside part could be depleted by spontaneous emission. Increasing w_p/w_g decreases n_{gf}/n_{gi}: The final inversion of population is low and a pulse laser with high energy is obtained. The expected values of the Q-switched laser pulse at various sizes w_g of the laser mode were also shown in Fig. 21.

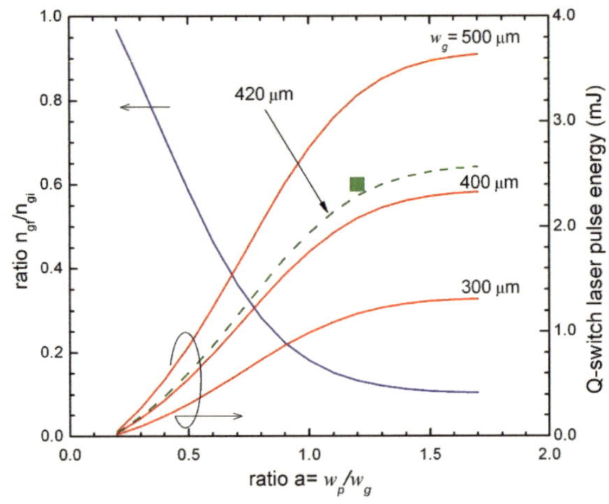

Fig. 21. Ratio n_{gf}/n_{gi} versus w_p/w_g and Q-switched laser pulse energy for various laser beam radii w_g. The pulse energy obtained from the 1.1-at.% Nd:YAG is given by the sign (■).

A plane-plane resonator operates due to thermal effects induced by optical pumping in the laser medium. The focal length f of Nd:YAG thermal lens can be evaluated by relation: $f= (\pi K_c w_p^2)/[P_h\cdot(dn/dT)]$ (Innocenzi et al., 1990). Nd:YAG has thermal conductivity K_c= 10.1 Wm^{-1}K^{-1} (Sato & Taira, 2006), thermal coefficient of the refraction index is dn/dT= 0.73×10^{-5} K^{-1}, while ~0.24 of the absorbed pump power is transformed into heat (P_h) under efficient laser emission at 1.06 μm. The average thermal lens of the 1.1-at.% Nd:YAG was evaluated as ~6.5 m, whereas value of w_g was determinate by an ABCD description of the resonator as ~420 μm. Sign of Fig. 21 is the experimental value of the laser pulse energy obtained from the 1.1-at.% Nd:YAG. Agreement with theoretical modeling is good, especially if uncertainties in evaluation of thermal focal lens or of other system parameters (such as losses L) are considered. The model can be improved by taking into account variation of pump beam radius w_p and of laser beam spot size w_g along the resonator length.

Energies of 2.03 mJ and 1.37 mJ were measured from the 1.5-at.% and 2.0-at.% Nd ceramics, respectively. The pump pulse energy increased from 27 mJ for the 1.1-at.% Nd:YAG to 33 mJ for the 1.5-at.% Nd:YAG, and to 32 mJ for the 2.0-at.% Nd:YAG. The decrease of the $^4F_{3/2}$ upper-level lifetime with Nd-doping level could be a reason for lower laser performances recorded with the highly-doped Nd:YAG compared with the 1.1-at.% Nd:YAG. The OCM transmission has also to be optimized for the highly-doped Nd:YAG ceramics.

Nd (at.%)	Average pulse energy (mJ)		Standard deviation (mJ)	
	Ox	Oz	Ox	Oz
1.1	2.36	2.34	0.02	0.06
1.5	2.03	2.03	0.03	0.03
2.0	1.33	1.34	0.05	0.02

Table 3. Average laser pulse energy and standard deviation measured along Ox and Oz axis of the Nd:YAG/Cr^{4+}:YAG ceramics media.

Very important for performances of the monolithic laser is the uniformity of the ceramic Nd:YAG/Cr^{4+}:YAG material. Table 3 presents average values of the laser pulse energy determined along Ox and Oz axes, estimated by scanning the medium with a 0.5-mm step. The laser pulse energies were very close to those measured at the media center. Moreover, standard deviation was small, below 3% for the 1.1-at.% Nd:YAG, and less than 4% for the highly-doped Nd:YAG. The results indicate a very good homogeneity as well as quality of the composite all-ceramics Nd:YAG/Cr^{4+}:YAG media, in spite of the high, 9-mm diameter. As an example, laser pulse energy measured from the highly-doped, 2.0-at.% Nd:YAG/Cr^{4+}:YAG medium is given in Fig. 22.

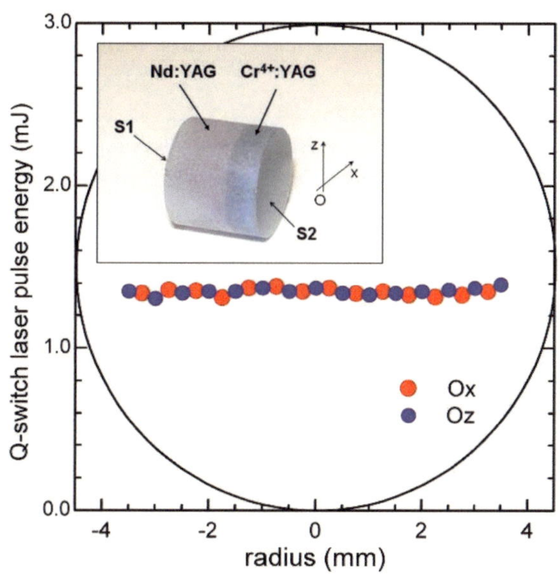

Fig. 22. Laser pulse energy measured along horizontal and vertical axis of the 2.0-at.% Nd:YAG/Cr^{4+}:YAG ceramics (inset shows the composite medium).

The three-beam output micro-laser (Fig. 23a) was realized with the composite, all-ceramics 1.1-at.% Nd:YAG/Cr^{4+}:YAG medium. The guiding line (Pavel et al., 2011b) was designed to assure a distance between a focusing point and the laser axis, φ_c of 4.5 mm and a depth of the focusing point inside the combustion chamber, b_c of 9 mm. Air breakdown is illustrated in Fig. 23b, and an automobile electrical spark plug is shown for comparison.

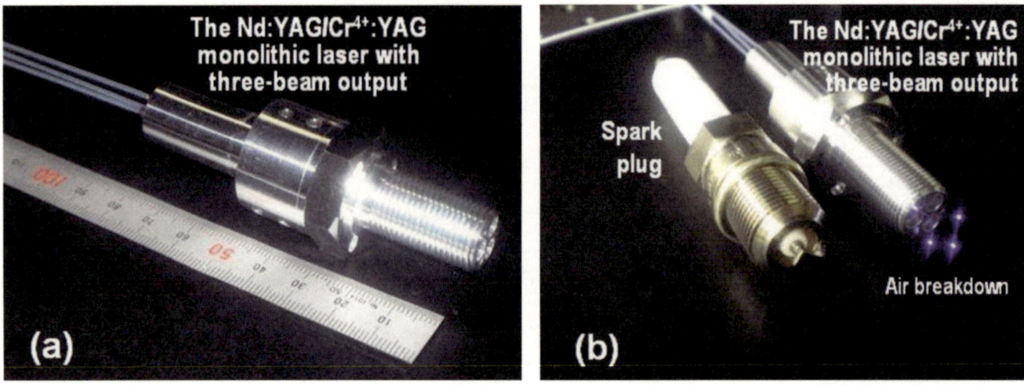

Fig. 23. (a) The passively Q-switched Nd:YAG/Cr^{4+}:YAG micro-laser with three-beam output is shown. (b) Air breakdown in three points is realized.

Various other characteristics of a Q-switched laser pulse, such us the delay time (i.e. the time between the moment when the pump pulse begins and the moment when the laser pulse develops), or the pulse jitter and standard deviation were determined. Figure 24 presents these parameters function of the pump pulse energy. As expected, delay time decreases with the increase of the pump pulse power. Therefore, the use of independent pumping lines allows control of the air breakdown timing by changing the pump energy of each line. Furthermore, real simultaneous ignition in all three points can be obtained by a small (less than 5%) tuning of the pump energy of each individual line. Time jitter is low (2.1 μs at the pump energy of 26.7 mJ and 1.0 μs at 32-mJ available energy of the pump pulse) and thus it would not have a negative impact on an automobile engine that is ignited by the laser. A second laser pulse was not observed. Nevertheless, increasing the pump pulse duration would enable obtaining of multiple laser pulses (Weinrotter et al., 2005; Tsunekane et al., 2010), which are useful for engine ignition especially if lean fuel-air mixtures are used.

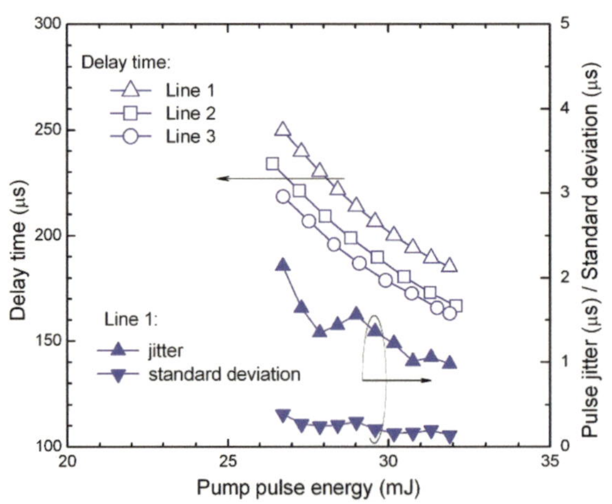

Fig. 24. Time delay of the Q-switched laser pulse, and time jitter and standard deviation versus pump pulse energy, for the repetition rate of 5 Hz (250-μs pump pulse duration).

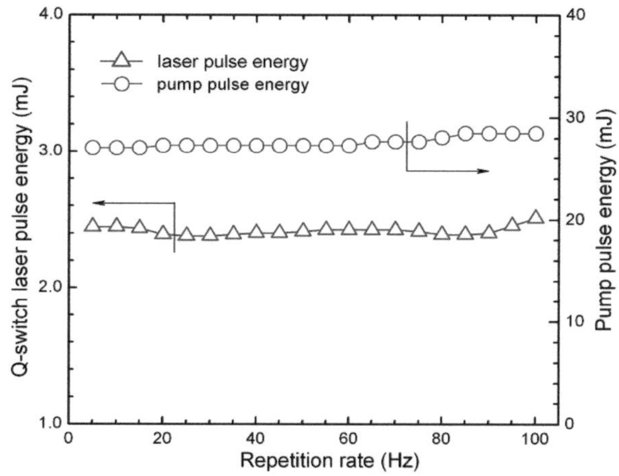

Fig. 25. Q-switched laser pulse energy and pump pulse energy versus pump repetition rate.

Previous experiments were performed at 5-Hz pump repetition, and temperature of the composite Nd:YAG/Cr^{4+}:YAG ceramics was not controlled. However, higher repetition rates are necessary for operation of a car engine, usually up to 60 Hz. Therefore, variation of laser pulse characteristics was investigated versus the pump repetition rate. Experiments concluded that an increase from 5 to 100 Hz of the pump repetition rate improved the laser pulse energy from 2.37 mJ to 2.51 mJ, i.e. by only a 6% fraction, as shown in Fig. 25. This change required a small (5 to 6%) increase of the minimal pump energy, from ~27 mJ at 5 Hz pump repetition rate to 28.4 mJ at the pump repetition rate of 100 Hz. Variation with temperature of the laser performances was behind this work purpose. Nevertheless, in prior papers (Tsunekane and Taira, 2009; Dascalu and Pavel, 2009; Pavel et al., 2010b) we have measured only a slight increase of the laser pulse energy when temperature of a Nd:YAG laser passively Q-switched by Cr^{4+}:YAG SA (build of discrete, single-crystals components) was increased up to 150°C. Future experiments would also consider testing of the laser to shock and vibration conditions that are similar to those experienced in a car engine.

4. Conclusion

A passively Q-switched Nd:YAG/Cr^{4+}:YAG giant-pulse emitting micro-laser with up to three-beam output has been realized. The device incorporates a composite, all-ceramics Nd:YAG/Cr^{4+}:YAG monolithic structure that was pumped by similar, independent lines. The laser size is comparable to that of an electrical spark plug, being the first demonstration of this kind of device to the best of our knowledge. Laser pulses with energy of ~2.4 mJ and 2.8-MW peak power at 5-Hz repetition rate were obtained from a 10-mm thick Nd:YAG/Cr^{4+}:YAG ceramics, just such as "giant micro-photonics". Increasing pump repetition rate up to 100 Hz improved the laser pulse energy by 6% and required only a 6% increase of the pump pulse energy compared with operation at 5 Hz. Pulse timing of the laser beams can by controlled by changing the pump energy of each individual line. On the other hand, simultaneous multi-point ignition is possible by less than 5% tuning of the pump energy of each individual pump line. These lasers will enable studies on the performances of internal combustion engines with multi-point ignition.

5. Acknowledgment

This work was financed by Japan Science and Technology Agency, and partially supported by DENSO Company, Japan. Permanent support from and fruitful discussions with Mr. Kenji Kanehara of Nippon Soken. Inc. is acknowledged.The authors thank Mr. Nobuo Mizutani of IMS Equipment Development Division for the help with the laser module design. N. Pavel acknowledges partial support of the Romanian Ministry of Education and Research, project 12106/01.10.2008.

6. References

Agnesi, A.; Dell'Acqua, S. & Reali, G. C. (1997). 1.5 Watt passively Q-switched diode-pumped cw Nd:YAG laser. *Opt. Commun.*, Vol. 133, No. 1-6, pp. 211-215.

An, J; Zhao, S. Z.; Li, G. Q.; Yang, K. J.; Sun, Y. M.; Li, D. C.; Wang, J. & Li, M. (2006). Doubly passively Q-switched intracavity-frequency-doubling Nd^{3+}:YAG/KTP green laser with GaAs and Cr^{4+}: YAG saturable absorbers. *Opt. Eng.*, Vol. 45, No.12, Art. No. 124202.

Aniolek, K. W.; Schmitt, R. L.; Kulp, T. J.; Richman, B. A.; Bisson, S.E. & Powers, P. E. (2000). Microlaser-pumped periodically poled lithium niobate optical parametric generator-optical parametric amplifier. *Opt. Lett.* Vol. 25, No. 8, pp. 557-559.

Bibeau, C.; Beach, R. J.; Mitchell, S. C.; Emanuel, M. A.; Skidmore, J.; Ebbers, C. A.; Sutton, S. B. & Jancaitis, K. S. (1998). High-average-power 1-μm performance and frequency conversion of a diode-end-pumped Yb:YAG laser. *IEEE J. Quantum Electron.*, Vol. 34, No. 10, pp. 2010-2019.

Cheng, K.; Zhao, S.; Li, Y.; Li, G.; Li, D., Yang, K.; Zhang, G.; Ge, H. & Yu, Z. (2011). Diode-pumped doubly passively Q-switched Nd:$LuVO_4$/KTP green laser with Cr^{4+}:YAG and GaAs saturable absorbers. *Opt. Commun.*, Vol. 284, No. 1, pp. 344-349.

Dascalu, T. & Pavel, N. (2009). High-temperature operation of a diode-pumped passively Q-switched Nd:YAG/Cr^{4+}:YAG laser. *Laser Phys.* Vol. 19, No. 11, pp. 2090-2095.

Degnan, J. (1995). Optimization of passively Q-switched lasers. *IEEE J. Quantum Electron.*, Vol. 31, No. 11, pp. 1890-1901.

Dong, J.; Shirakawa, A.; Takaichi, K.; Ueda, K.; Yagi, H.; Yanagitani, T. & Kaminskii, A.A. (2006). All-ceramic passively Q-switched Yb:YAG/Cr^{4+}:YAG microchip laser. *Electron. Lett.*, Vol. 42, No. 20, pp. 1154-1156.

Dong, J.; Ueda, K.; Shirakawa, A.; Yagi, H.; Yanagitani, T. & Kaminskii, A.A. (2007). Composite Yb:YAG/Cr^{4+}:YAG ceramics picosecond microchip lasers. *Opt. Express*, Vol. 15, No. 22, pp. 14516-14523.

Eimerl, D. (1987). High average power harmonic generation. *IEEE J. Quantum Electron.*, Vol. 23, No. 5, pp. 575-592.

Feng, Y.; Lu, J.; Takaichi, K.; Ueda, K.; Yagi, H.; Yanagitani, T. & Kaminskii, A.A. (2004). Passively *Q*-switched ceramic Nd^{3+}:YAG/Cr^{4+}:YAG lasers. *Appl. Opt.*, Vol. 43, No. 14, pp. 2944-2947.

Hanson, F. (1995). Improved laser performance at 946 and 473 nm from a composite Nd:$Y_3Al_5O_{12}$ rod. *Appl. Phys. Lett.*, Vol. 66, No. 26, pp. 3579-3551.

Honea, E. C.; Ebbers, C. A.; Beach, R. J.; Speth, J. A.; Skidmore, J. A.; Emanuel, M. A. & Payne, S. A. (1998). Analysis of an intracavity-doubled diode-pumped *Q*-switched Nd:YAG laser producing more than 100 W of power at 0.532 mm. *Opt. Lett.*, Vol. 23, No. 15, pp. 1203-1205.

Huang, S.; L.; Tsui; T.; Y., Wang, C. H. & Kao, F. J. (1999). Timing jitter reduction of a passively Q-switched laser, *Jpn. J. Appl. Phys.*, Vol. 38, Part 2, No. 3A, pp. L239-241.

Innocenzi, M. E.; Yura, H. T.; Fincher, C. L. & Fields, R. A. (1990). Thermal modeling of continuous-wave end-pumped solid-state lasers. *Appl. Phys. Lett.* Vol. 56, No. 19, pp. 1831-1833.

Jaspan, M. A.; Welford, D. & Russell, J. A. (2004). Passively Q-switched microlaser performance in the presence of pump-induced bleaching of the saturable absorber. *Appl. Opt.*, Vol. 43, No. 12, pp. 2555-2560.

Kajava, T. T & Gaeta, A. L. (1997). Intra-cavity frequency-doubling of a Nd:YAG laser passively *Q*-switched with GaAs . *Opt. Commun.*, Vol. 137, No. 1-2, pp. 93-97.

Kofler, H.; Tauer, J.; Tartar, G.; Iskra, K.; Klausner, J.; Herdin, G. & Wintner, E. (2007). An innovative solid-state laser for engine ignition. *Laser Phys. Lett.*, Vol. 4, No. 4, pp. 322-327.

Kroupa, G.; Franz, G. & Winkelhofer, E. (2009). Novel miniaturized high-energy Nd:YAG laser for spark ignition in internal combustion engines. *Opt. Eng.*, Vol. 48, No. 1, art. 014202 (5 pages).

Li, S. T.; Zhang, X. Y.; Wang, Q. P.; Li, P.; Chang, J.; Zhang, X. L. & Cong, Z. H. (2007). Modeling of Q-switched lasers with top-hat pump beam distribution. *Appl. Phys. B*, Vol. 88, No. 2, pp. 221-226.

Liu, J.; Yang, J. & He, J. (2004). Diode-pumped passively Q-switched intracavity frequency doubled Nd:$GdVO_4$/KTP green laser. *Opt. & Laser Techn.*, Vol. 36, No. 1, pp. 31-33.

Ma, J. X.; Alexander, D. R. & Poulain, D. E. (1998). Laser spark ignition and combustion characteristics of methane-air mixtures. *Comb. Flame,* Vol. 112, No. (4), pp. 492-506.

Morsy, M. H.; Ko, Y. S.; Chung, S. H. & Cho, P. (2001). Laser-induced two-point ignition of premixture with a single-shot laser. *Comb. Flame,* Vol. 125, No. 4, pp. 724-727.

Ostroumov, V. G.; Heine, F.; Kück, S; Huber, G; Mikhailov, V. A. & Shcherbakov, I. A. (1997). Intracavity frequency-doubled diode-pumped Nd: $LaSc_3(BO_3)_4$ lasers. *Appl. Phys. B,* Vol. 64, No. 3, pp. 301-3.5.

Paschotta, R. (2008). Encyclopedia of Laser Physics and Technology. Handbook/Reference Book. ISBN-10: 3-527-40828-2, ISBN-13: 978-3-527-40828-3; 2008 WILEY-VCH Verlag GmbH & Co. KGaA, Weinheim.

Pavel, N.; Saikawa, J. & Taira, T. (2001b). Diode end-pumped passively Q-switched Nd:YAG laser intra-cavity frequency doubled by LBO crystal. *Opt. Commun.*, Vol. 195, No. 1-4, pp. 233-240.

Pavel, N.; Saikawa, J.; Kurimura, S. & Taira, T. (2001a). High average power diode end-pumped composite Nd:YAG laser passively Q-switched by Cr^{4+}:YAG saturable absorber. *Jap. J. Appl. Phys.*, Vol. 40, Part. 1, No. 3A, pp. 1253-1259.

Pavel, N.; Tsunekane, M. & Taira, T. (2010a). High peak-power passively Q-switched all-ceramics Nd:YAG/Cr^{4+}:YAG lasers. Proceedings SPIE, Vol. 7469, Micro- to Nano-Photonics II - ROMOPTO 2009 Conference, 31 August - 03. Sept., 2009, Sibiu, Romania; paper 746903.

Pavel, N.; Tsunekane, M.; Kanehara, K. & Taira, T (2011a). Composite all-ceramics, passively Q-switched Nd:YAG/Cr^{4+}:YAG monolithic micro-laser with two-beam output for multi-point ignition. In: *CLEO 2011, Laser Science to Photonics Applications Conference,* Baltimore, Maryland, USA, 1-6 May, 2011, paper CMP1.

Pavel, N; Tsunekane, M. & Taira, T. (2010b). Enhancing performances of a passively Q-switched Nd:YAG/Cr^{4+}:YAG microlaser with a volume Bragg grating output coupler. *Opt. Lett.* Vol. 35, No. 10, pp. 1617-1619.

Pavel, N; Tsunekane, M. & Taira, T. (2011b). Passively Q-switched Nd:YAG/Cr^{4+}:YAG all-ceramics, composite, monolithic micro-lasers with multi-beam output for laser ignition. In: *CLEO Europe - EQEC 2011 Conference,* Münich, Germany, 22-26 May, 2011, paper CA7.1.

Phuoc, T. X. & White, F. P. (1999). Laser-induced spark ignition of CH_4/Air mixtures. *Comb. Flame,* Vol. 119, No. 3, pp. 203-216.

Phuoc, T. X. (2000). Single-point versus multi-point laser ignition: Experimental measurements of combustion times and pressures. *Comb. Flame,* Vol. 122, No. 4, pp. 508-510.

Sakai, H.; Kan, H. & Taira T., (2008). >1 MW peak power single-mode high-brightness passively Q-switched Nd^{3+}:YAG microchip laser. *Opt. Express,* Vol. 16, No. 24, pp. 19891-19899.

Sato, Y. & Taira, T. (2006). The studies of thermal conductivity in $GdVO_4$, YVO_4, and $Y_3Al_5O_{12}$ measured by quasi-one-dimensional flash method. *Opt. Express*, Vol. 14, No. 22, pp. 10528-10536.

Shimony, Y.; Burshtein, Z. & Kalisky, Y. (1995). Cr^{4+}:YAG as passive Q-switch and Brewster plate in a pulsed Nd-laser. *IEEE J. Quantum Electron.,* Vol. 31, No. 10, pp. 1738-1741.

Song, J.; Li, C.; Kim, N. S. & Ueda, K. I. (2000). Passively Q-switched diode-pumped continuous-wave Nd:YAG–Cr^{4+}:YAG laser with high peak power and high pulse energy. *Appl. Opt.*, Vol. 39, No. 27, pp. 4954-4958.

Taira, T. (2007). RE^{3+}-ion-deped YAG ceramic lasers. *IEEE J. Sel. Top. Quantum Electron.*, Vol. 13, No. 3, pp.798-809.

Tang, D. Y.; Ng, S. P.; Qin, L. J. & Meng, X. L. (2003). Deterministic chaos in a diode-pumped Nd:YAG laser passively Q switched by a Cr^{4+}:YAG crystal, *Opt. Lett.*, Vol. 28, No. 5, pp. 325-327.

Tsunekane, M. & Taira, T. (2009). Temperature and polarization dependences of Cr:YAG transmission for passive Q-switching. In: *Conference on Lasers and Electro-Optics/International Quantum Electronics Conference*, Baltimore, Maryland, USA, May 31 - June 5, 2009, paper JTuD8.

Tsunekane, M.; Inohara, T.; Ando, A.; Kanehara, K. & Taira, T. (2008). High peak power, passively Q-switched Cr:YAG/Nd:YAG micro-laser for ignition of engines. In: *Advanced Solid-State Photonics*, OSA Technical Digest Series (CD) (Optical Society of America, 2008); 27-30 January, 2008, Nara, Japan, paper MB4.

Tsunekane, M.; Inohara, T.; Ando, A.; Kido, N.; Kanehara, K. & Taira, T. (2010). High peak power, passively Q-switched microlaser for ignition of engines. *IEEE J. Quantum Electron.,* Vol. 46, No. 2, pp. 277-284.

Tsunekane, M.; Taguchi, N.; Kasamatsu, T. & Inaba, H. (1997). Analytical and experimental studies on the characteristics of composite solid-state laser rods in diode-end-pumped geometry. *IEEE J. Sel. Top. Quantum Electron.*, Vol. 3, No. 1, pp. 9-18.

Weinrotter, M.; Kopecek, H. & Wintner, E. (2005b). Laser ignition of engines. *Laser Phys.,* Vol. 15, No. 7, pp. 947-953.

Weinrotter, M.; Kopecek, H.; Tesch, M.; Wintner, E; Lackner, M. & Winter, F. (2005a). Laser ignition of ultra-lean methane/hydrogen/air mixtures at high temperature and pressure," *Exp. Therm. Fluid Science,* Vol. 29, No. 8, pp. 569-577.

Zayhowski, J.; Cook, C. C.; Wormhoudt, J. & J. H. Shorter, J. H. (2000). Passively Q-switched 214.8-nm Nd:YAG/Cr^{4+}:YAG microchip-laser system for the detection of NO. In *Advanced Solid State Lasers,* OSA Technical Digest Series (Optical Society of America, 2000), Davos, Switzerland, Feb. 2000, paper WB4.

Zayhowsky, J. J. & Dill III, C. (1994). Diode-pumped passively Q-switched picosecond microchip lasers. *Opt. Lett.*, Vol. 19, No. 18, pp. 1427-1429.

Zhang, X.; Zhao, S., Wang, Q., Zhang, Q., Sun, L. & Zhang, S. (1997). Optimization of Cr^{4+}-doped saturable-absorber Q-switched lasers, *IEEE J. Quantum Electron.,* Vol. 33, No. 12, pp.2286-2294.

Zhang, X.; Zhao, S.; Wang, Q.; Ozygus, B. & Weber, H. (2000). Modeling of passively Q-switched lasers. *J. Opt. Soc. Am. B.*, Vol. 17, No. 7, pp. 1166-1175.

5

Laser Pulses for Compton Scattering Light Sources

Sheldon S. Q. Wu[1], Miroslav Y. Shverdin[2], Felicie Albert[1] and Frederic V. Hartemann[1]

[1]*Lawrence Livermore National Laboratory*

[2]*AOSense, Inc.*

USA

1. Introduction

Nuclear Resonance Fluorescence (NRF) is an isotope specific process in which a nucleus, excited by gamma-rays, radiates high energy photons at a specific energy. This process has been well known for several decades, and has potential high impact applications in homeland security, nuclear waste assay, medical imaging and stockpile surveillance, among other areas of interest. Although several successful experiments have demonstrated NRF detection with broadband bremsstrahlung gamma-ray sources[1], NRF lines are more efficiently detected when excited by narrowband gamma-ray sources. Indeed, the effective width of these lines is on the order of $\Delta E / E \sim 10^{-6}$.

NRF lines are characterized by a very narrow line width and a strong absorption cross section. For actinides such as uranium, the NRF line width is due to the intrinsic line width Γ_0 connecting the excited state to the ground state and to the Doppler width,

$$\Delta = E_R \sqrt{\frac{kT_{eff}}{Mc^2}} \tag{1}$$

where E_R is the resonant energy (usually MeV), M the mass of the nucleus, k the Boltzmann constant, c the speed of light and T_{eff} the effective temperature of the material. This model is valid as long as $\Gamma_0 + \Delta \gg 2kT_D$ where T_D is the Debye temperature. Because in most cases $\Gamma_0 \ll 1$ eV, the total width is just determined by the thermal motion of the atoms. Debye temperatures for actinides are usually in the range 100-200 K. Within this context, the NRF absorption cross section near the resonant energy is[2]:

$$\sigma(E) = \pi^{3/2} \left(\frac{\lambda}{2\pi}\right)^2 \left(\frac{2J_j + 1}{2J_i + 1}\right) \frac{\Gamma_0}{\Delta} \exp\left[-\left(\frac{E - E_R}{\Delta}\right)^2\right] \tag{2}$$

where J_i and J_j are the total angular momentum for the ground and excited states respectively, and λ the radiation wavelength. Typically, strong M1 resonances at MeV energies are on the order of tens of meV wide with an absorption cross section around 10 barns.

Currently, Compton scattering is among the only physical processes capable of producing a narrow bandwidth radiation (below 1%) at gamma-ray energies, using state-of-the art accelerator and laser technologies. In Compton scattering light sources, a short laser pulse and a relativistic electron beam collide to yield tunable, monochromatic, polarized gamma-ray photons. Several projects have recently utilized Compton scattering to conduct NRF experiments: Duke University[2], Japan[3] and Lawrence Livermore National Laboratory (LLNL)[4]. In particular, LLNL's Thomson-Radiated Extreme X-rays (T-REX) project demonstrated isotope specific detection of low-Z and low density materials (^{7}Li) shielded behind high-Z and high-density materials (Pb, Al).

The T-REX device and the MEGa-ray machine currently in progress at LLNL are laser-based Compton sources that produce quasi-monochromatic, tunable, polarized gamma-rays via Compton scattering of energetic short duration laser pulses from high-brightness relativistic electron bunches. The detection of low-Z and low-density objects shielded by a high-Z dense material is a long-standing problem that has important applications ranging from homeland security and nonproliferation[5] to advanced biomedical imaging and paleontology. X-rays are sensitive to electron density, and x-ray radiography yields poor contrast in these situations. Within this context, NRF offers a unique approach to the so-called inverse density radiography problem. Since NRF is a process in which nuclei are excited by discrete high-energy (typically MeV) photons and subsequently re-emit gamma-rays at discrete energies determined by the structure of the nucleus, and the resonance structure is determined by the number of neutrons and protons present in the nucleus, NRF provides isotope-specific detection and imaging capability[6]. Thus Compton sources have a lot of potential high profile applications, since NRF can be used for nuclear waste assay and management, homeland security and stockpile stewardship, among others. Other processes and physics can be investigated with narrowband gamma-rays, such as near-threshold photofission, giant dipole resonances, and detailed studies of nuclear structure.

This chapter presents, primarily within the context of NRF-based applications, the theoretical and experimental aspects of a mono-energetic gamma-ray Compton source. In Section 2, we will introduce the fundamental concepts and basic physics involved in Compton scattering. In Section 3, we will discuss concepts necessary for understanding laser technology underlying high brightness Compton sources. Section 4 will cover amplification applications and review of some laser pulse measurement techniques. Nonlinear conversion will be discussed in Section 5. Finally, Section 6 contains descriptions of some simulation tools.

2. Mechanics of compton scattering

An electromagnetic wave that is incident upon a free charged particle will cause it to oscillate and so emit radiation. The accelerated motion of a charged particle will generate radiation in directions other than that of the incident wave. This process may be described as Compton scattering of the incident radiation. In this section, we will review the elementary physics of this interaction.

2.1 Overview

Consider an electromagnetic wave with the electric field $\mathbf{E} = \hat{\mathbf{e}} E_0 \cos(\mathbf{k} \cdot \mathbf{x} - \omega t)$ is incident upon a charged particle of charge e and mass m. Assuming that the resultant motion of the charge is nonrelativistic, the particle acceleration is $\ddot{\mathbf{x}} = \frac{e}{m}\mathbf{E}$. Such an accelerated motion will radiate energy at the same frequency as that of the incident wave. The power radiated per unit solid angle in some observation direction $\hat{\mathbf{q}}$ can be expressed in the cgs system of units as

$$\frac{dP}{d\Omega} = \frac{cE_0^{\ 2}}{4\pi}\left(\frac{e^2}{mc^2}\right)^2 (\hat{\mathbf{e}}\times\hat{\mathbf{q}})^2. \tag{3}$$

From a scattering perspective, it is convenient to introduce a scattering cross section, defined by

$$\frac{d\sigma}{d\Omega} = \frac{\text{Energy radiated/ unit solid angle/ unit time}}{\text{Incident energy flux/unit area/unit time}} \tag{4}$$

where the incident flux is $S = cE_0^{\ 2} / 4\pi$. Thus the scattering cross section is

$$\frac{d\sigma}{d\Omega} = \frac{1}{S}\frac{dP}{d\Omega} = r_0^{\ 2}(\hat{\mathbf{e}}\times\hat{\mathbf{q}})^2 \tag{5}$$

where $r_0 \equiv e^2 / mc^2$ is the classical electron radius. The result in Eq. (5) is valid for radiation polarized along a specific direction $\hat{\mathbf{e}}$. For unpolarized incident radiation, the scattering cross section is

$$\frac{d\sigma}{d\Omega} = \frac{r_0^{\ 2}}{2}\left(1+\cos^2\theta\right) \tag{6}$$

where θ is the scattering angle between the incident $\hat{\mathbf{k}}$ and scattered $\hat{\mathbf{q}}$ directions. The total scattering cross section obtained by integration over all solid angles is

$$\sigma_T = \frac{8\pi}{3}r_0^{\ 2} = 6.65\times 10^{-25} cm^2 \tag{7}$$

or better known as the Thomson cross section.

The Thomson cross section is only valid at low energies where the momentum transfer of the radiation to the particle can be ignored and the scattered waves has the same frequency as the incident waves. In 1923, Arthur Compton at Washington University in St. Louis found that the scattered radiation actually had a lower frequency than the incident waves. He further showed that if one adopts Einstein's ideas about the light quanta, then conservation laws of energy and momentum in fact led quantitatively to the observed frequency of the scattered waves. In other words, when the photon energy $\hbar\omega$ approaches the particle's rest energy mc^2, part of the initial photon momentum is transferred to the particle. This would become known as the Compton effect. For this effect to be neglected, we must have $\hbar\omega \ll mc^2$ in the frame where the charged particle is initially at rest. If this condition is violated, important modifications to Eq. (6) must be made due to the Compton effect. Quantum electrodynamical calculations give the modified scattering cross section for unpolarized radiation, in the initial rest frame of the charged particle,

$$\frac{d\sigma}{d\Omega} = \frac{r_0^2}{2}\left(\frac{\omega'}{\omega}\right)^2\left(\frac{\omega}{\omega'} + \frac{\omega'}{\omega} - \sin^2\theta\right) \tag{8}$$

where ω and ω' are the incident and scattered frequencies, respectively. They are related by the Compton formula:

$$\frac{\omega'}{\omega} = \frac{1}{1 + \frac{\hbar\omega}{mc^2}(1 - \cos\theta)} \tag{9}$$

The total cross section for scattering is then modified to

$$\sigma_{KN} = 2\pi r_0^2 \left\{ \frac{1+\varepsilon}{\varepsilon^3}\left[\frac{2\varepsilon(1+\varepsilon)}{1+2\varepsilon} - \ln(1+2\varepsilon)\right] + \frac{\ln(1+2\varepsilon)}{2\varepsilon} - \frac{1+3\varepsilon}{(1+2\varepsilon)^2} \right\} \tag{10}$$

where $\varepsilon = \hbar\omega / mc^2$ is the normalized energy of the incident light. This is called the Klein-Nishina cross section, first derived by Klein and Nishina in 1928. For $\varepsilon \ll 1$, Eq. (8) reduces to the Thomson cross section of Eq. (6). For scattering by high energy photons, the total cross section decreases from the Thomson value, as shown in Figure 1.

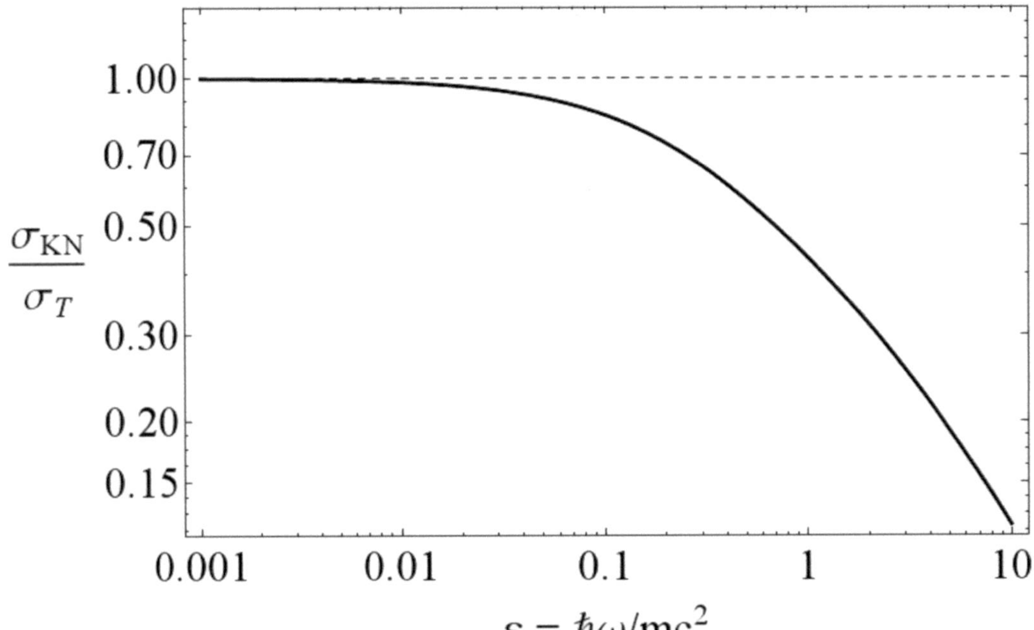

Fig. 1. Plot of the Klein-Nishina expression for the Compton scattering cross section. For small values of ε, the Compton cross section approaches the Thomson limit. At large values of ε, the Compton cross section approaches the asymptotic solution $\sigma_{KN} \sim \frac{3}{8}\frac{\sigma_T}{\varepsilon}(\ln 2\varepsilon + \frac{1}{2})$.

While the scattering expressions are simple in the frame where the particle is initially at rest, it is often desirable to do calculations in the laboratory frame where the electromagnetic fields are experimentally measured. To facilitate this, we will need generalizations of Eqs. (8) and (9) valid in an arbitrary inertial frame. The expressions are well known but their derivation is slightly involved, so we will simply quote the results[7]:

$$\frac{d\sigma}{d\Omega}=\frac{r_0{}^2 X}{2\gamma^2\left(1-\boldsymbol{\beta}\cdot\hat{\mathbf{k}}\right)^2}\left(\frac{\omega'}{\omega}\right)^2 \tag{11}$$

$$\frac{\omega'}{\omega}=\frac{1-\boldsymbol{\beta}\cdot\hat{\mathbf{k}}}{1-\boldsymbol{\beta}\cdot\hat{\mathbf{q}}+\frac{\varepsilon}{\gamma}\left(1-\hat{\mathbf{k}}\cdot\hat{\mathbf{q}}\right)} \tag{12}$$

where $\boldsymbol{\beta}$ is the initial particle velocity and $\gamma\equiv\frac{1}{\sqrt{1-\beta^2}}$. The spin-averaged relativistic invariant X can be expressed as

$$X=\frac{1}{2}\left(\frac{\kappa}{\kappa'}+\frac{\kappa'}{\kappa}\right)-1+2\left(\epsilon_\mu\pi^\mu-\frac{\epsilon_\mu p^\mu\pi_\nu p'^\nu}{\kappa}+\frac{\epsilon_\mu p'^\mu\pi_\nu p^\nu}{\kappa'}\right)^2 \tag{13}$$

where $\kappa=k_\mu p^\mu$, $\kappa'=q_\mu p^\mu$ are products of the 4-vectors $k^\mu=\hbar(\omega/c,\mathbf{k})$, $q^\mu=\hbar(\omega'/c,\mathbf{q})$ and $p^\mu=mc(\gamma,\gamma\boldsymbol{\beta})$; ϵ^μ and π^μ are the incident and scattered 4-polarizations, respectively; $p'_\mu=p_\mu+k_\mu-q_\mu$ is the 4-momentum of the charged particle after the scattering event. When the photon polarizations are not observed, this expression must be summed over the final and averaged over the initial polarizations, giving

$$\bar{X}=\frac{1}{2}\sum_{pol}X=\frac{\kappa}{\kappa'}+\frac{\kappa'}{\kappa}+2\left(\frac{m^2c^2}{\kappa}-\frac{m^2c^2}{\kappa'}\right)+\left(\frac{m^2c^2}{\kappa}-\frac{m^2c^2}{\kappa'}\right)^2 \tag{14}$$

The above describes the interaction of a single electron with an incident electromagnetic wave. For realistic laser-electron interactions, one needs to take into account the appropriate photon and electron phase space. The most useful expression to describe the source is typically the local differential brightness, which may be written in terms of the number of scattered photons, N, per unit volume per unit time per solid angle per energy interval[8]:

$$\frac{d^6N}{dq\,d\Omega\,d^3x\,dt}=\frac{d\sigma}{d\Omega}\delta(q-q_0)n_e n_\lambda\frac{p_\mu k^\mu}{\gamma m\hbar k}d^3\mathbf{p}\,d^3\mathbf{k} \tag{15}$$

where $n_e d^3\mathbf{p}$ and $n_\lambda d^3\mathbf{k}$ represent the electron beam phase space density and laser photon phase space density, respectively. The factor $\frac{p_\mu k^\mu}{\gamma m\hbar k}$ represents the relative velocity between the electron and photon and is equal to c for $\gamma=1$. The delta function in $q-q_0$ is required simply to preserve energy-momentum conservation relation expressed in Eq. (12).

Let us briefly apply some of theory discussed above to a realistic Compton source. Consider a Compton scattering geometry in Figure 2. Here, $\mathbf{p}=m_0c\mathbf{u}$. Let us assume $\varepsilon\ll 1$, it is evident from Eq. (12) that the scattered photon energy maximizes for on-axis scattering events, with $\varphi = \pi, \theta = 0$. For a relativistic electron with $\gamma \gg 1$, we may use the approximation, $\beta \approx 1 - \frac{1}{2\gamma^2}$, and find that Eq. (12) simplifies to $\frac{\omega'}{\omega} \approx 4\gamma^2$. In the case of highly relativistic electrons, the scattered radiation can be many orders of magnitude more energetic than the incident radiation! One can intuitively understand why this occurs. The counter-propagating incident radiation in the electron rest frame is up-shifted by a factor 2γ due to the relativistic Doppler effect. The scattered radiation is emitted at approximately the same energy in the electron frame. The Lorentz transformation back to the laboratory frame gives another factor of 2γ. Using typical experimental values for an interaction laser operating at 532 nm, $\hbar\omega = 2.33\,\text{eV}$, colliding head-on with a relativistic electron with $\gamma = 227$, we find that $\frac{\varepsilon / \gamma}{1-\beta} \approx 0.002$ which justifies neglecting the Compton effect to first order. More importantly, the scattered radiation will have energy $\hbar\omega' \approx 4\gamma^2\hbar\omega = 4.8 \times 10^5\,\text{eV}$ which is well within the gamma-ray regime. The electron energy is decreased by a corresponding amount.

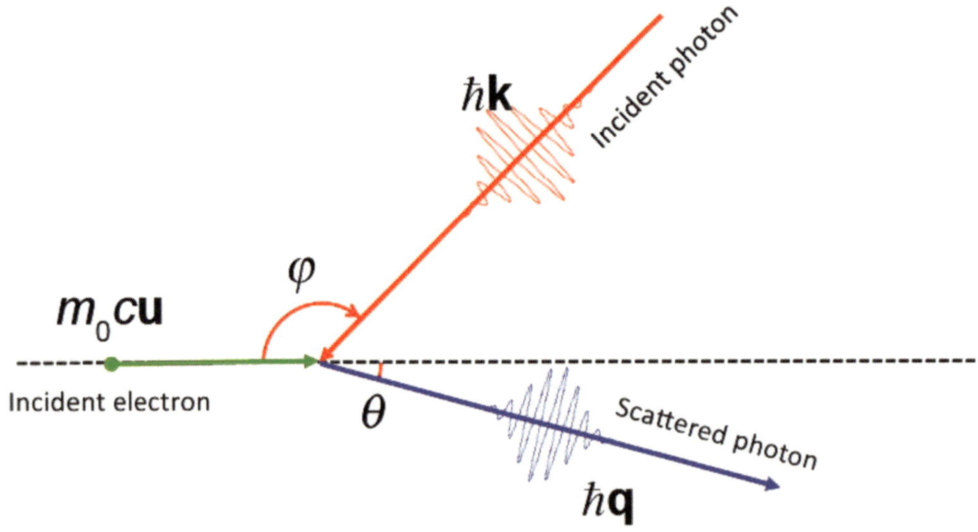

Fig. 2. Definition of the Compton scattering geometry in the case of a single incident electron.

2.2 Spectral broadening mechanisms

In the case of a weakly focused laser pulse (typical for Compton scattering sources), the coordinates of the wavevector (k_x, k_y, k_z) remain close to their initial values $(0, 0, k_0)$.[9] The exact nonlinear plane wave solution for the 4-velocity has been derived in earlier work[10,11]:

$$u_\mu = u_\mu^0 + A_\mu - k_\mu \frac{A_\nu (A^\nu + 2u_0^\nu)}{2k_\nu u_0^\nu} \tag{16}$$

where u_μ^0 is the initial 4-velocity and A_μ is the laser 4-potential measured in units of mc/e. For brevity, QED units are adopted for the remainder of this section: charge is measured in units of the electron charge *e*, mass in units of the electron rest mass m, length in units of the reduced electron Compton wavelength $\lambda_c \equiv \frac{\hbar}{mc}$, and time in units of λ_c / c.
By using the derived nonlinear 4-velocity in conjunction with the energy-momentum conservation relation, rewritten as:

$$u_\mu (k^\mu - q^\mu) = \lambda_c k_\mu q^\mu, \tag{17}$$

one immediately obtains:

$$\left(u_\mu^0 + A_\mu - k_\mu \frac{A_\nu A^\nu + 2u_\nu^0 A^\nu}{2u_\nu^0 k^\nu} \right)(k^\mu - q^\mu) = \lambda_c k_\mu q^\mu, \tag{18}$$

which after applying the Lorentz gauge condition, $k_\mu A^\mu = 0$, and the vacuum dispersion relation, $k_\mu k^\mu = 0$, simplifies to:

$$u_\mu^0 k^\mu - \left(u_\mu^0 - \frac{k_\mu}{2u_\nu^0 k^\nu} \langle A_\nu A^\nu \rangle \right) q^\mu = \lambda_c k_\mu q^\mu. \tag{19}$$

This new relation may be understood as a modified form of the Compton formula, now including the nonlinear ponderomotive force of the laser field. It is important to note that the general form of the laser potential is, for a linearly polarized plane wave, $(A_0 g(\phi)\cos(\phi), 0, 0)$, where ϕ is the phase and $g(\phi)$ the laser pulse envelope. Hence nonlinear effects are a direct consequence of the inhomogeneous nature of the laser electrical field.
In the geometry described in Figure 3, Eq. (12) can be rewritten using Eq. (19) as:

$$\frac{q}{k} = \frac{\gamma - u\cos(\epsilon + \varphi)}{\gamma - u\cos(\theta + \epsilon) + [1 - \cos(\varphi + \theta + \epsilon)] \left[\frac{\langle -A_\mu A^\mu \rangle}{2[\gamma - u\cos(\varphi + \epsilon)]} + \lambda_c k \right]}. \tag{20}$$

Here the small angle ϵ is different for each electron and represents the emittance of the electron beam. Note also that $\langle -A_\mu A^\mu \rangle$ is the nonlinear radiation pressure. By looking at the variation of q as a function of all the different parameters in Eq. (20), one finds for on-axis observation, $\theta = 0$:

$$\begin{aligned}
&\Delta q / q \propto \Delta k / k, \\
&\Delta q / q \propto -\frac{1}{4} \Delta \varphi^2, \\
&\Delta q / q \propto 2\Delta \gamma / \gamma, \\
&\Delta q / q \propto -\gamma^2 \Delta \epsilon^2, \\
&\Delta q / q \propto -\frac{\Delta A^2}{1 + A^2}.
\end{aligned} \tag{21}$$

While the gamma-ray spectral width depends directly on the electron and laser energy spreads, it is also strongly affected by the electron beam emittance because of the γ^2 factor. This provides a quick overview of the various sources of spectral broadening in a Compton scattering light source. Note that the negative variations are asymmetric broadening toward lower photon energies.

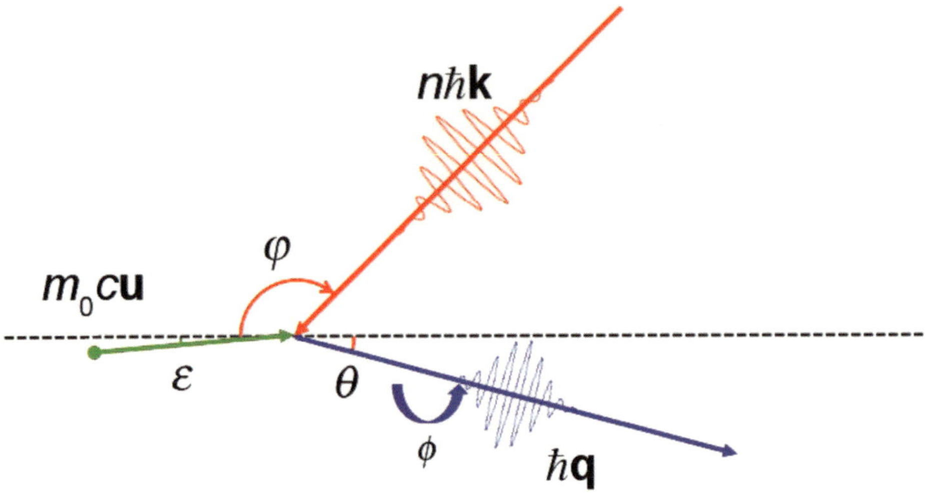

Fig. 3. Definition of the Compton scattering geometry in the case of an electron beam with some finite transverse emittance ϵ.

2.3 Compton scattering modeling

The Compton formula derived above provides a good approximation for the on-axis spectrum and within small angles ($\theta \ll 1/\gamma$) of radiation. In general, one has to take into account the differential cross section to derive the source brightness. In Compton scattering, the total number of photons scattered by an electron distribution is given by:

$$N = \int_{R^4} \sigma n_e n_\lambda \frac{u_\mu k^\mu}{\gamma k} d^4x. \tag{22}$$

In the case of a single electron, where $n_e = \delta[\mathbf{x}(\tau) - \mathbf{x}]$, the integral over space yields:

$$N = \int_{-\infty}^{+\infty} \sigma \frac{\kappa}{\lambda k} n_\lambda [x_\mu(\tau)] c dt. \tag{23}$$

By differentiating the equation above, the number of photons scattered per unit frequency and solid angle is derived, assuming that in the case of an uncorrelated incident photon phase space, corresponding to the Fourier transform limit, the phase space density takes the form of a product, $n_\lambda = n_\lambda(x_\mu)\tilde{n}_\lambda(k_\nu)$:

$$\frac{d^2N}{dq d\Omega} = \int_{-\infty}^{+\infty} \tilde{n}_\lambda(k)\delta(q - q_0) \int_{-\infty}^{+\infty} \frac{d\sigma}{d\Omega} \frac{\kappa}{\gamma k} n_\lambda [x_\mu(\tau)] c dt dk \tag{24}$$

The quantity $d\sigma / d\Omega$ is the differential cross-section in Eq. (11). In turn, the integral over k can be formally performed to obtain

$$\frac{d^2N}{dqd\Omega} = \frac{d\sigma}{d\Omega}\frac{\kappa}{\gamma k}\left[\frac{\tilde{n}_\lambda(k)}{|dq_0 / dk|}\right]_{k=k_p} \int_{-\infty}^{+\infty} n_\lambda[x_\mu(\tau)]cdt. \tag{25}$$

For a Gaussian laser pulse, the photon density in the Fourier domain is

$$\tilde{n}_\lambda = \frac{1}{\sqrt{\pi}\Delta k}\exp\left[-\left(\frac{k-k_0}{\Delta k}\right)^2\right] \tag{26}$$

and the incident photon density can be modeled analytically within the paraxial approximation, and in the case of a cylindrical focus:

$$n_\lambda(\mathbf{x},t) = \frac{N_\lambda}{(\pi / 2)^{3/2} w_0^2 c\Delta t}\frac{1}{1+(z/z_0)^2}\exp\left[-2\left(\frac{t-z/c}{\Delta t}\right)^2 - 2\frac{r^2}{w_0^2[1+(z/z_0)^2]}\right], \tag{27}$$

where N_λ is the total number of photons in the laser pulse, Δt the pulse duration, w_0 the $1/e^2$ focal radius and $z_0 = \pi w_0^2 / \lambda_0$ is the Rayleigh range. To evaluate the integral in Eq. (25), we replace the spatial coordinates by the ballistic electron trajectory:

$$\begin{aligned} x(t) &= x_0 + \frac{u_x}{\gamma}ct \\ y(t) &= y_0 + \frac{u_y}{\gamma}ct \\ z(t) &= z_0 + \frac{u_z}{\gamma}ct \\ r^2(t) &= x^2(t) + y^2(t) \end{aligned} \tag{28}$$

where we can divide x, y and r by w_0 and z and ct by z_0 to obtain the normalized quantities $\bar{x}$, $\bar{y}$, $\bar{z}$, $\bar{r}$ and $\bar{t}$. One finally obtains the expression:

$$\begin{aligned} \frac{d^2N}{dqd\Omega} &= \frac{1}{\sqrt{\pi}\Delta k}\left[\frac{d\sigma}{d\Omega}\frac{\kappa}{\gamma k}\frac{e^{-(k-k_0)^2/\Delta k^2}}{|\partial_k q_c(k)|}\right]_{k=k_p} \frac{N_\lambda}{\sqrt{\pi/2}^3 w_0^2 c\Delta t} \\ &\times \int_{-\infty}^{\infty}\frac{1}{1+\bar{z}^2}\exp\left[-2\left(\frac{z_0}{c\Delta t}\right)^2(\bar{t}-\bar{z})^2 - 2\frac{\bar{r}^2}{1+\bar{z}^2}\right]d\bar{t}. \end{aligned} \tag{29}$$

Example spectra for the photon source developed at LLNL are shown in Figure 4. It also shows the importance of considering recoil in the case of narrowband gamma-ray operation. The Klein-Nishina formalism presented above is well-suited to model recoil, but does not take into account nonlinear effects. Within the context of laser-plasma and laser electron interactions, nonlinear effects are neglected unless the normalized laser potential A approaches unity. $A = \sqrt{-A_\mu A^\mu}$ is most commonly described in practical units:

$A = 8.5 \times 10^{-10} \lambda_0 I^{1/2}$, where λ_0 is the laser wavelength measured in μm and I the laser intensity in units of W / cm^2. Typically, $A \approx 0.1$ or less for today's Compton sources. It has been shown that, despite being in a regime where $A^2 \ll 1$, nonlinear effects can strongly increase the width of the gamma-ray spectra[11].

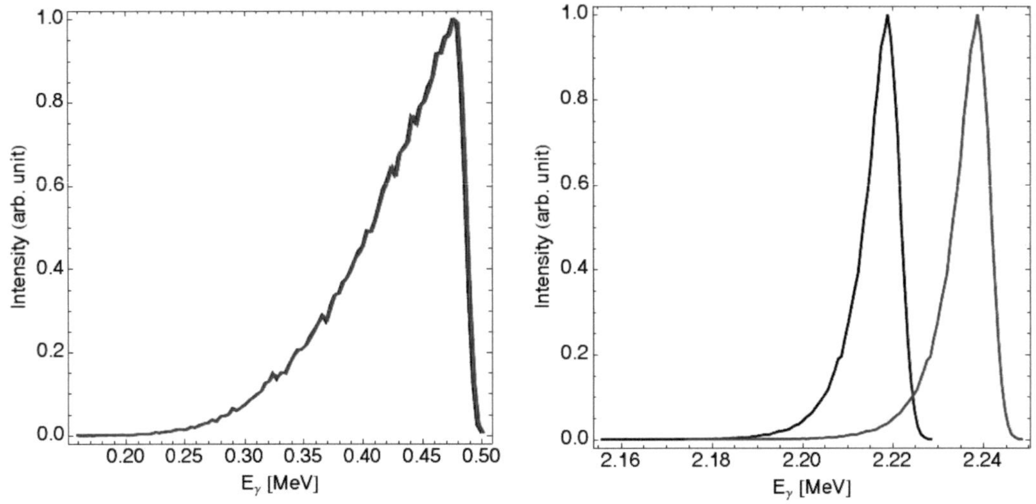

Fig. 4. Gamma-ray spectra in the case of Compton scattering (black) and Thomson scattering (blue). Left: T-REX source parameters (laser: 20 ps FWHM pulse duration, 532 nm wavelength, 34 μm rms spot size, 150 mJ; electron beam: 116 MeV, 40 μm rms spot size, 20 ps FWHM bunch length, 0.5 nC beam charge, 6 mm.mrad normalized emittance). Right: future Compton source parameters (laser 10 ps FWHM pulse duration, 532 nm wavelength, 12 μm rms spot size, 150 mJ; electron beam: 250 MeV, 15 μm rms spot size, 10 ps FWHM bunch length, 0.25 nC beam charge, 1 mm.mrad normalized emittance).

For completeness, we will briefly describe an alternative formalism based on classical radiation theory. Nonlinear spectra can also be calculated from the electron trajectories by using the covariant radiation formula that describes the number of photons scattered per unit frequency and solid angle:

$$\frac{d^2N}{dqd\Omega} = \frac{\alpha}{4\pi^2 q} \left| \mathbf{q} \times \int_{-\infty}^{+\infty} \mathbf{u}(\tau) e^{-iq_\nu x^\nu(\tau)} cd\tau \right|^2 \tag{30}$$

where α is the fine structure constant, $q^\mu = (q, \mathbf{q})$ the scattered 4-wavenumber and $u^\mu = dx^\mu / cd\tau = (\gamma, \mathbf{u})$ the electron 4-velocity along its trajectory. Eq. (17) can be generalized to describe the scattering process for nonlinear, three-dimensional, and correlated incident radiation. On the other hand, Eq. (25) properly satisfies energy-momentum conservation, therefore accounting for recoil, and also includes QED corrections when using the Klein-Nishina differential scattering cross-section. The two approaches yield complementary information, but are mutually incompatible, and coincide only for Fourier transform-limited (uncorrelated) plane waves in the linear regime and in the limit where $\lambda_c \to 0$. Returning to Eq. (24) and comparing it with Eq. (30), it is clear that the quantities $d\sigma / d\Omega$ and $|\mathbf{q} \times \mathbf{u}|^2$

play equivalent roles, and the Dirac delta-function $\delta(q-q_0)$ is related to the Fourier transform in the case of an infinite plane wave. Using Eq. (30), important modifications to the radiation spectrum under strong optical fields using the classical radiation formula have been predicted. The interested reader may peruse Refs. 9 and 11 for more details.

3. Laser technology

This chapter describes some of the main concepts necessary for understanding laser technology underlying high brightness, Compton scattering based mono-energetic gamma-ray sources. Using the recently commissioned T-REX device at LLNL as a model[12,13], the two main components are a low emittance, low energy spread, high charge electron beam and a high intensity, narrow bandwidth, counter-propagating laser focused into the interaction region where the Compton scattering will occur. A simplified schematic of the Compton-scattering source is shown in Figure 5. The laser systems involved are among the current state-of-the art.

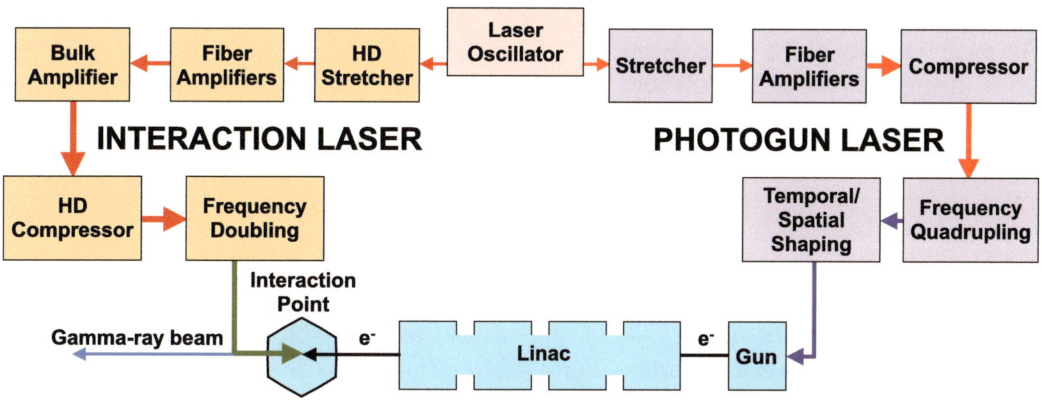

Fig. 5. Block diagram of the LLNL MEGa-ray Compton source with details of the laser systems.

An ultrashort photogun laser facilitates generation of a high charge, low emittance electron beam. It delivers spatially and temporally shaped UV pulses to RF photocathode to generate electrons with a desired phase-space distribution by the photoelectric effect. Precisely synchronized to the RF phase of the linear accelerator, the generated electrons are then accelerated to relativistic velocities. The arrival of the electrons at the interaction point is timed to the arrival of the interaction laser photons. The interaction laser is a joule-class, 10 ps chirped pulse amplification system. A common fiber oscillator, operating at 40 MHz repetition rate, serves as a reference clock for the GHz-scale RF system of the accelerator, seeds both laser systems and facilitates sub-picosecond synchronization between the interaction laser, the photogun laser, and the RF phase. Typical parameters for the laser system are listed in Table 1.

To help understand the function and the construction of these laser systems, this chapter we will provide a brief overview of the key relevant concepts, including mode-locking, chirped-pulse amplification, nonlinear frequency generation, laser diagnostics and pulse characterization techniques. We will also describe the most common laser types involved in generation of energetic ultrashort pulses: solid state Nd:YAG and Titanium:Sapphire, and fiber-based Yb:YAG.

Parameters	Oscillator	Photogun Laser	Interaction Laser
Repetition Rate	40.8 MHz	10-120 Hz	10-120 Hz
Wavelength	1 μm	263 nm	532 nm or 1 μm
Pulse Energy	100 pJ	30-50 μJ	150 mJ - 800 mJ
Pulse Duration	1 ps	2 or 15 ps	10 ps
Spot size on target	n/a	1-2 mm	20-40 μm RMS

Table 1. Summary of key laser parameters for T-REX and MEGa-ray Compton scattering light source.

3.1 Chirped pulse amplification

Chirped pulse amplification, invented over two decades ago[14], allow generation of highly energetic ultrashort pulses. The key concept behind CPA is to increase the pulse duration prior to amplification, thereby reducing the peak intensity during the amplification process. The peak intensity of the pulse determines the onset of various nonlinear processes, such as self-focusing, self-phase modulation, multi-photon ionization that lead to pulse break-up and damage to the amplifying medium. A measure of nonlinear phase accumulation, the "B-integral", is defined by:

$$\phi_{NL} = \frac{2\pi}{\lambda}\int_{-\infty}^{\infty} n_2 I(z)dz \tag{31}$$

where $I(z)$ is a position dependent pulse intensity and n_2 is the material dependent nonlinear refractive index ($n = n_0 + n_2 I$). ϕ_{NL} should normally be kept below 2-3 to avoid significant pulse and beam distortion. For a desired final pulse intensity, we minimize the value of the accumulated B-integral by increasing the beam diameter and the chirped pulse duration. After amplification, the stretched pulse is recompressed to its near transform limited duration. The four stages of the CPA process are illustrated in Figure 6.

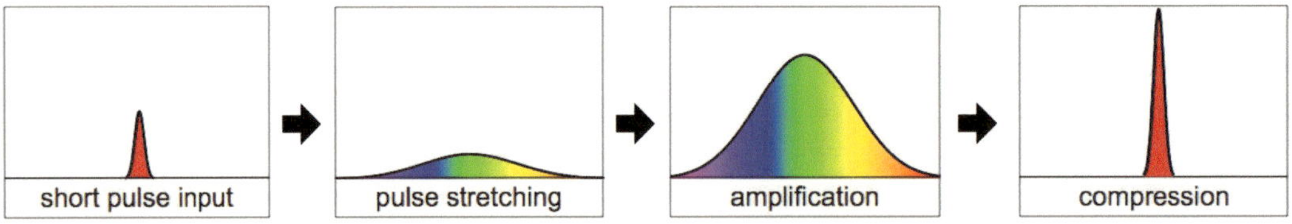

Fig. 6. Basic CPA scheme: ultrashort seed pulse is stretched, amplified, and recompressed.

The seed pulse in any CPA system is produced by an ultrashort oscillator. Typically, the pulse energy from a fiber oscillator is in the 10 sec of pico-Joule range, and 10 sec of nJ from a solid-state oscillator. The oscillator pulse is then stretched in time and amplified. Depending on the system, pulse stretching can occur in a bulk grating stretcher, prism-based stretcher, chirped fiber Bragg grating, or chirped volume Bragg grating. Stretching and amplification can occur in several stages depending on the final energy. Initial pre-amplifiers provide very high gain (up to 30 dB or more is possible) to boost the pulse energy to mJ levels. Power amplifiers then provide lower gain but higher output energy. Many amplifier stages can be multiplexed. State of the art custom systems can produce over 1 PW of peak power, corresponding to for example 40 J in 40 fs.[15,16] Many implementations of the chirped pulse amplification have been demonstrated. Typically, amplifiers consist of high bandwidth laser gain medium, such as Ti:Sapphire or Yb:YAG. The laser crystal is pumped by an external source to produce population inversion prior to the arrival of the seed pulse. The seed pulse then stimulates emission and grows in energy.

In another amplification scheme, called optical parametric chirped pulse amplification (OPCPA), no energy is stored in the medium. Instead, gain is provided through a nonlinear difference frequency generation process. A very high quality, energetic narrowband pump laser is coincident on a nonlinear crystal at the same time as the chirped seed pulse. The seed pulse then grows in energy through a parametric three-photon mixing process. The advantage of this scheme is that no energy is stored in the medium and the nonlinear crystal is much thinner compared to a laser crystal. Hence, thermal management and nonlinear phase accumulation issues are mitigated in OPCPA. A key difficulty to implementing OPCPA, is that the pump laser has to be of very high quality.

After the amplification stage, pulses are recompressed in a pulse compressor. For joule class final pulses, recompression occurs exclusively in grating based compressors. Detailed dispersion balance is required to produce a high fidelity final pulse. Details of pulse stretchers and compressors are provided in the following section.

3.2 Pulse stretching and compressing

The action of various dispersive elements is best described in the spectral domain. Given a time dependent electric field, $E(t) = A(t)e^{i\omega_0 t}$, where we factor out the carrier-frequency term, its frequency domain is the Fourier transform:

$$\mathcal{A}(\omega) = \int_{-\infty}^{\infty} A(t)e^{-i\omega t}dt = \sqrt{I_s(\omega)}e^{i\phi(\omega)}. \tag{32}$$

Here, we explicitly separate the real spectral amplitude $\sqrt{I_s(\omega)}$, and phase $\phi(\omega)$. We define spectral intensity $I_s(\omega) \equiv |\mathcal{A}(\omega)|^2$, and temporal intensity $I(t) \equiv |A(t)|^2$ We can Taylor expand the phase of the pulse envelope $\mathcal{A}(\omega)$ as

$$\phi(\omega) = \sum_{n=0}^{\infty} \frac{1}{n!}\phi^{(n)}(0)\omega^n \tag{33}$$

where $\phi^{(n)}(0) \equiv \frac{d^n\phi(\omega)}{d\phi^n}$ evaluated at $\omega = 0$. The $n = 1$ term is the group delay, corresponding to the time shift of the pulse; terms $n \geq 2$ are responsible for pulse dispersion. The terms $\phi^{(n)}(0)$, where $n = 2$ and 3, are defined as group delay dispersion (GDD) and third order dispersion (TOD), respectively. For an unchirped Gaussian pulse, $E(t) = E_0 e^{-\frac{t^2}{\tau_0^2}}$, we can analytically calculate its chirped duration, τ_f assuming a purely quadratic dispersion ($\phi^{(n)}(0) = 0$ for $n > 2$).

$$\tau_f = \tau_0 \sqrt{1 + 16 \frac{GDD^2}{\tau_0^4}} \tag{34}$$

The transform limited pulse duration is inversely proportional to the pulse bandwidth. For a transform limited Gaussian, the time-bandwidth product (FWHM intensity duration [sec] × FWHM spectral intensity bandwidth [Hz]) is $\frac{2}{\pi}\log 2 \approx 0.44$. From Eq. (34), for large stretch ratios, the amount of chirp, GDD, needed to stretch from the transform limit, τ_0, to the final duration, τ_f, is proportional to τ_0.

In our CPA system, a near transform limited pulse is chirped from 200 fs (photogun laser) and 10 ps (interaction laser) to a few nanosecond duration. Options for such large dispersion stretchers include chirped fiber Bragg gratings (CFBG), chirped volume Bragg gratings (CVBG), and bulk grating based stretchers. Prism based stretchers do not provide sufficient dispersion for our bandwidths. The main attraction of CFBG and CVBG is their extremely compact size and no need for alignment. Both CFBG and CVBG have shown great recent promise but still have unresolved issues relating to group delay ripple that affect recompressed pulse fidelity[17]. High pulse contrast systems typically utilize reflective grating stretchers and compressors which provide a smooth dispersion profile. Grating stretchers and compressors are guided by the grating equation which relates the angle of incidence ψ and the angle of diffraction ϕ measured with respect to grating normal for a ray at wavelength λ,

$$\sin\psi + \sin\phi = \frac{m\lambda}{d} \tag{35}$$

where d is the groove spacing and m is an integer specifying the diffraction order. Grating stretchers and compressors achieve large optical path differences versus wavelength due to their high angular resolution[18,19].

A stretcher imparts a positive pulse chirp (longer wavelengths lead the shorter wavelengths in time) and a compressor imparts a negative pulse chirp. The sign of the chirp is important, because other materials in the system, such as transport fibers, lenses, and amplifiers, have positive dispersion. A pulse with a negative initial chirp could become partially recompressed during amplification and damage the gain medium.

The dispersion needs to be balanced in the CPA chain to recompress the pulse to its transform limit. The total group delay, GD_{total} versus wavelength for the system must be a constant at the output, or

$$\begin{aligned} GD_{total}(\lambda) = GD_{stretcher}(\lambda) + GD_{compressor}(\lambda) + GD_{fiber}(\lambda) + \\ + GD_{oscillator}(\lambda) + GD_{amplifier}(\lambda) + GD_{material}(\lambda) = C_1 \end{aligned} \tag{36}$$

where C_1 is an arbitrary constant. From Eq. (33), $\phi(\omega) = \int_{-\infty}^{\infty} GD(\omega) d\omega$. Any frequency dependence in the total group delay will degrade pulse fidelity.

Several approaches to achieving dispersion balance exist. In the Taylor expansion (Eq. (33)) to the phase of a well behaved dispersive elements, when $n < m$ the contribution of a term n to the total phase is much larger than of a term m, or $\left|\phi^{(n)}(0)\Delta\omega^n / n!\right| \gg \left|\phi^{(m)}(0)\Delta\omega^m / m!\right|$, where $\Delta\omega$ is the pulse bandwidth. One approach to achieve dispersion balance is to calculate the total GDD, TOD, and higher order terms for the system and attempt to zero them. For example, when terms up to the 4th order are zero, the system is quintic limited. This term-by-term cancellation approach can become problematic when successive terms in the Taylor expansion do not decrease rapidly. It is often preferable to minimize the residual group delay, GD_{RMS}, over the pulse bandwidth,

$$GD_{RMS} = \sqrt{\frac{\int_{-\infty}^{\infty}\left(GD(\omega)-\overline{GD}\right)^2 |\mathcal{E}(\omega)|^2 \, d\omega}{\int_{\infty}^{\infty}|\mathcal{E}(\omega)|^2 \, d\omega}} \tag{37}$$

where $\overline{GD}$ is the mean group delay defined as

$$\overline{GD} \equiv \frac{\int_{-\infty}^{\infty} GD(\omega)|\mathcal{E}(\omega)|^2 \, d\omega}{\int_{-\infty}^{\infty}|\mathcal{E}(\omega)|^2 \, d\omega}. \tag{38}$$

A sample pulse stretcher is shown in Figure 7. The stretcher consists of a single grating, a lens of focal length *f*, a retro-reflecting folding mirror placed *f* away from the lens and a vertical roof mirror. The folding mirror simplifies stretcher configuration and alignment by eliminating a second grating and lens pair. A vertical roof mirror double passes the beam through the stretcher and takes out the spatial chirp. The total dispersion of the stretcher is determined by the distance from the grating to the lens, L_f. When $L_f = f$, the path lengths at all wavelengths are equal and the net dispersion is zero. When $L_f > f$, the chirp becomes negative, same as in the compressor. In a stretcher, $L_f < f$ producing a positive chirp. We can calculate the dispersion, or $GD(\omega) = n_\omega L(\omega)/c$, where $L(\omega)$ is the frequency dependent propagation distance and n_ω is the frequency dependent refractive index, using various techniques.

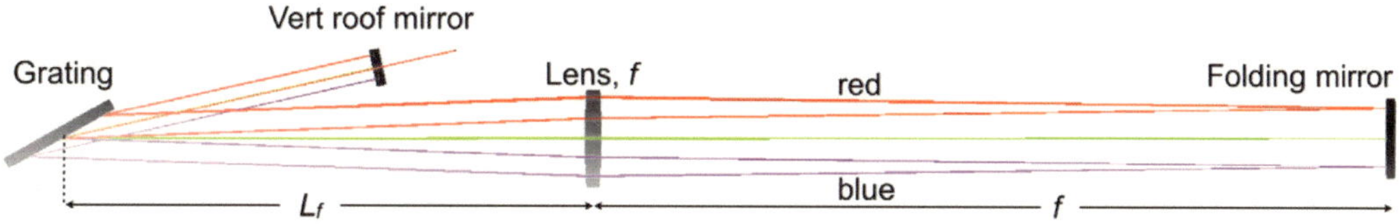

Fig. 7. Grating and lens based stretcher is folded with a flat mirror. The distance from the grating to the focal length of the lens determine the sign and magnitude of the chirp.

Here, we give a compact equation for $GDD(\omega) = \frac{\partial GD(\omega)}{\partial \omega}$, from which all other dispersion terms can be determined. Assuming an aberration free stretcher in the thin lens approximate, $GDD(\omega)$ is then given by

$$GDD(\omega) = 2(f - L_f)\frac{\lambda}{\pi c^2}\frac{(\sin\phi + \sin\psi)^2}{\cos^2\phi} \tag{39}$$

where the diffraction angle, ϕ is a function of ω. The aberration free approximation is important, because a real lens introduces both chromatic and geometric beam aberrations which modify higher order dispersion terms from those derived from Eq. (39).

A compressor, which is typically a final component in the CPA system, undoes all of the system dispersion. The compressor used on the photogun laser is shown in Figure 8. Compared to the stretcher, the lens is absent. The magnitude of the negative chirp is now controlled by the distance from the horizontal roof to the grating. The horizontal roof, here folds the compressor geometry and eliminates a second grating. As in the stretcher of Figure 7, the compressor's vertical roof mirror double passes the beam and removes the spatial chirp. Because a compressor has no curved optics, it introduces no geometric or chromatic aberrations. The dispersion of a real compressor is very precisely described by the Treacy formula or its equivalent given by

$$GDD(\omega) = -L_1 \frac{\lambda}{\pi c^2} \frac{(\sin\phi + \sin\psi)^2}{\cos^2\phi} \quad (40)$$

The actual stretcher design for MEGa-ray photogun laser is shown in Figure 9. The all-reflective Offner design uses a convex and a concave mirror to form an imaging telescope with magnification of -1 and relay planes at the radius of curvature (ROC) center of the concave mirror[20]. The grating is placed inside the ROC to impart a positive chirp. A vertical

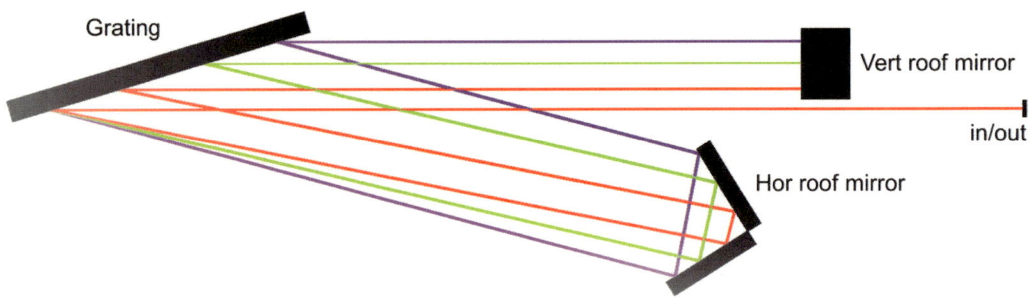

Fig. 8. Folded grating compressor. The distance from the horizontal roof mirror to the grating determines the magnitude of the negative chirp.

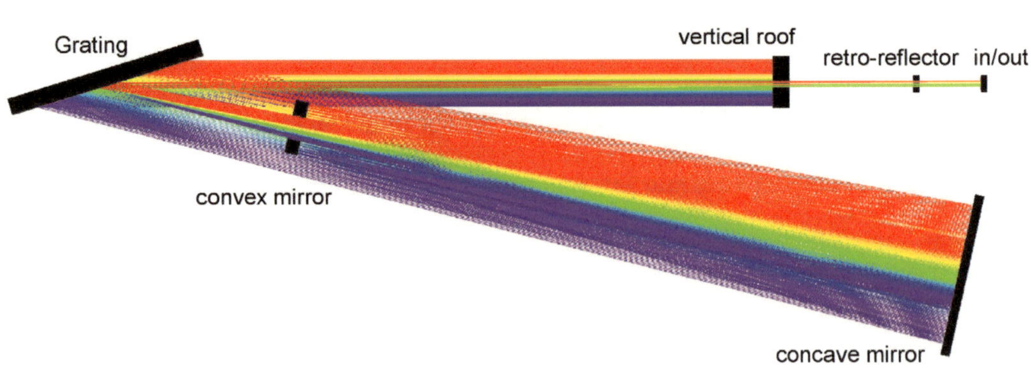

Fig. 9. Raytraced design of the Offner stretcher for the photogun laser.

roof mirror folds the stretcher geometry and eliminates the second grating. The Offner stretcher is compact and has low chromatic and geometric aberrations. Consequently, its dispersion profile is nearly aberration free, meaning the GDD and higher order dispersion terms are closely predicted by Eq. (39).

3.3 Ultrashort oscillators

Essential to building a high energy ultrashort laser system is a highly stable, robust oscillator. The oscillator serves as a front end, seeding an amplifier chain. Modelocking process generates pulses with duration from few picoseconds down to few femtoseconds. Modelocking is at the heart of all ultrashort laser systems. Some more exotic nonlinear techniques have generated few femtoseconds pulses without utilizing modelocking[21,22]; however, these are still at a laboratory research stage. The basic idea of modelocking consists of adjusting the relative phases of all the cavity laser modes to be in-phase. The phase adjustment occurs either passively or actively.
Consider a simple two mirror cavity of length L. Confining our discussion to TEM_{00} modes, the allowed mode spacing in the frequency domain is $\Delta\nu = c / 2n_g L$, where c is the speed of light and n_g is the group refractive index, $n_g = n(\omega) + \omega \frac{\partial n}{\partial \omega}$. $\Delta\nu$ is also known as the free spectral range (FSR). For higher order modes, the frequency spacing is slightly modified because the Guoy phase shift increases with mode number. While the mode spacing is typically in the few GHz to 100s of MHz range, the total bandwidth for 100 fs pulse ≈10 THz. This results in ≈10000 laser modes in the cavity. In the time domain, total electric field can be expressed as

$$E(t) = \frac{1}{2}\sum_{q=1}^{N} E_q \exp(i\omega_q t + i\phi_q) + c.c. \tag{41}$$

where the sum is taken over all N modes, and ϕ is the phase associated with each mode. When all the phases are equal, the pulse intensity is a periodic function with pulse to pulses spacing given by $1/\Delta\nu$. Assuming Gaussian pulses, the pulse width becomes $\Delta t = 0.44 / N\Delta\nu$, where Δt is the $1/e^2$ pulse duration and $N\Delta\nu$ is equivalent to $1/e^2$ pulse bandwidth.
The idea behind modelocking consists of introducing a variable cavity loss condition. The cavity loss is engineered so that the pulse with the highest intensity will experience the smallest loss. The modelocking mechanism can be either passive or active. In active modelocking, an active element, such as an acousto-optic or an electro-optic modulator periodically modulates the cavity loss. In the time domain, the modulator acts as a shutter which lets light through only when it is open. Using a fast modulator synchronized to the cavity roundtrip time, we can obtain pulses as short as a few picoseconds. This type of mechanism is known as amplitude modulation (AM) modelocking.
Active modelocking can also occur by modulating the phase of the light with an electro-optic phase modulator. The idea, here, is that light passing through a phase modulator acquires sidebands at upshifted and downshifted frequencies. When the frequency shifted sidebands are outside the gain bandwidth of the laser medium, they are attenuated after many roundtrips. When the phase modulator is synchronized with the cavity roundtrip

time, only the sidebands which pass through the modulator at zero frequency shift will be amplified. Therefore, a short pulse experiences the largest gain. For a homogeneous laser, the primary mechanism that limits pulse duration of an actively modelocked laser is gain narrowing. The pulse duration, τ_p can be calculated for a steady state case by[23]

$$\tau_p \approx 0.45\left(\frac{G}{\Delta_m}\right)^{1/4} \times \left(\frac{1}{\nu_m \Delta\nu_g}\right)^{1/2} \tag{42}$$

where G is the roundtrip gain, Δ_m is the modulation depth, ν_m is the modulation frequency, and $\Delta\nu_g$ is the FWHM gain bandwidth. From this equation, the highest modulation frequency limits the final pulse duration. Pulses shorter than a few picoseconds are difficult to generate using this technique because of the limit on the driving electronics.

In contrast to active modelocking, passive modelocking does not involve external modulation, and can produce pulses as short as ≈5 fs. The basic idea, here, is to use a saturable absorber inside the cavity. The saturable absorber has a loss which decreases with intensity. Hence, the highest intensity pulses experience the largest cavity gain. A saturable absorber can also be described as a very fast shutter, much faster than an actively driven device. Implementation of the saturable absorber takes several forms. Bulk lasers typically rely on either a semiconductor saturable absorber mirror (SESAM) or on Kerr-lens modelocking (KLM). A SESAM is a multi-layer semiconductor device which reflects light due to Bragg reflections off the periodic structure. A thin absorbing top layer saturates with intensity, increasing the mirror reflectivity. SESAM structures are used on various types of diode pumped ultrashort laser operating in the 1 µm and longer spectral range.

The shortest pulses to date are generated with Ti:Sapphire systems and rely on Kerr-lens modelocking. Kerr lensing is a third-order nonlinear effect due to intensity dependence of the refractive index, $n = n_0 + n_2 I$, where I is the pulse intensity. The intensity dependence of the refractive index leads to beam self-focusing when $n_2 > 0$. The nonlinear medium acts as a positive lens. Inside an oscillator, this implies that the spatial mode of a short pulse is smaller than the spatial mode of the longer pulse. By introducing a small aperture, we can filter out the larger spatial modes. The modelocked pulse will then experience the largest gain in the resonator.

Finally, ultrashort fiber lasers have generated sub 100 fs pulses. Fibers lasers are highly portable, reliable, relatively insensitive to external perturbations, provide long term hands free operation, and can scale to average powers above 1 kW in a diffraction limited beam and CW operation[24]. For many applications, ultrashort fiber oscillators are replacing traditional solid-state systems. For example, on T-REX we employ a 10 W Yb:doped mode-locked oscillator which, when compressed, produces 250 pJ, sub 100 fs, near transform limited pulses at 40.8 MHz repetition rate with a full bandwidth from 1035 nm to 1085 nm. The oscillator, based on a design developed at Cornell[25], fits on a rack mounted, 12"×12" breadboard (see Figure 10). An example of the oscillator spectrum from the LLNL fiber oscillator is shown in Figure 11. The passively modelocked fiber oscillator is based on a type of saturable absorber process called nonlinear polarization evolution[26]. Nonlinear polarization evolution is a nonlinear process also based on the intensity dependence of the refractive index. Inside a non polarization maintaining fiber, two orthogonally polarized

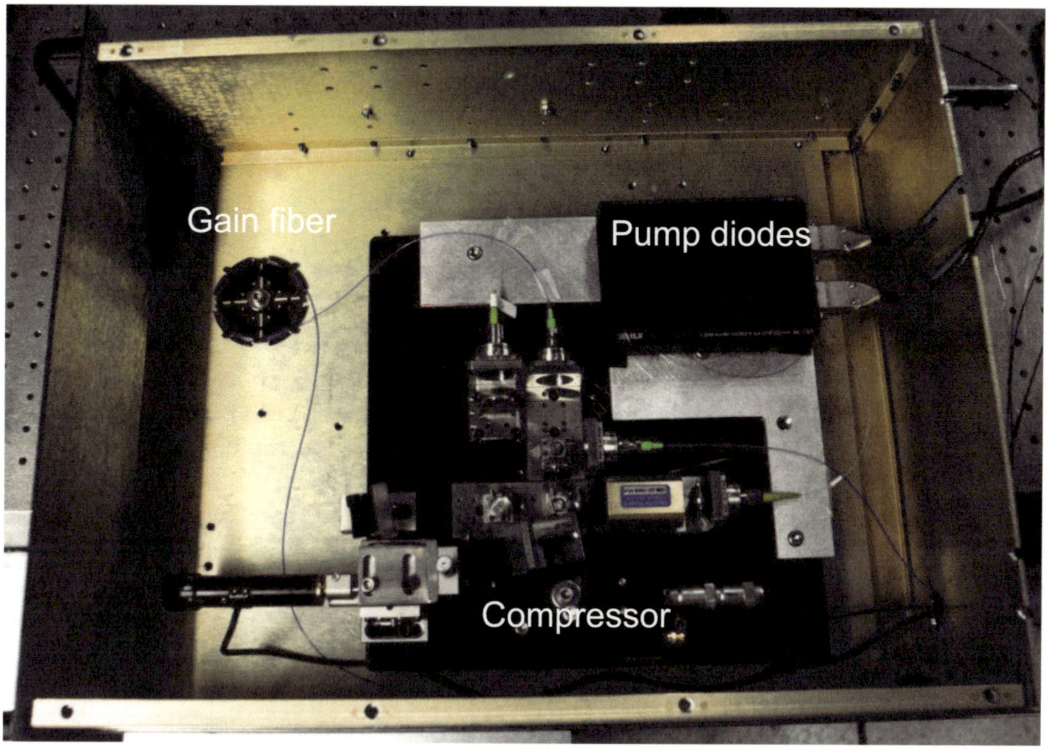

Fig. 10. The ultrashort Yb^{+} doped fiber oscillator fits on a 12"x12" breadboard. When recompressed, the anomalous dispersion oscillator generates 100 fs pulses.

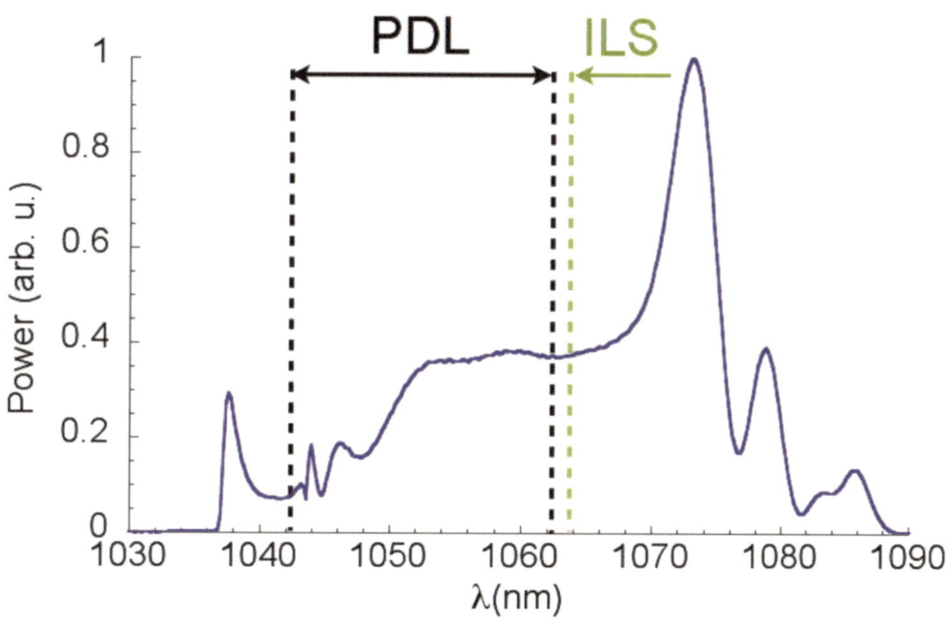

Fig. 11. Experimental pulse spectrum of the MEGa-ray oscillator of Figure 10. The same oscillator seeds both the interaction and the photogun lasers.

electric fields will experience different intensity dependent phase shifts, rotating the final polarization state. When a polarizer is placed at the output, the polarization rotation is converted to amplitude modulation. By properly selecting the length of the gain fiber and adjusting the input and output polarization state, we can produce a modelocked pulse. The fiber oscillator cavity also contains a pulse compressor which disperses the circulating pulse. The pulse compressor partially compensates for fiber dispersion. The output modelocked pulse from a fiber oscillator is typically slightly chirped.

4. Amplification of laser pulses

A laser amplifier coherently boosts the energy of the input pulse. In a standard master oscillator power amplifier (MOPA) configuration, a picojoule scale seed pulse from the oscillator can be amplified to Joule scale and above. The most extreme example is the laser at the National Ignition Facility, where a few nanojoule oscillator is amplified to a final pulse energy of 10 kJ.[27] The action of the amplifier is illustrated in Figure 12. An input pulse with intensity I_{in} is amplified to output intensity I_{out}. An amplifier can be characterized by its small signal (undepleted) gain, G_0, saturation fluence, J_{sat}, and its frequency dependent lineshape function, $g(\nu)$. In general, we need to solve the laser rate equations to determine the output pulse shape, spectral distribution, and energy. For a homogeneous gain medium, when (1) the temporal pulse variation and the population inversion is sufficiently slow (compared to T_2 coherence time) to justify rate equation analysis, and (2) transient effects relating to spontaneous emission and pumping occur on a time scale much longer than the pulse duration, an analytical solution can be obtained using a Frantz-Nodvik approach[28]. Eq. (43) is derived for a monochromatic input pulse[29].

$$I_{out}(t) = I_{in}(t) \times \left[1 - \left(1 - G_0^{-1}\right) e^{-J_{in}(t)/J_{sat}(t)}\right]^{-1} \tag{43}$$

Here, $J_{in}(t) \equiv \int_0^t I_{in}(t')dt'$ is the instantaneous fluence. Eq. (43) can be modified to describe amplification of a chirped pulse. This involves determining the frequency dependence of the small gain, $G_0(\nu)$ and the saturation fluence, $J_{sat}(\nu)$, both of which depend on the emission cross-section $\sigma_e(\nu)$. For a four-level system, $J_{sat}(\nu) = h\nu / \sigma_e(\nu)$, and $G_0 = \exp[N\sigma_e(\nu)]$, where N is the population inversion. If the instantaneous pulse frequency, $\bar{\nu} = \frac{1}{2\pi}\frac{d\phi(t)}{dt}$ is a monotonic function of t, we can define time as a function of frequency, $t(\nu)$. In this instantaneous frequency approach, we can rewrite Eq. (43) as a function of ν. Various aspects of chirped pulse amplification, such as total output energy, gain narrowing, square pulse distortion, and spectral sculpting can be analytically calculated using the modified form of Eq. (43).

A major constraint on the ultrashort pulse amplifier is that the gain medium must be sufficiently broad to support the required pulse duration. To illustrate the operation of the ultrashort oscillator combined with an amplifier, we use the T-REX laser system. A single fiber oscillator seeds both, the photogun laser and the interaction laser systems. An experimental oscillator spectrum showing the bandpass for the photogun and the interaction lasers is shown in Figure 11. Two different chains of fiber Yb:doped fiber amplifiers tuned for peak gain at 1053 nm for PDL and 1064 nm for ILS, boost the seed pulse to ≈100 μJ/pulse. The seed is chirped to a few nanosecond duration prior to amplification. Each fiber amplifier stage provides 20 dB of gain. The repetition rate is gradually stepped down to 10 kHz with acousto-optic modulators (AOM), inserted in between fiber amplifiers. The AOMs also remove out of band amplified spontaneous emission (ASE), preserving pulse fidelity.

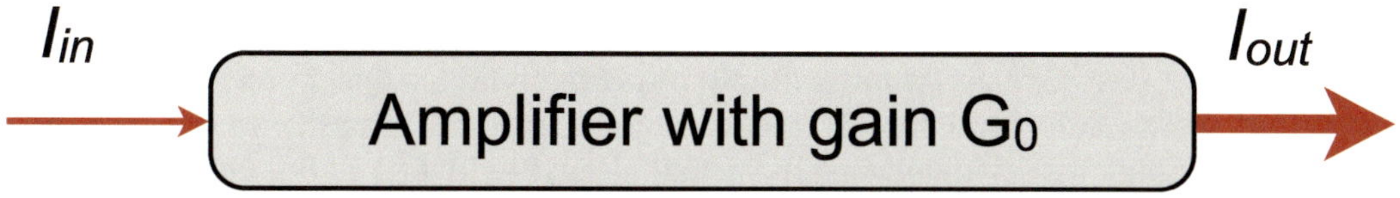

Fig. 12. Basic scheme for pulse amplification. A gain medium with stored energy coherently amplifies the input pulse.

The total energy from the fiber amplifier is limited by the total stored energy and by the accumulated nonlinear phase in Eq. (31). Several initial amplifier stages consist of telecom-type, 6.6 μm core polarization maintaining (PM) fibers. The output from this core sized fiber is limited to ≈1μJ. Next, a series of large mode field diameter, 29 μm core photonic band-gap (PBG) fibers accommodate pulse energies up to 100 μJ. We are currently developing even larger core (80 μm) PBG fiber rod based amplifiers to attain over 1 mJ per pulse at the output. A major challenge with PBG and other large core fibers is careful control of the refractive index uniformity to prevent generation of higher order spatial modes, which degrade beam quality.

4.1 Introduction to FROG

Current electronic bandwidths preclude direct measurement of pulses shorter than ≈50 psec. A wide variety of indirect characterization techniques exist for short pulses produced by mode-locked lasers. Almost exclusively, these methods rely on a nonlinear interaction between two or more pulses. A very common measurement technique is pulse intensity autocorrelation. Here, a signal is measured as a function of delay between the pulse and a replica of the pulse,

$$A(\tau) = \int_{-\infty}^{\infty} I(t)I(t-\tau)dt \,. \tag{44}$$

While this method is easy to implement it does not define the actual pulse shape. Different pulse shapes can result in the same autocorrelation trace. Intensity autocorrelation provides a rough measure of the pulse duration, when we assume that the pulse intensity envelope is relatively smooth in time and when spatio-temporal aberrations are absent. To fully characterize a pulse, we must determine the actual electric field profile in time and space $E(t,x,y,z)$. An extension of the intensity autocorrelation is interferometric autocorrelation, which involves splitting and recombining the pulse and its replica inside an interferometer. Interferometric stability of the measurement allows the user to resolve the electric field fringes in the measurement. Here the the interferometric autocorrelation signal, A_{IFAC} is given by

$$A_{IFAC} = \int_{-\infty}^{\infty} |E(t) + E(t-\tau)|^2 \, dt \tag{45}$$

Interferometric autocorrelation carries some of the phase information of the field and more accurately resolves the pulse intensity profile. However, no study has shown that the phase information can be fully recovered. Another disadvantage is that interferometric autocorrelation is time consuming to implement and necessarily involves many laser shot sequences.

Since the 1990s, several techniques have emerged which characterize both the intensity and the phase of the pulse. One of the best known techniques is Frequency Resolved Optical Grating (FROG)[30]. In its most simple implementation, a spectrometer is added to the output of the autocorrelator. By measuring the spectral intensity as a function of relative pulse delay we can determine the actual electric field $E(t)$ of the pulse. The phase information is recovered by applying an iterative algorithm to the measured pulse spectrogram. FROG can be implemented using a variety of nonlinearities. The most common is the 2nd harmonic FROG, where a nonlinear crystal is used in a sum-frequency configuration. Mathematically, it determines

$$I_{FROG}^{SHG}(\omega,\tau) = \left| \int_{-\infty}^{\infty} E(t)E(t-\tau)\exp(-i\omega t)dt \right|^2 \tag{46}$$

which is a 2D spectrogram plotting delay versus frequency. Several inversion algorithms exist to process the FROG data and extract pulse intensity and phase. Conceptually, the first pulse $E(t)$ is a probe, and its replica $E(t-\tau)$ is a gate. In a cross-correlation measurement, the gate pulse should be much shorter than the probe pulse. In practice, however, a shorter pulse is not available and in many cases not required.

Many different implementations of the FROG measurement have been implemented utilizing the 3rd order $\chi^{(3)}$ nonlinearity. These include Polarization Gated (PG-FROG), where the nonlinearity is the nonlinear polarization rotation of the probe pulse when mixed with a gate pulse; third harmonic generation (THG-FROG), where the measured signal results from the four-wave mixing process; self-diffraction (SD-FROG) where the signal results when two of the overlapping beams set-up a grating which diffracts portions of both beams along new k-vectors directions. There are various advantages and disadvantages associated with each FROG implementation relating to signal sensitivity, apparatus complexity, potential ambiguities, dynamic range, and ease of analysis. Generally, SHG FROG is best suited for characterization of oscillators and other low intensity pulses. One disadvantage of SHG FROG is a time reversal ambiguity between $E(t)$ and $E(-t)$. Both pulses produce identical FROG spectrograms. Hence we need to perform an additional procedure, such as inserting a dispersive element into the beam, to remove this ambiguity. Third order FROG automatically removes the time reversal ambiguity. It also produces FROG spectrograms that more intuitively correspond to actual pulse shapes.

FROG measurement can be implemented in a multi-shot configuration, where measurements are performed sequentially and then analyzed, or in a single-shot configuration. In a single-shot configuration, complete pulse information is gathered in a single laser shot. Properly optimized software can provide the pulse shape and phase nearly instantaneously. Similar to single-shot autocorrelation, single shot FROG measurement interferes one portion of the beam with another portion of the beam. Similar to a streak camera, one of the spatial axes becomes a time axis and a complete spectrogram (spectral intensity versus time and wavelength) is recorded during a single shot. The single-shot FROG techniques are sensitive to spatio-temporal aberrations. Spatio-temporal aberrations refer to temporal differences as a function of position on the beam. A common example includes pulse front tilt, where the pulse arrival time changes across the beam aperture. This can be observed when an ultrashort pulse come off a diffraction grating or passes through a prism. Another spatio-temporal aberration is spatial chirp, where the pulse spectrum varies across the beam aperture. All types of spatio-temporal aberrations can be visualized by representing the pulse as a 3D map, where for example, the spectral intensity is plotted versus position and frequency[31]. A straight forward single shot FROG implementation using GRENOUILLE can measure both of these types of spatio-temporal aberrations[32].

Finally, FROG technique is highly robust. Because the pulse information is overdefined, analysis of the recorded FROG traces can uncover certain systematic errors during the measurement process and the measured data set can be checked for self-consistency. The retrieval process is highly visual, as the retrieved spectrogram is easily compared to the recorded spectrogram. While every pulse has a spectrogram, not every spectrogram has an associated electric field. Improperly recorded spectrograms will not result in a retrievable pulse shape. FROG technique is extremely powerful and produces complete electric field information for both simple and ultra-complex shapes. FROG can be implemented for a variety of laser wavelengths from the UV to the IR, and across a wide range of pulse duration, from few femtosecond to 10's of picoseconds.

4.2 Chirped Pulse Amplification with narrowband pulses

In CPA, the gain bandwidth of the amplifying medium is typically broad enough to minimize gain narrowing of the seed pulse. On the photogun laser, Yb doped glass has a wide bandwidth spanning from 1000 nm to 1150 nm, and is well suited for amplifying 100 fs pulses. On the interaction laser, however, bulk amplification is in Nd:YAG, which has a narrow bandwidth of 120 GHz.

Nd:YAG technology is a workhorse of high average power lasers. The basic idea is that the gain for a Nd:YAG crystal rod is provided by either flashlamps or laser diodes. Previously, joule-class Nd:YAG amplifiers were flashlamp pumped; recently, however, diode pumped joule-class systems have been introduced commercially. Flashlamp pumping is inefficient compared to diode pumping and causes significant thermal lensing and thermal birefringence in the crystal. As a result, flashlamp pumped rod systems are limited to approximately 10-20 W. Diode pumping significantly reduces the total pump energy, enabling operation at higher repetition rates and higher average power. The diode pumped Nd:YAG based laser system for the next generation gamma-ray source at LLNL is designed to operate at nearly 100 W of average power. Even higher average powers are obtained with slab Nd:YAG amplifiers. Compared to rod geometry, slab geometry is much more efficient at removing the heat from the system. The beam aspect ratio can also be scaled more favorable to reduce unwanted nonlinear effects. However, slab rod geometries are much more difficult to engineer and are not currently commercially viable.

Here, we describe an implementation of CPA with Nd:YAG amplifiers. Due to the limited Nd:YAG bandwidth, the seed pulses gain narrow to sub nanometer bandwidths after amplification, sufficient to support 5-10 ps pulse durations. Traditional two-grating stretchers and compressors cannot provide adequate dispersion in a table-top footprint. We will describe novel hyper-dispersion technology that we developed for CPA with sub-nanometer bandwidth pulses[33]. The meter-scale stretcher and compressor pair achieve 10x greater dispersion compared to standard two-grating designs. Previously, D. Fittinghoff *et al.* suggested hyper-dispersion compressor arrangements[34]. F. J. Duarte described a conceptually similar hyper-dispersion arrangement for a prism-based compressor[35].

We utilize commercial Nd:YAG amplifiers for two reasons: (1) Nd:YAG technology is extremely mature, relatively inexpensive, and provides high signal gain; (2) nominally 10 ps transform limited laser pulses are well-suited for narrowband gamma-ray generation. Employing a hyper-dispersion stretcher and compressor pair, we generated 750 mJ pulses at 1064 nm with 0.2 nm bandwidth, compressed to near transform limited duration of 8 ps.

The nearly unfolded version of the hyper-dispersion compressor is shown in Figure 13, with a retro-mirror replacing gratings 5-8. Compared to standard Treacy design, this compressor contains two additional gratings (G2 and G3). The orientation of G2 is anti-parallel to G1: the rays dispersed at G1, are further dispersed at G2. This anti-parallel arrangement enables angular dispersion, $d\theta / d\lambda$, which is greater than possible with a single grating. The orientation of G3 and G4 is parallel to, respectively, G2 and G1. G3 undoes the angular dispersion of G2 and G4 undoes the angular dispersion of G1 producing a collimated, spatially chirped beam at the retro-mirror. After retro-reflection, the spatial chirp is removed after four more grating reflections. The number of grating reflections (8), is twice that in a Treacy compressor. High diffraction efficiency gratings are essential for high throughput efficiency. We utilize multi-layer-dielectric (MLD) gratings developed at LLNL with achievable diffraction efficiency >99%.[36] The magnitude of the negative chirp is controlled by varying L_1, the optical path length of the central ray between G1 and G2 and L_2, the optical path length between G2 and G3.

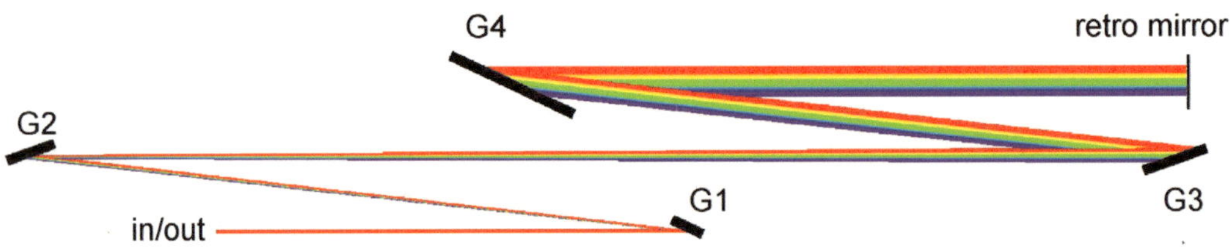

Fig. 13. Unfolded version of the hyper-dispersion compressor with anti-parallel gratings.

An analytical formula for group delay dispersion (GDD) as a function of wavelength for the compressor can be derived using the Kostenbauder formalism[37,38]:

$$GDD = -\frac{\lambda}{\pi c^2}\frac{(\sin\phi + \sin\psi)^2}{\cos^4\phi} \times \left[2L_1\cos^2\phi + L_2(\cos\phi + \cos\psi)^2\right] \tag{47}$$

Here ψ is the angle of incidence and ϕ is the angle of diffraction of the central ray of wavelength λ at the first grating measured with respect to grating normal. We assume that the groove density of gratings 1 and 2 is the same. Note that the expression for GDD reduces to that of the standard two-grating compressor when $L_2 = 0$. Higher order dispersion terms can be derived by noting that ϕ is a function of wavelength.

Compressor design can be folded to reduce the total number of gratings and simplify compressor alignment. The compressor shown in Figure 14 has been designed for the MEGa-ray machine and is similar to the experimental design on T-REX. The compressor consists of two 40×20 cm multi-layer dielectric (MLD) gratings arranged anti-parallel to each other, a vertical roof mirror (RM), and a series of two periscopes and a horizontal roof mirror. These six mirrors set the height and the position of the reflected beam on the gratings and invert the beam in the plane of diffraction.

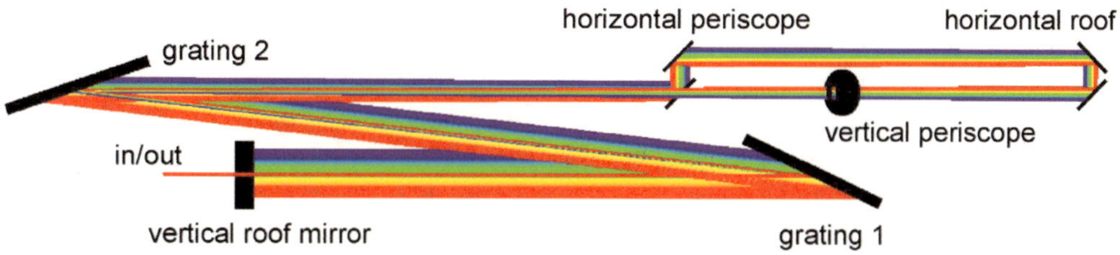

Fig. 14. The compact hyper-dispersion compressor consisting of two 1740 grooves/mm gratings has a footprint of 3x0.7 m and group delay dispersion of -4300 ps^2/rad.

The beam is incident at Littrow angle minus 3 degrees (64.8°) on the 1740 grooves/mm gratings. The vertical roof mirror here is equivalent to the retro-mirror in Figure 13. The beam undergoes a total of 8 grating reflections and 16 mirror reflections in the compressor. The total beam path of the central ray is 22 m. On T-REX, the MLD gratings had diffraction efficiency above 97%, enabling an overall compressor throughput efficiency of 60%. Here, the magnitude of the chirp is tuned by translating the horizontal roof mirror. The compressor, with its relatively compact footprint of 3×0.7 m provides $GDD = -4300$ ps^2/rad, or a pulse dispersion of 7000 ps/nm. This chirps the incident 0.2 nm bandwidth pulse to 3 ns. The two grating separation in a standard Treacy compressor with the same dispersion would be 32 m.

The hyper-dispersion stretcher design is conceptually similar. The unfolded version is shown in Figure 15. Compared to the standard Martinez stretcher, the hyper-dispersion design contains two extra mirrors G1 and G4, arranged anti-parallel to G2 and G3. We define L_f as the path length from G2 to the first lens. Then the total ray path length from G1 to the first lens, $L_1 + L_f$, must be smaller than the lens focal length, f to produce a positive chirp. The magnitude of the chirp is controlled by varying the value of $L_1 + L_f - f$.

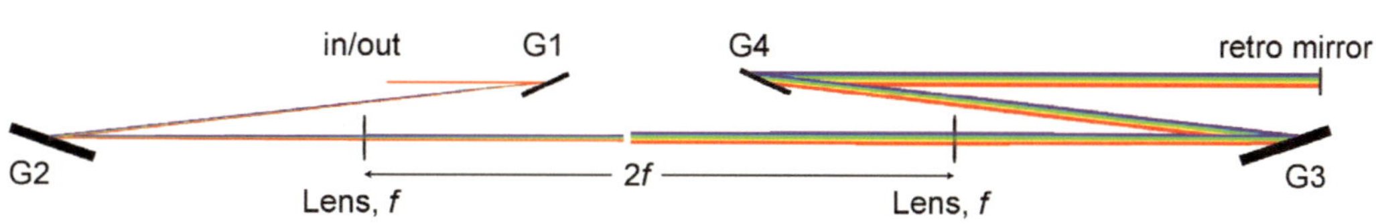

Fig. 15. Unfolded hyper-dispersion stretcher utilizes four gratings, as opposed to two gratings in the standard Martinez design.

We modify Eq. (47) to derive the GDD formula for the aberration free hyper-dispersion stretcher shown in Figure 15:

$$GDD = -\frac{\lambda}{\pi c^2}\frac{(\sin\phi + \sin\psi)^2}{\cos^4\phi} \times \left[L_1\cos^2\phi + (L_f - f)(\cos\phi + \cos\psi)^2\right] \quad (48)$$

The folded CAD version of the stretcher built for T-REX is shown in Figure 16. For high fidelity pulse recompression, the stretcher is designed with a nearly equal and opposite chirp compared to that of the compressor. The small difference accounts for other dispersive elements in the system. We again use two large 1740 grooves/mm MLD gratings, with footprints of 20×10 cm and 35×15 cm. The beam is incident at the same Littrow angle less 3° as in the compressor. A large, 175 mm diameter, $f = 3099$ mm lens accommodates the large footprint of the spatially chirped beam. A folding retro-mirror is placed *f* away from the lens, forming a 2-*f* telescope seen in the unfolded version of Figure 15. The beam height changes through off-center incidence on the lens. After 4 grating reflections, the beam is incident on the vertical roof mirror, which is equivalent to the retro-mirror in Figure 15. The two 45° mirrors fold the beam path, rendering a more compact footprint. After 8 grating reflections, the compressed pulse arrives and 2nd pass retro-mirror. The beam is then retro-reflected through the stretcher, undergoing a total of 16 grating reflections. We double pass through the stretcher to double the total pulse chirp. Beam clipping on the lens prevents reducing the lens to G2 distance to match compressor dispersion in a single pass.

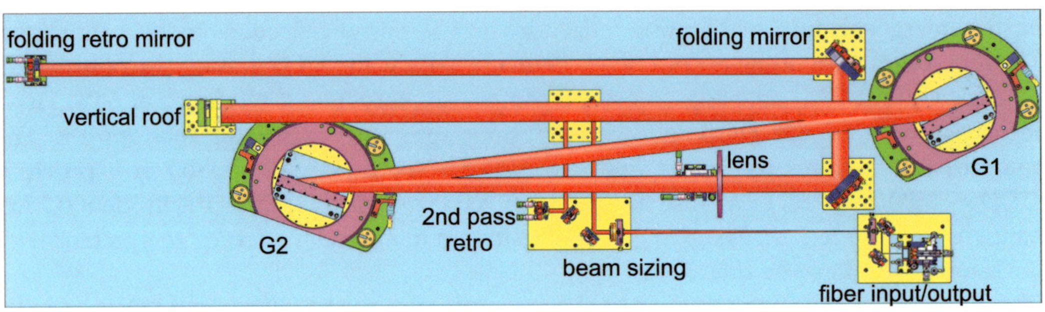

Fig. 16. The compact hyper-dispersion stretcher matches the GDD of the compressor.

Chromatic and geometric lens aberrations modify higher order dispersion terms in the stretcher, requiring raytracing for more accurate computation. We use a commercial ray-tracing software (FRED by Photon Engineering, LLC) to compute ray paths in the stretcher and in the compressor.

From raytrace analysis, the GDD for the stretcher is 4300 ps^2/rad, and the TOD/GDD ratio is -115 fs, at the 1064 nm central wavelength; for the compressor, the TOD/GDD ratio is -84 fs. The TOD mismatch would result in a 3% reduction in the temporal Strehl ratio of the compressed pulse. We can match the GDD and the TOD of the stretcher/compressor pair by a 1° increase of the angle of incidence on gratings 1 and 2 in the stretcher.

We employed the hyper-dispersion stretcher-compressor pair in our interaction laser. Commercial Q-switched bulk Nd:YAG laser heads amplified stretched pulses from the fiber chain to 1.3 J, with 800 mJ remaining after pulse recompression. We characterize the compressed pulse temporal profile using multi-shot second harmonic generation (SHG) frequency resolved optical gating (FROG) technique[39,40].

In the measurement, we use a 0.01 nm resolution 1 m spectrometer (McPherson Model 2061) to resolve the narrow bandwidth pulse spectrum at the output of a background free SHG auto-correlator. The measured field of the FROG spectrogram, $\sqrt{I(\omega,\tau)}$, is shown in Figure 17. Numerical processing then symmetrizes the trace and removes spurious background and noise. The FROG algorithm converges to the spectrogram shown in Figure 17. The FROG algorithm discretizes the measurement into a 512×512 array. The FROG error between the measured and the converged calculated profile, defined as,

$$\Delta_{FROG} \equiv \sqrt{\frac{1}{N^2}\sum_{i,j=1}^{N}\left[I_{FROG}^{meas}(\omega_i,\tau_j) - I_{FROG}^{calculated}(\omega_i,\tau_j)\right]^2} \tag{49}$$

with $N = 512$ as the array dimension, is $\Delta_{FROG} = 5.3\times10^{-3}$.

The intensity profile corresponding to FROG spectrum of Figure 17 is shown in Figure 18. The pulse is slightly asymmetric and contains a small pre-pulse, caused by a small residual TOD mismatch. We calculate that the FWHM is 8.3 ps, with 84% of the pulse energy contained in the 20 ps wide bin indicated by the dashed box, and the temporal Strehl ratio is 0.78. The temporal strehl is the ratio of the peak measured intensity, and the peak transform-limited intensity for the measured spectral profile. The temporal waveform on the logarithmic scale of Figure 18 shows a post pulse at 160 ps, 100 dB lower than the main pulse. This post-pulse causes the satellite wings and the fringing in the FROG measurement. Frequency doubling will further improve the pulse contrast.

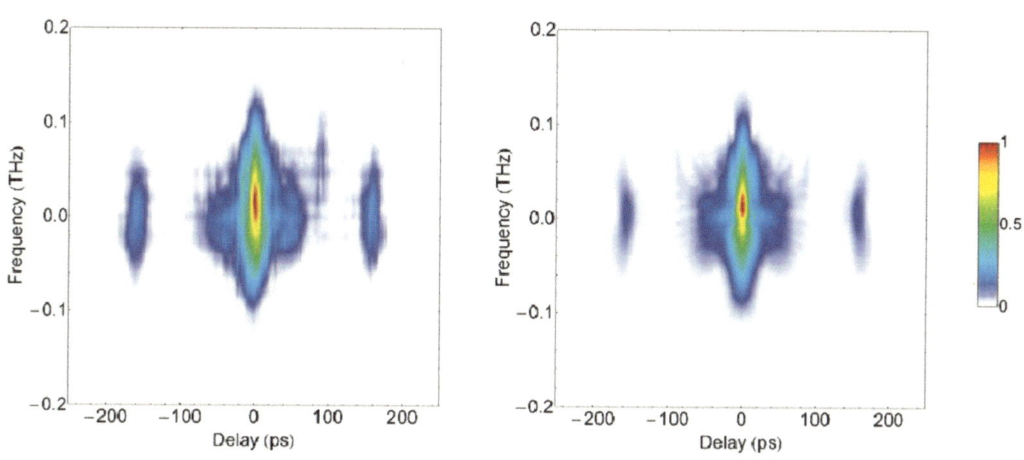

(a) (b)

Fig. 17. Experimental measurement of the 800 mJ pulse duration. (a) Field of the experimentally measured FROG spectrogram. (b) Lowest error field obtained by a FROG algorithm using a 512x512 discretization grid.

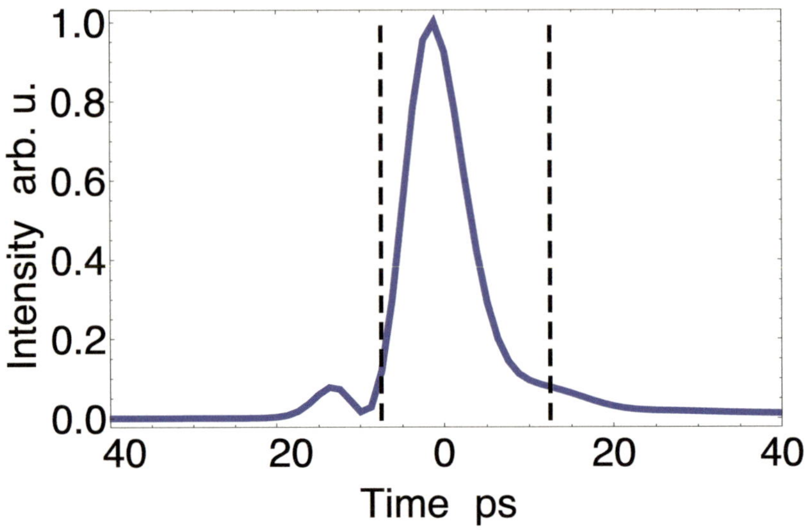

(a)

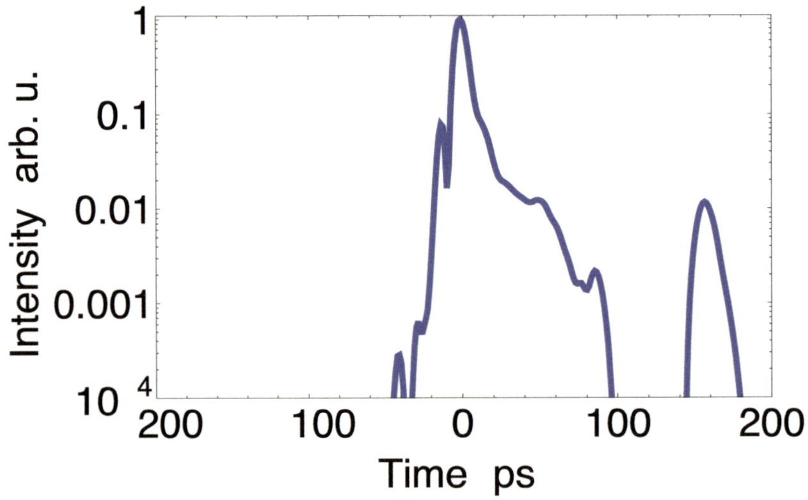

(b)

Fig. 18. Temporal pulse intensity obtained by analyzing a numerically processed FROG spectrogram on the linear scale (a) and log scale (b). FWHM of the pulse duration is 8.3 ps, and 84% of the energy is contained in the 20 ps bin (dashed box).

The FROG technique also measures the pulse spectrum, as shown in Figure 19. Comparing the FROG measured spectrum with the direct IR spectral measurement performed with an $f = 1$ m spectrometer indicates good agreement.

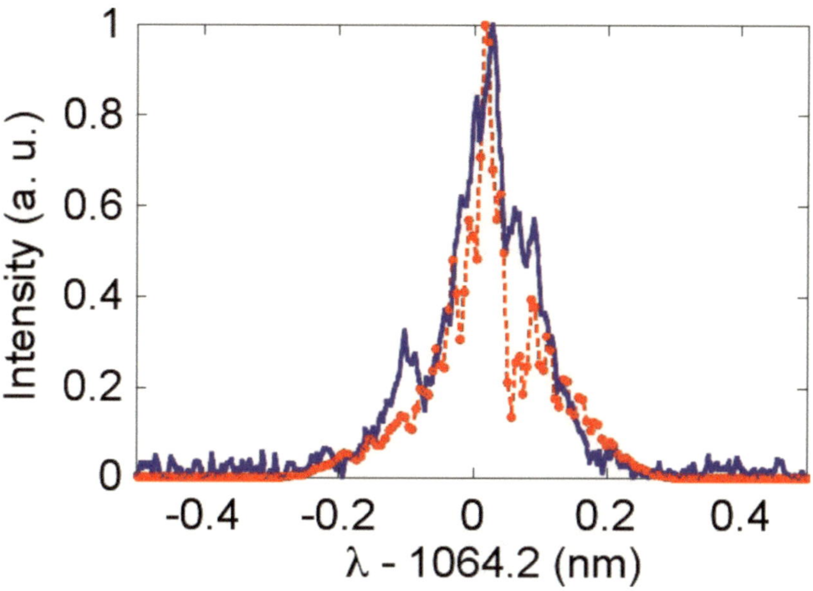

Fig. 19. Pulse spectrum from FROG (red dots) and spectrometer (solid line) measurements.

5. Nonlinear optics and frequency conversion

In linear optics, the response of the dielectric medium is not modified by the applied electric field of the laser field. However, at high intensities provided by ultrashort laser pulses (GW/cm^2), the applied electric field becomes comparable to the interatomic electric field. In linear optics, we assume that the electron performs linear harmonic motion about the nucleus. At the high field intensities, this model is no longer valid as the electron is driven further away from the nucleus. In the frequency domain, nonlinear motion corresponds to generation of new frequencies at some multiple of the applied frequency. Some examples of the nonlinear processes include sum and difference frequency generation, Raman scattering, Brillouin scattering, high harmonic generation, four-wave mixing, optical parametric oscillation and nonlinear polarization rotation.

More formally, the response of a dielectric medium to an applied electric field can be described by an induced polarization,

$$\mathbf{P} = \epsilon_0 \chi^{(1)} \mathbf{E} + \epsilon_0 \chi^{(2)} \mathbf{E}\mathbf{E} + \ldots \tag{50}$$

where ϵ_0 is the free space permittivity, $\chi^{(n)}$ is the nth order susceptibility, and $\mathbf{E}$ is the electric field. Because $\chi^{(1)} \gg \chi^{(2)}$ for an off-resonant medium, higher order terms become important only when the applied electric field is sufficiently high. Nonlinear processes are generally classified by the order n of the $\chi^{(n)}$. Second harmonic generation, and sum/difference generation are $\chi^{(2)}$ processes and are generally much stronger than $\chi^{(3)}$ processes such as four-wave mixing. However, in media lacking inversion symmetry such as amorphous glass, or gas $\chi^{(2)} = 0$ and sum/difference generation is not observed.

Sum/difference frequency conversion is a practical method for efficient, high power generation at wavelengths not accessible by common laser sources. Here, we describe frequency doubling. In a $\chi^{(2)}$ process, **EE** term produces excitation at twice the fundamental frequency. Let $\mathbf{E} = \mathbf{A}\cos\omega t$, then $\mathbf{E}^2 = \frac{\mathbf{A}^2}{2}(1+\cos 2\omega t)$. The magnitude of the nonlinear susceptibility varies with the applied frequency and depends on the electronic level structure of the material. Under well-optimized conditions harmonic efficiencies can exceed 80%. When selecting an appropriate nonlinear crystal, we consider various application dependent factors such as the magnitude of the nonlinear coefficient, acceptance bandwidth, absorption, thermal acceptance, thermal conductivity, walk-off angle, damage threshold, and maximum clear aperture. For pulse durations in the 200 fs to 10 ps range and for the fundamental wavelength ≈1 µm, beta barium borate (BBO) is an excellent candidate for 2ω, 3ω, and 4ω generation. The main drawback, is that the largest clear crystal aperture is ≈20 mm which limits its use to low pulse energies (<10-100 mJ). For higher pulse energies, deuterated and non-deuterated potassium dihydrogen phosphate (DKDP and KDP), lithium triborate (LBO) and yttrium calcium oxyborate (YCOB) can be grown to much larger apertures. YCOB is particularly attractive for its high average power handling, high damage threshold, and large effective nonlinearity[41]. For frequency doubling, typical required laser intensities are in the 100 MW/cm² to 10 GW/cm² range.

The crystal must be cut along an appropriate plane to allow phase matching and to maximize the effective nonlinear coefficient d_{eff} which is related to $\chi^{(2)}$ and the crystal orientation. The interacting waves at ω and 2ω acquire different phases, $\phi(\omega) = k_\omega z = n_\omega \omega z / c$ and $\phi(2\omega) = k_{2\omega} z = 2n_{2\omega}\omega z / c$ as they propagate along the crystal in z direction. An interaction is phase matched when $k_{2\omega} = 2k_\omega$. A uniaxial crystal contains two

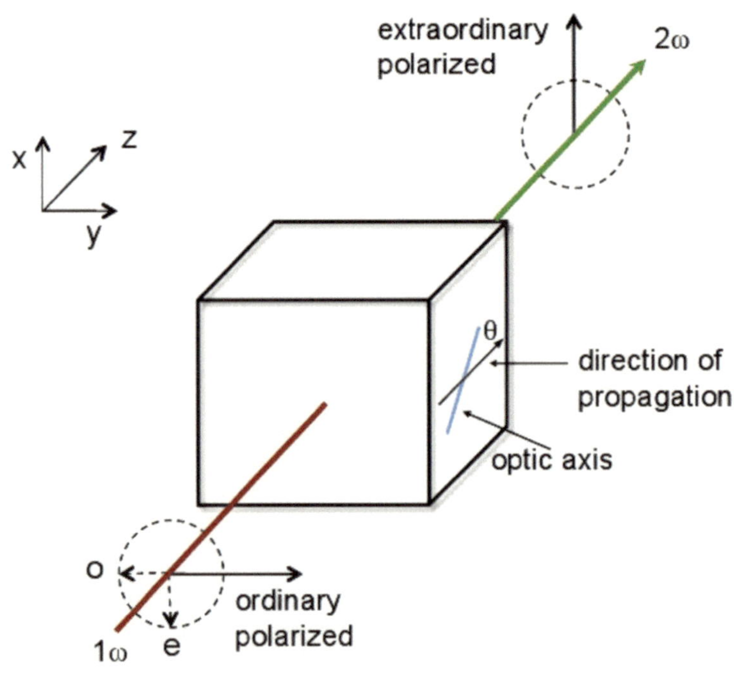

Fig. 20. Frequency doubling with OOE type phase-matching in a uniaxial nonlinear crystal. θ is the phase-matching angle between the optic axis and the propagation direction.

polarization eigenvectors, one parallel to the optic axis (the axis of rotation symmetry) and one perpendicular to it. An electric field inside the crystal contains a component perpendicular to the optic axis (ordinary polarization) and a component in the plane defined by the optic axis and the direction of propagation (extraordinary polarization), as illustrated in Figure 20. The refractive index of the extraordinary polarization, n^e, varies with θ, the angle between the direction of propagation and the optic axis; the ordinary refractive index, n^o has no angular dependence. In the example shown in Figure 20, the crystal is rotated along the y-axis until $n^e_{2\omega}(\theta) = n^o_{\omega}$. The illustrated phase matching condition, where both incident photons have the same polarization is known as type I phase matching. In Type II phase matching, the incident field has both an ordinary and an extraordinary polarization component.

Coupled Eqs. (51) and (52), given in SI units, describe Type I 2ω generation process relevant for 200 fs - 10 ps duration pulses. Here, we make a plane wave approximation, justified when we are not focusing into the crystal, and when the crystal is sufficiently thin to ignore beam walk-off effects. We also ignore pulse dispersion in the crystal, justified for our pulse bandwidth and crystal thickness. We can account for two-photon absorption, which becomes important for 4ω generation in BBO, by adding $\beta|A_{2\omega}|^2 A_{2\omega}$ term to the left hand side of Eq. (52).

$$\frac{\partial A_{\omega}}{\partial z} + \frac{1}{v_{g,\omega}}\frac{\partial A_{\omega}}{\partial t} + \frac{\alpha_{\omega}}{2}A_{\omega} = i\frac{2\omega}{n(\omega)c}d_{eff}A^*_{\omega}A_{2\omega}\exp(-i\Delta kz) \tag{51}$$

$$\frac{\partial A_{2\omega}}{\partial z} + \frac{1}{v_{g,2\omega}}\frac{\partial A_{2\omega}}{\partial t} + \frac{\alpha_{2\omega}}{2}A_{2\omega} = i\frac{2\omega}{n(2\omega)c}d_{eff}A^2_{\omega}\exp(i\Delta kz) \tag{52}$$

We obtain an analytical solution assuming qausi-CW pulse duration, which eliminates the time dependent terms, and a low conversion efficiency, or a constant A_{ω}. The efficiency of 2ω harmonic generation, $\eta_{2\omega} = I_{2\omega} / I_{\omega}$, reduces to:

$$\eta_{2\omega} = \frac{8\pi^2 d^2_{eff} L^2 I_{\omega}}{\epsilon_0 n^2_{\omega} n_{2\omega} c\lambda^2_{\omega}}\frac{\sin^2(\Delta kL/2)}{(\Delta kL/2)^2} \tag{53}$$

As an example, in our laser systems, we implement frequency conversion on both, the photogun and the interaction laser systems. On T-REX, we generate the 4th harmonic of the fundamental frequency by cascading two BBO crystals. The first, 1 mm thick crystal cut for Type I phase matching, frequency doubles the incident pulse from 1053 nm to 527 nm. The second 0.45 mm thick BBO crystal cut for Type I phase matching, frequency doubles 527 nm pusle to 263 nm. The overall conversion efficiency from IR to UV is 10%, yielding 100 μJ at 263 nm. Here, frequency conversion is primarily limited by two-photon absorption in the UV and the group velocity mismatch (GVM) between the 2ω and 4ω pulses. GVM results in temporal walk-off of the pulse envelopes and, in the frequency domain, is equivalent to the acceptance bandwidth.

On the interaction laser, we frequency double the high energy pulses to increase the final gamma-ray energy. On T-REX we use a large aperture (30x30 mm) 6 mm thick DKDP crystal to frequency double 800 mJ pulse from 1064 nm to 532 nm with up to 40% conversion efficiency. Here, the pulse bandwidth is relatively narrow (≈0.2 nm) and group velocity walk-off is insignificant. The conversion efficiency is primarily limited by beam quality and temporal pulse shape. Generated 532 nm pulse energy is plotted versus the compressed input pulse energy in Figure 21. At maximum IR energy, the conversion efficiency unexpectedly decreases. This may indicate onset of crystal damage, degradation in pulse quality, or an increase in phase mismatch.

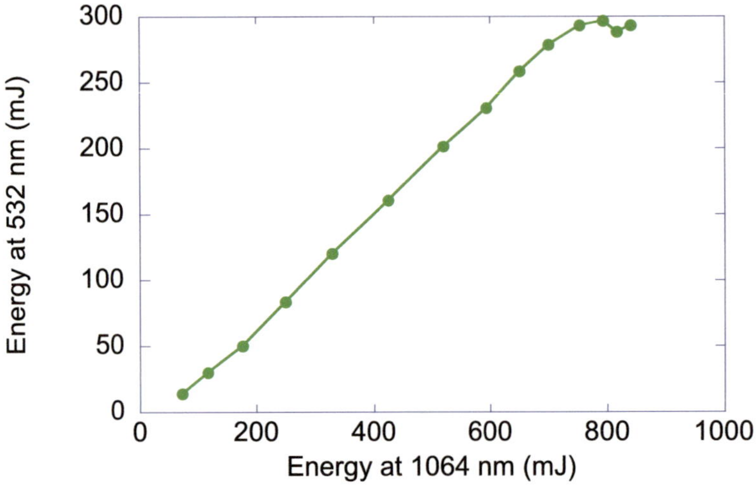

Fig. 21. Frequency doubling of the 10 ps T-REX ILS laser with a peak efficiency of 40%.

6. Simulation tools

A wide selection of proprietary, commercial and open source simulation tools exist for various aspects of laser design. Some are narrow in scope; others are optimized for system level design. The brief overview given here is not comprehensive and simply exposes the reader to some of the available choices. As with any modeling tool, its use requires in-depth familiarity with the subject and a good intuition of the expected results. The simulation should serve as a design guide and should be checked against experimental results. In general, simulation tools require verification, validation, and benchmarking. When verifying a code, we check that the underlying equations of the physical model properly describe the phenomena being studied. For example, a nonlinear frequency conversion code designed for nanosecond laser pulses may not properly describe frequency conversion with femtosecond pulses because it excludes pulse dispersion effects.

Benchmarking a simulation involves comparing the results from several different codes. This may be particularly important when modeling a new concept or utilizing a new code. Finally, validating a code involves comparing the results of the simulation to a real experiment. Ideally, the code can be validated for a certain range of bounding parameters. Code validation, when possible, is perhaps the most important aspect of ensuring the precision and accuracy of a particular modeling tool.

Various system level simulation codes can be divided into ray-tracing codes and physical optics codes. Ray-tracing assumes that the light wave can be modeled as a large number of

1-dimensional rays which propagate through various interfaces according to Snell's law. In their basic implementation, ray-tracing does not account for various wave effects such as diffraction or interference. The advantage of ray-tracing is that it is very fast and efficient for dealing with complicated interfaces. Ray-tracing can produce accurate results when dealing with large beams and features much larger than the light's wavelength, where wave-like effects can be disregarded. Ray-tracing can further be sub-divided into sequential and non-sequential variety. Sequential ray-tracing assumes that the order of each interface is predetermined. An example of sequential ray-tracing is a system involving a light source, a sequence of lenses, and an imaging plane. In non sequential ray-tracing, the sequence of the interfaces is determined during the raytrace, as the code tracks the position, angle and direction of each ray. An individual surface may be struck by the ray multiple times. An example of a nonsequential raytrace is a folded pulse compressor. While non-sequential raytracing approximates a real system better than a sequential raytrace, the performance is significantly slower. Most of the major codes include both sequential and non-sequential raytracing modes.

Raytracing codes are generally optimized for lens design in imaging applications. The user can select existing lenses from an extensive lens catalogue and design new lenses by specifying surface curvature and lens material. Program feature allow for rapid optimization of various system parameters, such as wavefront aberration or RMS spot size, as well as tolerance analysis. Some of the well known programs include ZEMAX, OSLO, and Code V. These codes also have advanced features allowing non-sequential analysis, as well as coherent physical optics propagation.

Ray-tracing can accurately calculate coherent wave-like effects by utilizing gaussian decomposition. Here, a beam is decomposed into a summation of TEM_{00} modes. Each mode is described by four waves that represent its waist and several additional rays that describe beam divergence. The rays are then propagated by geometrical optics but retain both the phase and intensity information of the beam. The Gaussian decomposition algorithm can accurately model beam propagation in both the near and the far-fields. An example of such code is FRED, which performs coherent beam propagation for a wide variety of optical elements, such as lenses, gratings, mirrors, and prisms.

As an alternative to ray-tracing, physical optics codes treat the wave properties of light by solving some simplified form of the nonlinear Schrodinger's equation typically using fast Fourier transform methods. A major advantage of the physical optics codes is their ability to model the time dependent light properties. These codes can often model nonlinear light properties, such as frequency conversion, and Raman scattering. The Fourier based codes are typically far better at simulating far field light propagation, coherent wave effects, and complex non-geometric optical elements. A disadvantage of the physical optics codes is that they do not handle light propagation through irregular refractive interfaces. Examples of sophisticated physical optics codes include PROP, developed at LLNL, MIRO, developed by CEA, and its commercial variant Commod Pro. Another script language based commercial code is GLAD.

In our experience, we find that physical optics based codes are ideal for top level system design as well as setting various system requirements, such as beam size, intensity, time duration, etc. Once the top level system design is complete, ray-trace codes are ideal for precise specification of system components, such as the lens focal lengths, inter-component distances, and optical element's aperture. A code such as FRED is particularly attractive because it allows for easy calculation of coherent beam effects and can be used with various CAD packages.

7. Conclusion

We have presented a brief overview of the fundamental physics and applications of Compton-scattering based compact mono-energetic gamma-ray sources, emphasizing the recently commissioned 2nd generation T-REX device, and the currently under construction, 3rd generation MEGa-ray Compton scattering light source at LLNL. We have also detailed the underlying laser technology and described several technological breakthroughs which enable development of Compton sources with the highest peak brightness in a compact footprint. Finally, we anticipate continued interest and applications in scientific and technological frontiers, in particular, in the field of nuclear photonics, for high energy Compton scattering light sources.

We are grateful to S.G. Anderson, D.J. Gibson, R.A. Marsh, M. Messerly, C.A. Ebbers, C.W. Siders, C.P.J. Barty, and the entire MEGa-ray team at Lawrence Livermore National Laboratory for their support and many useful discussions. This work was performed under the auspices of the U.S. Department of Energy by University of California, Lawrence Livermore National Laboratory under Contract W-7405-ENG-48.

8. References

[1] W. Bertozzi, J. A. Caggiano, W. K. Hensley, M. S. Johnson, S. E. Korbly, R. J. Ledoux, D. P. McNabb, E. B. Norman, W. H. Park, and G. A. Warren, Phys. Rev. C, 78, 041601(R) (2008).

[2] C.A. Hagmann, J. M. Hall, M. S. Johnson, D. P. McNabb, J. H. Kelley, C. Huibregtse, E. Kwan, G. Rusev, and A. P. Tonchev, J. Appl. Phys. 106, 084901 (2009).

[3] N. Kikuzawa, R. Hajima, N. Nishimori, E. Minehara, T. Hayakawa, T. Shizuma, H. Toyokawa, and H. Ohgaki, Appl. Phys. Express 2, 036502 (2009).

[4] F. Albert, S. G. Anderson, G. A. Anderson, S. M. Betts, D. J. Gibson, C. A. Hagmann, J. Hall, M. S. Johnson, M. J. Messerly, V. A. Semenov, M. Y. Shverdin, A. M. Tremaine, F. V. Hartemann, C. W. Siders, D. P. McNabb, and C. P. J. Barty, "Isotope-specific detection of low-density materials with laser-based monoenergetic gamma-rays," Opt. Lett. 35, 354-356 (2010)

[5] J. Pruet, D. P. McNabb, C. A. Hagmann, F. V. Hartemann, and C. P. J. Barty, J. Appl. Phys. 99, 123102 (2006).

[6] U. Kneissl, H. M. Pitz, and A. Zilges, Prog. Part. Nucl. Phys. 37, 349 (1996).

[7] J. M. Jauch, F. Rohrlich, *The Theory of Photons and Electrons*, Springer-Verlag (1976)

[8] F. V. Hartemann, W. J. Brown, D. J. Gibson, S. G. Anderson, A. M. Tremaine, P. T. Springer, A. J. Wootton, E. P. Hartouni, C. P. J. Barty, "High-energy scaling of Compton scattering light sources", Phys. Rev. ST Accel. Beams 8, 100702 (2005)

[9] F. Albert, S. G. Anderson, D. J. Gibson, R. A. Marsh, C. W. Siders, C. P. J. Barty, F. V. Hartemann, "Three-dimensional theory of weakly nonlinear Compton scattering", Phys. Plasmas 18, 013108 (2011)

[10] J.W. Meyer, "Covariant Classical Motion of Electron in a Laser Beam", Phys. Rev. D 3, 621–622 (1971)

[11] F.V. Hartemann, F. Albert, C.W. Siders and C.P.J. Barty, Phys. Rev. Lett., 105, 130801 (2010)

[12] F. Albert, S. G. Anderson, D. J. Gibson, C. A. Hagmann, M. S. Johnson, M. Messerly, V. Semenov, M. Y. Shverdin, B. Rusnak, A. M. Tremaine, F. V. Hartemann, C. W.

Siders, D. P. McNabb, C. P. J. Barty, "Characterization and applications of a tunable, laser-based, MeV-class Compton-scattering γ-ray source", Phys. Rev. ST Accel. Beams 13, 070704 (2010)
[13] D.J. Gibson, F. Albert, S.G. Anderson, S.M. Betts, M.J. Messerly, H.H. Phan, V.A. Semenov, M.Y. Shverdin, A.M. Tremaine, F.V. Hartemann, C.W. Siders, D.P. McNabb, and C.P.J. Barty, Phys. Rev. ST Accel. Beams 13, 070703 (2010).
[14] D. Strickland and G. Mourou, "Compression of amplified chirped optical pulses," Opt. Commun. 56(3), 219–221 (1985).
[15] M. D. Perry, G. Mourou, "Terawatt to petawatt subpicosecond lasers", Science 264: 917–924 (1994).
[16] S. Karsch, Z. Major, J. Fülöp, I. Ahmad, T. Wang, A. Henig, S. Kruber, R. Weingartner, M. Siebold, J. Hein, C. Wandt, S. Klingebiel, J. Osterhoff, R. Hörlein, and F. Krausz, "The Petawatt Field Synthesizer: A New Approach to Ultrahigh Field Generation," in Advanced Solid-State Photonics, OSA Technical Digest Series (CD) (Optical Society of America, 2008), paper WF1.
[17] Michael Sumetsky, Benjamin Eggleton, and C. de Sterke, "Theory of group delay ripple generated by chirped fiber gratings," Opt. Express 10, 332-340 (2002)
[18] E. B. Treacy, "Optical pulse compression with diffraction gratings", IEEE J. Quantum Electron., vol. 5, no. 9, p.454 , 1969.
[19] O. E. Martinez, "3000 times grating compressor with positive group velocity dispersion: Application to fiber compensation in 1.3-1.6 μm region," Quantum Electronics, IEEE Journal of , vol.23, no.1, pp. 59- 64, Jan 1987.
[20] G. Cheriaux, P. Rousseau, F. Salin, J. P. Chambaret, Barry Walker, and L. F. Dimauro, "Aberration-free stretcher design for ultrashort-pulse amplification," Opt. Lett. 21, 414-416 (1996)
[21] M. Y. Shverdin, D. R. Walker, D. D. Yavuz, G. Y. Yin, S. E. Harris, "Generation of a Single-Cycle Optical Pulse", Phys. Rev. Lett. 94, 033904 (2005)
[22] N. Zhavoronkov, G. Korn, "Generation of Single Intense Short Optical Pulses by Ultrafast Molecular Phase Modulation", Phys. Rev. Lett. 88, 203901 (2002)
[23] A. Siegman, Lasers, University Science Books, Sausality, CA (1986)
[24] Y. Jeong, J. Sahu, D. Payne, and J. Nilsson, "Ytterbium-doped large-core fiber laser with 1.36 kW continuous-wave output power," Opt. Express 12, 6088-6092 (2004)
[25] F. Ö. Ilday, J. R. Buckley, H. Lim, F. W. Wise, and W. G. Clark, "Generation of 50-fs, 5-nJ pulses at 1.03 μm from a wave-breaking-free fiber laser," Opt. Lett. 28, 1365-1367 (2003)
[26] M. E. Fermann, M. J. Andrejco, Y. Silberberg, and M. L. Stock, "Passive mode locking by using nonlinear polarization evolution in a polarization-maintaining erbium-doped fiber," Opt. Lett. 18, 894-896 (1993)
[27] G. H. Miller, E. I. Moses, C. R. Wuest, "The national ignition facility: enabling fusion ignition for the 21st century", Nuclear Fusion 44(12): S228 (2004).
[28] L. M. Frantz and J. S. Nodvik, "Theory of pulse propagation in a laser amplifier," J. Appl. Phys. 34, 2346–2349 (1963).
[29] T. Planchon, F. Burgy, J. P. Rousseau, and J. P. Chambaret, "3D Modeling of amplification processes in CPA laser amplifiers", Appl. Phys. B, Photophys. Laser Chem., vol. 80, no. 6, pp.661 - 667 , 2005.

[30] R. Trebino, Frequency-Resolved Optical Gating: TheMeasurement of Ultrashort Laser Pulses, Kluwer Academic Publishers, Boston, MA (2000)

[31] Selcuk Akturk, Xun Gu, Pablo Gabolde, and Rick Trebino, "The general theory of first-order spatio-temporal distortions of Gaussian pulses and beams," Opt. Express 13, 8642-8661 (2005)

[32] Patrick O'Shea, Mark Kimmel, Xun Gu, and Rick Trebino, "Highly simplified device for ultrashort-pulse measurement," Opt. Lett. 26, 932-934 (2001)

[33] M. Y. Shverdin, F. Albert, S. G. Anderson, S. M. Betts, D. J. Gibson, M. J. Messerly, F. V. Hartemann, C. W. Siders, and C. P. J. Barty, "Chirped-pulse amplification with narrowband pulses," Opt. Lett. 35, 2478-2480 (2010)

[34] D. N. Fittinghoff, W. A. Molander, and C. P. J. Barty, "Hyperdispersion grating arrangements for compact pulse compressors and expanders," in Frontiers in Optics, OSA Technical Digest Series (Optical Society of America, 2004), paper FThL5.

[35] F. J. Duarte, "Generalized multiple-prism dispersion theory for pulse compression in ultrafast dye lasers," Opt. Quantum Electron. 19, 223-229 (1987).

[36] M. D. Perry, R. D. Boyd, J. A. Britten, D. Decker, B. W. Shore, C. Shannon, and E. Shults, "High-efficiency multilayer dielectric diffraction gratings," Opt. Lett. 20, 940-942 (1995)

[37] A. G. Kostenbauder, "Ray-pulse matrices: A rational treatment for dispersive optical systems", IEEE J. Quantum Elect. 26: 1148–1157 (1990)

[38] Qiang Lin, Shaomin Wang, Javier Alda, and Eusebio Bernabeu, "Transformation of pulsed nonideal beams in a four-dimension domain," Opt. Lett. 18, 669-671 (1993)

[39] Daniel J. Kane and Rick Trebino, "Single-shot measurement of the intensity and phase of an arbitrary ultrashort pulse by using frequency-resolved optical gating," Opt. Lett. 18, 823-825 (1993)

[40] Daniel J. Kane, A. J. Taylor, Rick Trebino, and K. W. DeLong, "Single-shot measurement of the intensity and phase of a femtosecond UV laser pulse with frequency-resolved optical gating," Opt. Lett. 19, 1061-1063 (1994)

[41] Zhi M. Liao, Igor Jovanovic, Chris A. Ebbers, Yiting Fei, and Bruce Chai, "Energy and average power scalable optical parametric chirped-pulse amplification in yttrium calcium oxyborate," Opt. Lett. 31, 1277-1279 (2006)

6

Frequency-Tunable Coherent THz-Wave Pulse Generation Using Two Cr:Forsterite Lasers with One Nd:YAG Laser Pumping and Applications for Non-Destructive THz Inspection

Tadao Tanabe and Yutaka Oyama
Department of Materials Science, Graduate School of Engineering, Tohoku University
Japan

1. Introduction

The development of various original semiconductor devices, which act as electromagnetic wave generators in the THz region of the spectrum, has long been an active area of research in our group. The terahertz (THz) region, which lies between the microwave and infrared regions, offers a wealth of untapped potential. In most cases, the devices are based upon the utilization of THz lattice vibrations in compound semiconductors (*e.g.*, GaP and GaAs), which has recently become an important technology behind frequency-sweepable coherent THz-wave sources, following the invention of the semiconductor laser by Nishizawa in 1957 [1]. In 1963, Nishizawa was the first to predict the utility of the phonon and molecular vibration in semiconductors for optical communication and THz-wave generation. However, an important frequency gap between the microwave and optical frequencies remained, and presently is referred to as the THz region. In 1983, the semiconductor Raman laser was realized, which relied on the longitudinal optical phonon (LO phonon) mode of a GaP crystal. This work highlighted the generation of a 12-THz wave with a peak power as high as 3 W [2]. The output power of the Raman laser was increased by a phonon enhancement effect within the waveguide-structured GaP [3].

A high-power frequency-tunable THz-wave was generated via excitation of phonon-polaritons mode in GaP [4–8]. The frequency range was approximately 0.3–7.5 THz, in which the peak power was greater than 100 mW over most of the tunable region [5–7]. The generated THz-wave power and frequency regions have been shown to depend on carrier densities within the GaP crystals.

Furthermore, THz signal generators have been developed with various functions. Generation of narrow-linewidth THz waves has very useful applications in the fields of high-resolution spectroscopy, optical communications and *in-situ* security screening. The CW THz waves are generated from GaP by using semiconductor diode lasers. The linewidth is about 4 MHz. A 30 cm-long portable THz-wave generator is constructed using two Cr:Forsterite lasers pumped using a single Nd:YAG laser.

In this chapter, we review the photonic approaches of THz-wave sources and highlight the principles and performance of these THz-wave generating devices. Developments in THz technology allow spectroscopic investigation of low-energy excitations of macromolecules, such as molecular rotations, hydrogen bonding, and intermolecular interactions, with a broad frequency range and high resolution. Indeed, low-energy excitations are believed to be critical in understanding the complex behavior of biological molecules, cells, and tissues. For industrial applications, we have recently developed THz diagnosis technologies. Deformation of polyethylene can be monitored using polarized THz spectroscopy. The THz diagnoses the inside of object even with covered by materials.

THz wave has unique properties with high transparency for non-polarized materials such as with radio waves and easy handling as in the case of light. THz waves are expected to be a promising frequency for the non-destructive diagnosis of the interior of non-polarized materials [9-14]. For the THz inspection method, the spatial resolution is higher compared with that of conventional microwave techniques.

The energy of a THz wave is as low as room temperature, for example, whose energy level corresponds to hydrogen bonding, van der Waals interactions and free carrier absorption. Although the transparency of a THz wave is not as great as that of X-ray, a THz wave is sensitive to soft materials such as hydrate as well as to wet conditions. To date, our group has utilized THz waves for inspections of diffused water and defects in timber and concrete blocks using a 0.2 THz generator [15]. The phase transition of liquid crystal has been investigated based on molecular interactions using a GaP THz signal generator [16]. Polarization THz measurements are helpful for THz non-destructive diagnosis of the tensile strain in deformed UHMWPE [17]. Use of THz waves in conjunction with X-ray and γ-ray measurements shows promise for analysis macro-structures in organic materials and polymers. Recently, compact sized THz wave generators have been developed for practical use [18]. For single-frequency coherent THz waves, a geometrical optical design can be applied. A THz beam spot can be controlled to be as small as the wavelength of a THz wave. Such THz sources are suitable for spectroscopic imaging with spatial high-resolution.

As one application of the THz diagnosis, the study has focused on the interior copper conductors covered with insulating polyethylene. THz reflectivity of the copper surfaces was investigated by using the GaP THz wave generator. Surface evaluation of copper with various conditions was performed using THz diffused reflection spectroscopy. Copper is a basic metal. In particular, copper cables are used for a wide variety of electric components. Confirmation of copper conductors covered with plastics is essential for a social safety. A suitable way to evaluate the deterioration of interior copper conductors has not yet been fully established. A non-destructive method for the diagnosis of electric cables would be of value for quality evaluation in use of it.

2. THz-wave generation from semiconductors

Widely frequency-tunable high-power THz waves have been generated from GaP by pumping with a Q-switched Nd:YAG laser and an OPO, or two Cr:Forsterite lasers [4–7]. THz waves were generated in the frequency range from 0.3 to 7.5 THz using difference frequency generation (DFG) via the excitation of phonon–polaritons in GaP. Indeed, this process converts energy very efficiently, and resulted in a THz wave with an energy of 9 nJ/pulse (peak power of 1.5 W). Furthermore, frequency-tunable CW THz waves were generated by enhancing the power density of incident beams from semiconductor lasers [8].

The THz-wave output power (P_{THz}) can be increased by exploiting the inverse proportion to the beam spot size, S, based on the follow equation:

$$P_{THz} = \frac{A}{S} \cdot P_1 \cdot P_2 , \tag{1}$$

where A is the coefficient for generating THz waves from a GaP crystal pumped under non-collinear phase-matching conditions. Note that A is estimated to be 0.4×10^{-13} W^{-1} cm^2 [15, 16], while S is the spatial overlap of the cross-sectional areas of the pump and signal beams, and P_1 and P_2 are the effective powers of the pump and signal beams, respectively.

The two lasers used for the DFG of THz waves via excitation of the phonon–polariton mode in GaP crystals were a 1.064 μm Nd:YAG source and a β-BaB_2O_4-based OPO system. They were set up in a non-collinear configuration with a very small angle between the two beams, and the GaP crystals were positioned as depicted in Fig. 1 [4–7]. The wavelength of the pump beam was varied between 1.035 and 1.062 μm, which corresponded to generated THz-wave frequencies between 8 and 0.5 THz. The THz-wave energy can be collected with parabolic reflectors and determined using a pyroelectric DTGS detector operating at room temperature or a liquid-helium-cooled Si bolometer.

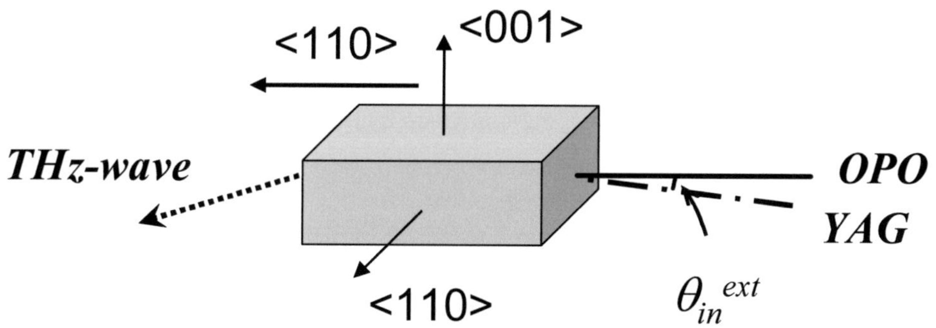

Fig. 1. Schematic of the YAG and OPO beam paths within the GaP crystal.

The THz-wave generation using two Cr:Forsterite lasers (Cr:F) was very similar to the OPO-based system. The pump and signal beams were combined using a cubic polarizer placed on a rotating platform on top of a linear stage, which automatically produced a very small angle between the two beams to fulfill the phase-matching condition and achieve a spatial overlap of the incoming beams. The non-collinear phase-matching was satisfied under the following condition [4, 6]. The calculated angle between the pump and signal light beams (θ_{in}) inside the GaP crystal matched quite well with respect to the experimental results, as long as $\theta_{in} << 1$:

$$\frac{\theta_{in}}{\nu} = \sqrt{\frac{2(\Delta q / q)(n_I^2 / n_L \cdot n_S)}{\nu_L \cdot \nu_S}} , \tag{2}$$

where ν_L, ν_S, and ν are the frequencies of the pump, signal, and THz waves, respectively; $(\Delta q/q)$ is the relative deviation of the wave-vector phonon–polaritons in the GaP crystal; n_I, n_L and n_S are the refractive indexes of the THz wave, pump, and signal beams, respectively;

and Δq is the deviation of q from the exact phase-matching value in the collinear configuration, which can be described as

$$\left(\frac{\Delta q}{q}\right)=\frac{n_I - n_{eff}}{n_I}. \tag{3}$$

In the case of the collinear interaction, the difference between the pump and signal wave numbers can be described by the following equations:

$$k_L - k_S = k = \frac{n_{eff}\cdot 2\pi\nu}{c},\ n_{eff} = n_S + \nu_L\cdot\frac{\delta n}{\delta\nu} \tag{4}$$

where k_L and k_S are the wave numbers of the pump and signal beams, respectively, and $\delta n/\delta\nu$ is the frequency dispersion of the refractive index. Moreover, the dispersion relationship of phonon–polaritons in GaP can be given by the following equation:

$$\left(\frac{cq}{2\pi\nu}\right)^2 = n_I^2 = \frac{\left(\varepsilon_S\nu_0^2 - \varepsilon_\infty\nu^2\right)}{\left(\nu_0^2-\nu^2\right)}, \tag{5}$$

where ε_s and ε_∞ are the static and optical frequency dielectric constants, respectively (ε_s = 11.15, ε_∞ = 9.20), and ν_0 = 11.01 THz is the pure transverse optical phonon frequency [19]. In general, this equation holds true as long as the THz frequency, ν, is not nearly equal to ν_0. The internal angle external to the GaP, θ_{in}^{ext}, is given by the relationship

$$\theta_{in} \approx \frac{\theta_{in}^{ext}}{n_S}, \tag{6}$$

where θ_{in} is calculated to be 0.1° at 2 THz; therefore, the two near-IR beams overlapped sufficiently to interact with each other. The polarization of the pump and signal beams was adjusted to be in the <001> and $<1\bar{1}0>$ directions, respectively. Figure 2 illustrates the relationship between θ_{in}^{ext} and the THz-wave frequency at which the maximum output power was obtained. The slope of the curve declined noticeably, which was reflected in the measured dispersion curve of the phonon–polariton branch of GaP. The θ_{in}^{ext} increased with the wavelength of the pump and signal beams, while the propagation direction of the THz wave was related to $(\Delta q/q)$. The output direction of the THz wave depends on the wavelength of the pump and signal beams, as well as the THz frequency. The angle of the propagation direction of the THz wave inside the crystal, θ_I, is given as

$$\sin\left(\theta_I - \theta_{in}\right) = \left(\frac{k_S}{q}\right)\sin\theta_{in},\ \text{with } \theta_I >> \theta_{in}. \tag{7}$$

From this, we obtain

$$\sin\theta_I \approx \sqrt{\left(\frac{2\Delta q}{q}\right)\cdot\left(\frac{\nu_S}{\nu_L}\right)\cdot\left(\frac{n_S}{n_L}\right)} \approx \sqrt{\frac{2\Delta q}{q}}. \tag{8}$$

The THz-wave direction was compensated with a pair of off-axis parabolic reflectors, where one of them was carefully moved on a translation stage. The GaP crystal was rotated at a lower frequency in the Cr:F-based system when compared to the OPO-based source to prevent total internal reflection of the THz wave.

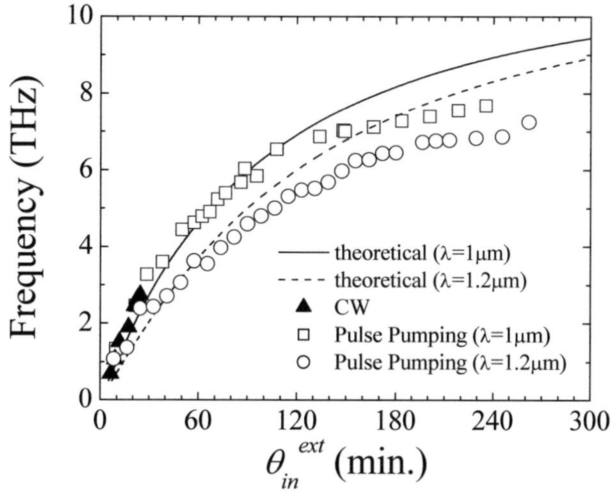

Fig. 2. The relationship between θ_{in}^{ext} and the THz-wave frequency at which the maximum output power was obtained in CW and pulse pumping. The solid line represents the calculated relationship described in more detail within the text.

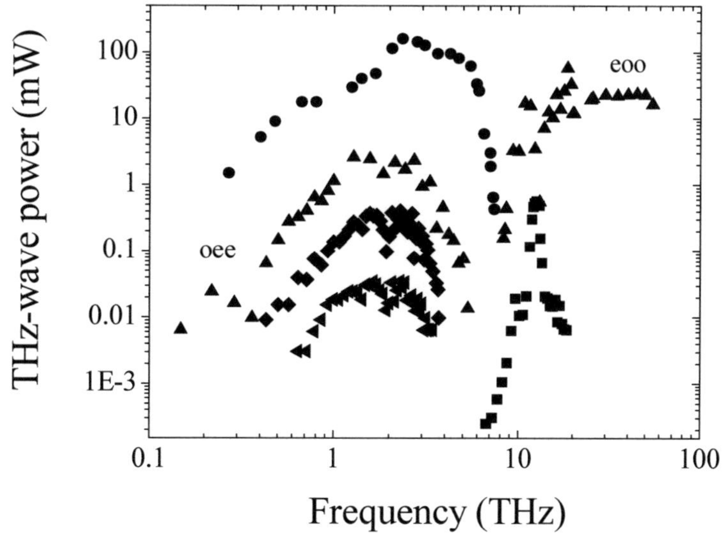

Fig. 3. Frequency dependence of the THz-wave output power from various semiconductor crystals. Note: the pump and signal energy was 3 mJ before incidence to the crystal: GaP (•), GaSe (▲), CdSe (■), $ZnGeP_2$-oee (◆), and $ZnGeP_2$-eoo (◂).

Figure 3 shows the frequency dependence of the maximum THz-wave output power at various θ_{in}^{ext}. The pulse energies of the YAG and OPO were both attenuated to 3 mJ before incidence to the GaP crystal. The THz-wave output power remained stable at approximately 100 mW over a wide frequency range (2.0–5.2 THz). The total tunable frequency range was 0.3–7.5 THz. In cases when the THz-wave frequencies were greater than 5.5 THz ($\theta_{in}^{ext} > 70°$), a much higher power was obtained by rotating the GaP crystal in order to prevent total internal reflection. The THz-wave output power increased linearly with the pump and signal beam energy. Note that the frequency bandwidth is equivalent to the pump and signal beams.

Figure 4 shows the THz-wave power generated as a function of the measured pump beam energy using the Cr:F-based system. It can be seen that the THz-wave power was nearly proportional to the pump energy. When the pump and signal energies were 11.4 and 11.6 mJ, respectively, the THz power increased to 1.5 W for a 10-mm-long crystal without causing surface damage.

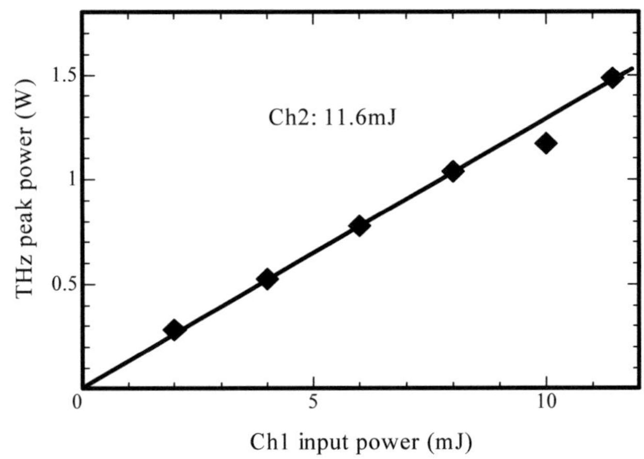

Fig. 4. Generated THz-wave power as a function of the pump beam (Ch1) energy/pulse in the 10-mm-long GaP crystal using a Cr:F source system. The signal beam (Ch2) energy/pulse was 11.6 mJ.

The continuous wave (CW) single-frequency THz waves were generated in a widely frequency-tunable pumping source consisting of an external cavity laser diode (ECLD) and a laser diode (LD)-pumped Nd:YAG laser combined with an ytterbium-doped fiber amplifier (FA) [8]. The estimated THz-wave output peak was 50 pW. In the automatic measurement of transmission spectroscopy, the wavelength of the ECLD was swept from 1.0538 to 1.0541 μm with external cavity fine-tuning. The THz frequency can be shifted according to the wavelength of ECLD. The THz-wave generation efficiency is related to the THz absorption coefficient in GaP crystals. High-power THz-wave output requires a stoichiometric control of the GaP to reduce the THz absorption due to free carriers and phonons in crystal.

In the waveguide structure, efficient THz-wave generation was achieved. THz-wave generation was demonstrated in a GaP waveguide with the same size as the wavelength of THz-wave under a collinear phase-matching condition. The conversion efficiencies of THz wave generation from the rod-type waveguides were estimated. Interestingly, higher conversion efficiencies were achieved as the waveguide size decreased. For example, in the case in which the waveguide cross section was 200 μm × 160 μm, the conversion efficiency increased to 7.6×10^{-12} W^{-1}. This value was an order of magnitude greater than that in bulk GaP crystals (7.4×10^{-13} W^{-1}).

In addition to the GaP crystal measurements, we also generated frequency-tunable coherent THz-waves using GaSe, $ZnGeP_2$, and CdSe semiconductors based upon difference-frequency generation. GaSe had a high second-order nonlinear optical (NLO) coefficient (d_{22} = 54 pm/V) [20]. Furthermore, GaSe crystals have merit because they can be used to construct a simple THz-wave generation system, since collinear phase-matched DFG eliminates the complexity of angle-tuning in both the input and output beams. THz waves were generated in a wide frequency range from the THz to the mid-infrared region as shown in Fig. 3. Note that Fig. 3 shows the frequency dependence of the THz-wave output power from each semiconductor crystal at various PM angles using the YAG and OPO-based sources. Upon closer inspection of Fig. 6, it is apparent that the combination of GaP (0.3–7.5 THz) and GaSe (10–100 THz) had the widest tunable range with the highest power.

In practical applications, THz-wave generation systems have been used for spectroscopic measurement and THz imaging. A portable THz-wave generator is necessary for practical use, by which it can be moved closer to sample for THz sensing in the field. For example, in organic and inorganic crystal fabrication processes, crystalline defects can be detected using a THz spectrometer. For this motivation, using only one small 30-cm-long Nd:YAG laser and two Cr:Forsterite crystals, we constructed two Cr:Forsterite lasers pumped with the YAG laser and generated THz waves with the compact device. We investigated the pulse duration and delay time to realize Cr:Forsterite lasers which are suitable for use as the THz-wave generator because it requires overlapping of two Cr:Forsterite laser pulses both temporally and spatially for DFG in GaP.

A Cr:Forsterite (Cr:Mg_2SiO_4) laser is a solid state laser that is tunable between 1130 and 1370 nm. The laser properties have been investigated, leading to the CW and mode-locked pulse operations. A Cr:Forsterite crystal has Cr^{4+} in the tetrahedrally coordinated Si^{4+} site, which acts as the lasing ion. Crystal growth processes induce impurities such as Cr^{3+} and Cr^{2+}. Those impurities can be decreased by annealing. Two Cr:Forsterite crystals (Cr:F-1, Cr:F-2) with different crystal properties were used for this study: Cr:F-1 is dark blue and Cr:F-2 is dark green. Respectively, they are a rectangular parallelepiped (-R) and a Brewster-cut crystal (-B) of 5 mm × 5 mm (cross-section) × 10 mm (length). Transmittance spectra of Cr:F-1 and Cr:F-2 were measured in the NIR region at room temperature and the absorption coefficient were estimated, respectively. The absorption peaks at 550, 660, 740, and 1060 nm are attributed to the Cr^{4+}. In contrast, Cr^{3+} has absorption at 474 and 665 nm. At around 700 nm, the Cr^{3+} absorption is dominant compared to that of the Cr^{4+}. The 732-nm absorption is considerably higher than that at 1064 nm, but the slope efficiency excited at 732 nm is less than 1064 nm. Consequently, a Nd:YAG laser was used for pumping Cr:Forsterite. The figure of merit (FOM) of Cr:F-1 and Cr:F-2 are 8.0 and 5.9, respectively. A Q-switched Nd:YAG laser with 10-ns pulse duration at 10 ~ 30Hz repetition was used to pump the Cr:Forsterite crystals. Selective frequency cavity oscillator systems were constructed using the Cr:F-1 and Cr:F-2. The Cr:Forsterite laser characteristics of the pulse duration and delay time were measured in terms of the Cr:Forsterite laser energy and cavity length. A THz-wave generation system was constructed using two Cr:Forsterite crystals and one Q-switched Nd:YAG laser. A plane-plane cavity oscillator was constructed for measurement of the slope efficiency in the Cr:Forsterite laser. The output coupler is of 6% transmittance around 1.2 μm. The Cr:Forsterite laser energies using Cr:F-1-B, Cr:F-1-R, and Cr:F-2-B were measured according to changes in the Nd:YAG laser energy. Cr:F-1-B has the highest slope-efficiency of 8.4%, with the lasing threshold of 2.2 mJ/mm^2. Slope efficiencies and lasing thresholds are respectively 6.8%, 2.0 mJ/mm^2 in Cr:F-1-R and 4.8%, and 4.6 mJ/mm^2 in Cr:F-2-B. These results suggest a relation of the slope efficiency to FOM and the crystal shape.

It is necessary to select a frequency for THz-wave generation. The cavity oscillator using a diffraction grating can select a frequency. The selective-frequency cavity was adjusted so that the first-order reflected light is directed to the direction of incident beam. The cavity can oscillate at 1250 nm when the incident beam angle is 31.2° normal to the grating of 830 lines/mm. The selective frequency Cr:Forsterite laser can be outputted from zero-order light. The pulse duration and delay time were optimized as a function of Cr:Forsterite laser energy and cavity length. The laser crystal was Cr:F-1-R. The delay time was measured as the time from the 50% rising edge of the Nd:YAG laser to the 50% rising edge of the Cr:Forsterite laser using photodiodes and an oscilloscope. The 200 points were recorded at the same optical condition; each result was plotted and fitted to a Gaussian function for measurement of the pulse duration and delay time.

Figure 5 shows the Cr:Forsterite laser characteristics of the pulse duration and delay time when the cavity length was fixed at 20 cm. The Nd:YAG pumping laser energy is changed to 66 mJ, 80 mJ, and 95 mJ. Changing the optical alignment at each Nd:YAG laser pumping

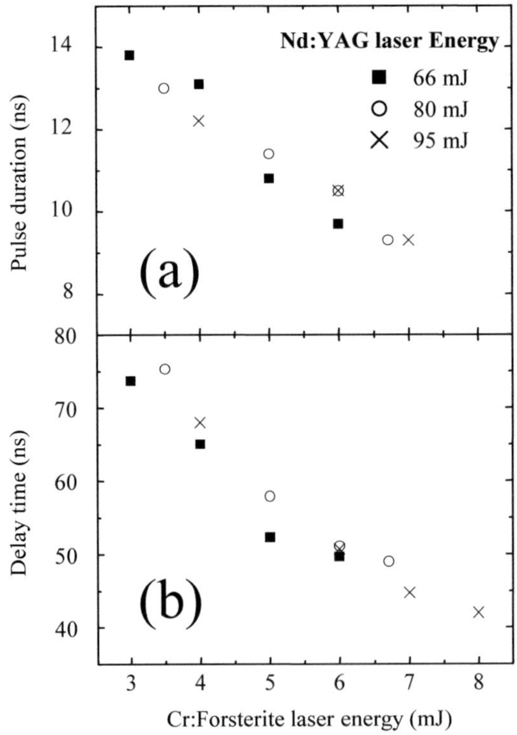

Fig. 5. Cr:Forsterite laser characteristics of the pulse duration (a) and delay time (b) at Cr:Forsterite laser energy of 3–8 mJ. The cavity length was fixed to 20 cm. Nd:YAG laser power was changed to 66 mJ, 80 mJ, and 95 mJ.

energy changed the Cr:Forsterite laser energy to maximum. Both the pulse duration and delay time are decreased when the Cr:Forsterite laser energy increases.
Figure 6 shows Cr:Forsterite laser characteristics of the pulse duration and delay time for the case in which the cavity length was changed from 12 cm to 30 cm at Cr:Forsterite laser energy of 3 mJ, 4 mJ, and 5 mJ. Both the pulse duration and delay time are increased as a function of the cavity length. The result of pulse duration at any cavity length can be understood according to the photon lifetime, which is

$$\tau = \frac{2L}{c(1-R)} \tag{9}$$

where τ is the photon lifetime, L is the cavity length, c is the velocity of light, and R is the transmittance of an output coupler. Therefore, the cavity length produces a long photon lifetime and long pulse duration.

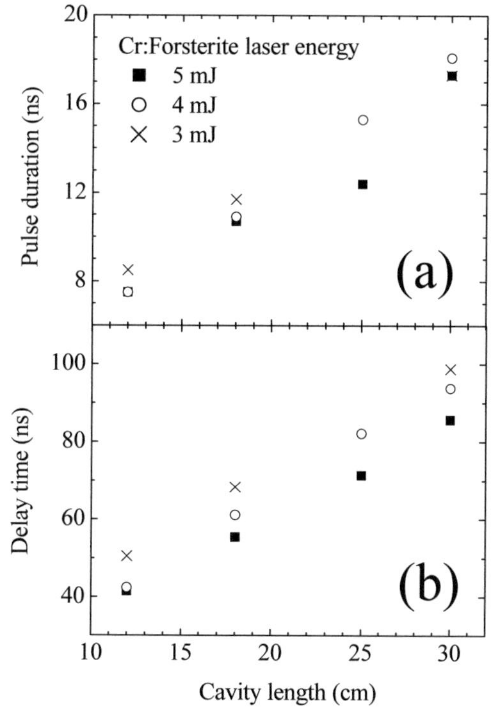

Fig. 6. Cr:Forsterite laser characteristics of the pulse duration (a) and delay time (b) with the cavity length of 12–30 cm. The Cr:Forsterite laser energy was 3 mJ, 4 mJ, and 5 mJ.

According to these results, the delay time can be controlled according to the Cr:Forsterite laser energy and cavity length. The Cr:Forsterite laser energy is controllable according to the conversion efficiency and pumping energy. The conversion efficiency is changed by optical alignment. Each pumping energy of the two Cr:Forsterite lasers can be changed using a pair of a polarizer and a half-wavelength plate, even if single Nd:YAG laser is used. The method can generate a higher-powered THz wave because the two laser conversion efficiencies are kept high.

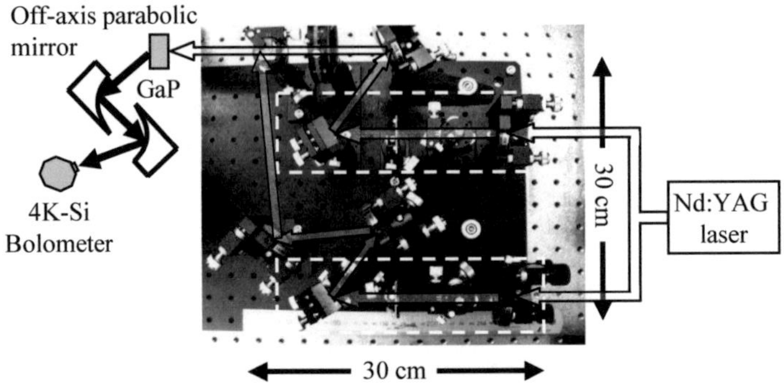

Fig. 7. Optical setup of THz-wave generation with two Cr:Forsterite lasers. The THz-wave detector was a liquid-He cooled silicon bolometer.

Figure 7 shows that two Cr:Forsterite laser oscillators with a 15-cm cavity for THz-wave generation were constructed to be 30 cm × 30 cm. The 30 cm-length Nd:YAG laser can be put diagonally under this Cr:Forsterite laser system. The two different frequency lasers were generated respectively using Cr:F-1-R and Cr:F-2-R with single Nd:YAG pumping of 95 mJ and 105 mJ. The Cr:Forsterite laser using Cr:F-1-R crystal energy is 5.0 mJ; the frequency is tuned to around 1238.4 nm. The Cr:Forsterite laser using Cr:F-2-R crystal energy is 1.5 mJ and the frequency is 1223.5 nm. The two pulses are overlapped temporally and spatially at the GaP crystal surface. The pulse timing was tuned using the conversion efficiency of Cr:Forsterite laser. The THz wave was generated with energy of 4.7 pJ around 2.95 THz. Both the Cr:Forsterite laser line widths were measured as less 0.07 nm using a spectrum analyzer. The linewidth of the generated THz wave is estimated as less than 30 GHz.

Changing the conversion efficiency and pump energy of a Cr:Forsterite laser controlled temporal overlap of two Cr:Forsterite laser pulses. A THz wave was generated using two Cr:Forsterite lasers pumped using a single Nd:YAG laser. The Cr:Forsterite laser system was built as 30-cm square for portable use in the field.

3. Polarized THz spectroscopy for uniaxially deformed ultra high molecular weight polyethylene

One research objective is developing THz evaluation technique of uniaxially deformed polyethylene. Polarized THz spectra have been measured for the deformed polyethylene at room temperature. The THz pulse has a linear polarization.

Ultra high molecular weight polyethylene (UHMWPE) has strength against impact force, friction, chemical attack, and coldness. Therefore this material has been used for many industrial applications, for example gears, gaskets and artificial joints. Several non-destructive tests for UHMWPE have been developed already. The conventional methods are XRD, FT-IR, Raman spectroscopy. These methods should be improved for easy and safe test of mechanical deformation of bulk polymer. The molecular vibration modes of polyethylene are well known. The 14 vibration modes of polyethylene are divided into two types: carbon - carbon vibration and CH_2 plane vibration. In THz spectra of UHMWPE, absorption band is appeared around 2.2 THz, which is assigned to B_{1u} translational lattice vibration mode [21, 22]. THz wave has high transparency for polymers comparing to mid-infrared. Furthermore, it is harmless and easy to use it. Until now, chemical degradation diagnosis of UHMWPE was reported with THz Time Domain Spectroscopy [23]. THz spectra have been obtained for γ-ray irradiated UHMWPEs with or without vitamin E doping. Vitamin E has anti-degradation effect on UHMWPE.

The UHMWPE plates were deformed at room temperature. One day after deformation, the polarized spectra were obtained with every strain. The measurement frequency is swept from1.5 to 3 THz with 15 GHz step. It takes about 10 minutes for measurement. The 2.2 THz absorption band is due to B_{1u} lattice translational vibration mode of PE. For the spectra with the polarized direction of THz wave parallel to the deformed direction, the absorbance decreases drastically as a function of strain. For the spectra with the THz wave direction perpendicular to the deformed direction, the absorbance decreases gradually in the perpendicular direction.

The integral absorption intensities of the 2.2 THz band are plotted as a function of strain in Figure 8(a). In the parallel direction, up to 25% strain, absorption intensity decreases rapidly. Over 25% strain, the intensity decreases slowly. The degrees of orientation are estimated from XRD. In figure 8(b), the THz absorption intensities in parallel direction are plotted against the degrees of orientation from XRD. This relative is appeared linearity. This result indicates the correlative between THz spectroscopy and XRD. These results suggest the absorption intensity decrease of parallel polarization is caused by orientation.

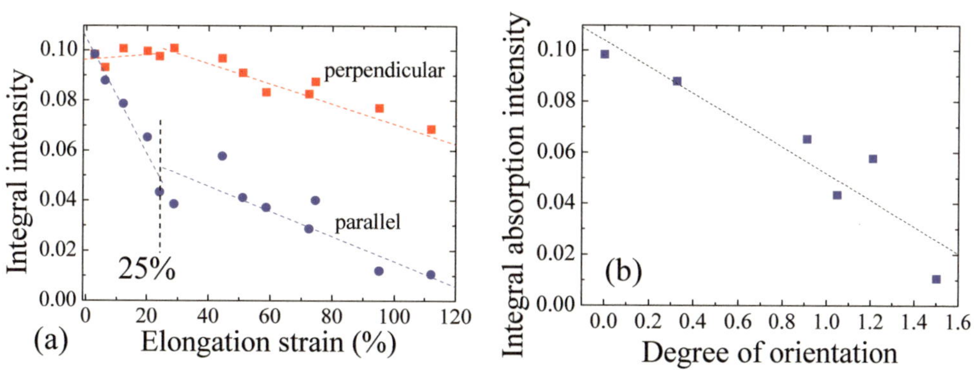

Fig. 8. (a) The integrated absorption intensities of the B_{1u} band as a function of the degree of elongation strain. Incident THz wave was polarized parallel (●) and perpendicular (■) to the axial direction of the deformation. (b) THz absorption intensity vs. degree of orientation. THz wave electric polarization direction is perpendicular to the deformation direction. Degree of orientation was evaluated by XRD measurements.

These results are discussed with lamellar model. Dependent of THz spectral feature on the polarization direction as well as strain amount is explained. THz absorption intensities are inversely proportional to the degree of orientation. When lamellas are oriented along c-axis, the dipole moment direction of B_{1u} vibration mode is perpendicular to c-axis. Then, the B_{1u} absorption band cannot be observed with THz wave polarization direction to the c-axis. According to the IR selection rule, absorption intensity decreases in the parallel polarization spectrum.

For *in-situ* spectroscopic measurements of UHMWPE during uniaxial deformation, the UHMWPE plate was strained at every 10 minutes with 1 % step. It is appeared that the red

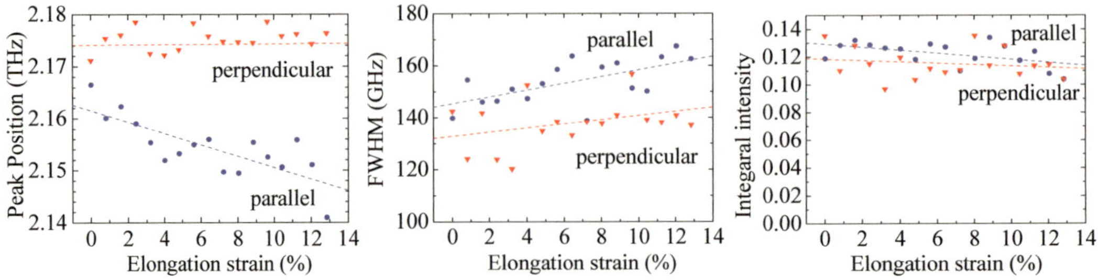

Fig. 9. The absorption peak position, FWHM, and integral absorption intensity of the B_{1u} band as a function of the degree of elongation strain.

shift and widening of B_{1u} absorption band as a function of strain. For perpendicular polarized spectra, the clear peak shift and widening are unclear. For the 2.2 THz band, the absorption peak position, full width at half maximum (FWHM) and integral absorption intensity are plotted in Figure 9 against the elongation strains, respectively. The peak position clearly shifts to lower frequency in the parallel polarized *in-situ* spectra. The peak shift rate is -7.6 % / strain. For absorption bands in higher frequency regions, Richard Wool and co-researchers reported that the peak shift rates of the CH_2 rocking B_{1u} band at 730 cm^{-1} and the CH_2 B_{2u} bending band at 1472 cm^{-1} are -3.0 % and -0.6 % / strain, respectively [24]. The relation between the peak shift rate and peak position is appeared linearity in Figure 10.

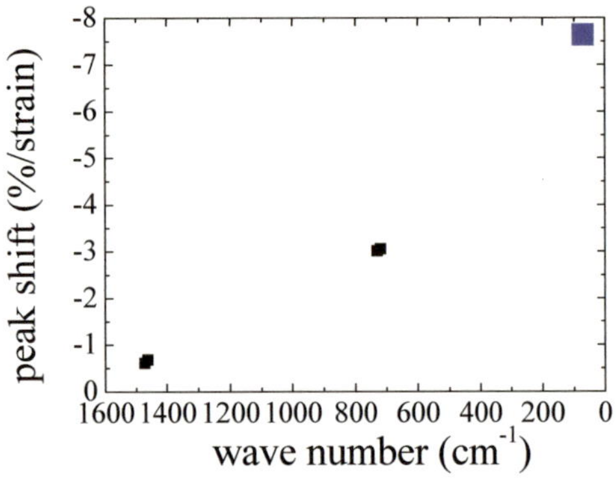

Fig. 10. The relation between peak shift rate and peak position of the B_{1u} band during uniaxial deformation.

Polarized THz wave spectra were obtained for evaluation of polymer chains in mechanically extensional-deformed UHMWPEs. THz absorption band was seen around 2.2 THz which is due to the B_{1u} translational lattice vibration mode. For the deformed UHMWPE, the dichroism is appeared in the 2.2 THz band intensity and peak position. The absorption intensity is smaller in spectra with the THz wave electric polarization direction parallel to the deformation direction than that with the polarization direction perpendicular. The peak position shift and absorption intensity is dependent on the amount of elongation strain, respectively. Based on these results, it is suggested that the THz nondestructive diagnosis of the tensile strain in deformed UHMWPE is possible based on the dichroism of B_{1u} band intensities on polarized THz wave spectroscopy.

4. THz reflection spectroscopy for metal conductor surfaces covered with insulating polyethylene

The THz wave has a lot of characteristics, for example, high permeability for non polar materials, less-invasive, high reflectance on metal surfaces and safety. Our one research target is a THz application for monitoring of metal surface corrosion. Even a metal surface has a corrosion protective covering for practical uses, THz wave is transparent for covering polymers. Here at the metal surface THz wave is absorbed due to corrosion products. This application has many merits as following, high spatial resolution comparing to that of conventional microwave techniques, non-contact, non-destructive and safety.

Until now, the THz spectral data base has been developed mainly for organic materials. That data for metal and inorganic materials is under construction. However, of course, for THz non-destructive applications, the THz spectral data of metal compounds is very important. THz characteristics of metal compounds are essential. We have focused on copper for the non-destructive THz inspection of electric power lines. At present, inspection method of electric cupper cable is based on pealing of cover with visual observation. For prevention of unexpected break down, the non-destructive inspection is desired as a THz application. The non-destructive THz inspection can be applied even for live power lines. Thus THz spectra have been measured for inorganic chemicals of copper, such as Cu_2O, CuO, $Cu(COO)_2 \cdot 1/2H_2O$. THz reflection spectrum of natural cupper oxides on an electric line surface was measured for practical applications. The difference of spectral features between standard copper chemicals and natural copper oxides has been discussed.

For sample preparation, Cu_2O, CuO and Copper oxalate were purchased with 99 % purity. These samples were formed into pellets with PE powder under various concentrations for transmittance THz spectral measurements. The PE powder has high permeability for THz wave. Sample and PE powder are mixed and pressed, then a 1 mm thickness PE pellet was prepared with the diameter of 20 mm. In order to escape the interference in THz spectral measurements, the pellet is formed with wedge shape.

For the THz spectral measurement, THz wave is generated from GaP based on non linear optical effect. We use Nd:YAG laser as a pumping light. Two near-infrared (NIR) beams are from Cr:Forsterite lasers, and we use two detectors of DTGS in a double beam configuration. Two NIR laser beams from Cr:Forsterite lasers are introduced to GaP, then THz wave is generated via the difference frequency generation method. The THz-wave path was purged with dry air to eliminate water vapour absorptions. Absorbance is calculated by Lambert-Beer law. For practical applications, the diffused reflection spectrum of natural copper oxide was measured. In reflection spectral measurements, 4K cooled Si bolometer was used as a THz detector. Samples are copper electric line with oxidized surface and not oxidized surface. THz wave induced to sample surfaces are separated into two directions. One THz wave is reflected at sample surfaces and the other is transmitted and reflected at the interface between an oxide film and a copper substance.

Figure 11(a) shows THz spectra of Cu_2O with various concentrations. One sharp absorption peak is appeared at 4.43 THz. Over wide frequency region from 2.2 to 4.2 THz, the broad band is observed. Following the background correction, these absorption bonds are split with a Gaussian function fitting. For each band the intensity are plotted against sample concentrations in a log-log graph, as shown in Figure 11(b). It is apparent that the each integrated intensity increase as a function of sample concentration.

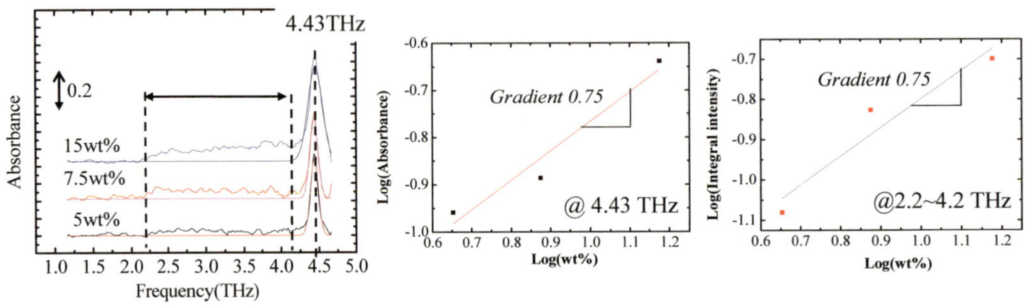

Fig. 11. THz spectra of Cu_2O with various concentrations at room temperature.

THz spectra of CuO are shown in Figure 12(a). A broad band is observed. The integrated intensity is plotted in Figure 12(b) as a function of sample concentrations in a log-log graph. This intensity increases against the concentration. The broad band is assigned to CuO. THz spectral feature of CuO is difference from that of Cu_2O.
Figure 13 shows THz spectra of copper oxalate. An asymmetric band is observed around 3.3 THz. The band is split into 3 peaks with a Gaussian function fitting. The peak positions are 2.7, 3.4, and 3.5 THz, respectively. For bands at 2.7 and 3.4THz, the intensities are increased as a function of sample concentration. That means the 2.7 and 3.4 THz bands are assigned to copper oxalate. But, for 3.5 THz band, the intensity is near constant regardless of sample concentrations. The 3.5 THz band is not due to the sample.

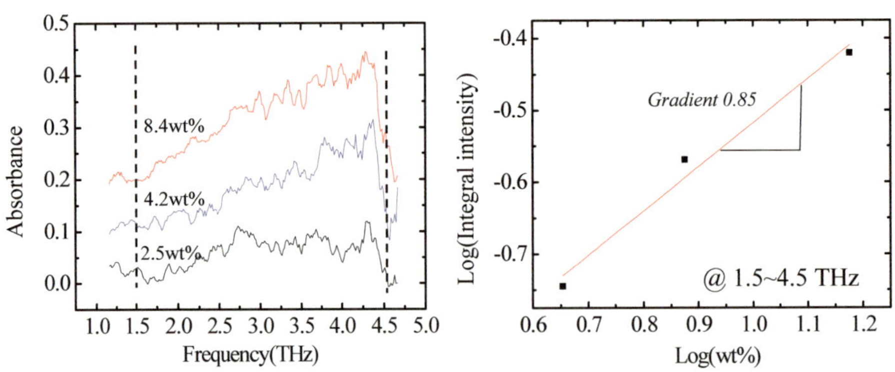

Fig. 12. THz spectra of CuO with various concentrations at room temperature.

Diffused reflection THz spectra of copper surface covered with oxide and without oxides were shown in Figure 14(a), respectively. In the region from 1.3~2.5 THz, the reflectance is difference in each spectrum. The THz reflectance from the no oxide surface is larger than that from the oxide surface, which is attributable to the very thin surface oxide layer, even though the oxide layer thickness is very thin compared with the wavelength of THz wave. The difference is plotted at each frequency as an absorption spectrum of natural oxide, shown in Figure 14(b). The absorption band is split into 3 peaks with a Gaussian function fitting. The positions are 1.51, 1.77 and 1.95 THz, respectively. Every peak position is not in agreement with that of standard copper chemicals. This is due to the difference of composition ratio for Cu and O between the natural copper oxide and standard copper chemicals. The depth profiles of relative concentration are obtained on CuO, Cu and O by secondary ion mass spectroscopy (SIMS). Cs^+ primary ions at an energy of 5 keV were used. The ion incidence angle is 60 ° relative to the surface normal. The ion current was 100 nA and the beam was scanned across areas of 300 μm × 300 μm. To reduce the crater effect, the analysis beam with square of 90 μm was positioned in the center of sputter scan area. CuO was used to monitor the distribution of oxygen. The monitoring CuO^- cluster leads to change of dynamic range with half order of magnitude around 3 μm. The depth profile indicates the thickness of the oxide layer was 3 μm. The O^- yield is saturated so that the oxygen distribution is nearly uniform. Matrix effect affect on the Cu^- yield. The difference of THz reflectivity is helpful for a non-destructive evaluation of the corrosion of metal surface. This result is one of killer applications of THz wave and greatly contributes in the field of non-destructive inspection for corroded metals, even covered with an insulating polyethylene.

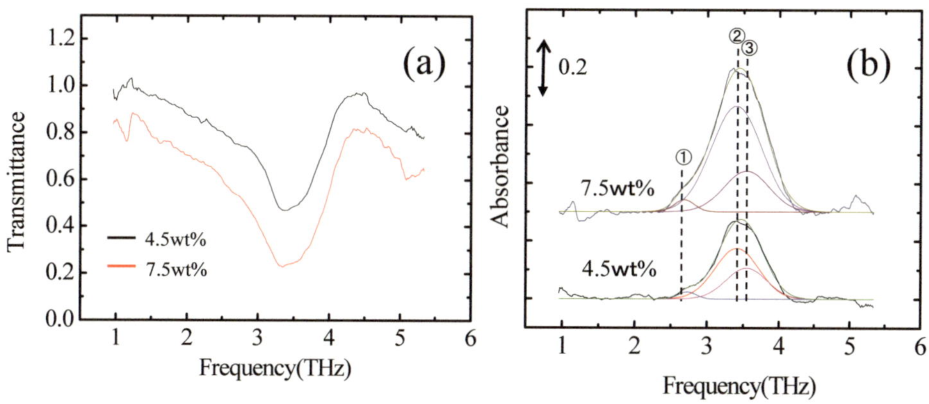

Fig. 13. THz spectra of $Cu(COO)_2 \cdot 1/2H_2O$ with various concentrations at room temperature.

For THz reflection spectroscopy, very thin surface oxide layer reveals serious difference on THz reflectivity at a specific frequency even when layer thickness is very thin compared with the wavelength of THz wave. This phenomenon contributes to detect the corroded surface of metals for non-destructive evaluation of corrosion. To realize such an evaluation, THz wave is the best, due to its high permeability for insulators and sensitive change of reflection intensity from metal surface even covered with insulators.

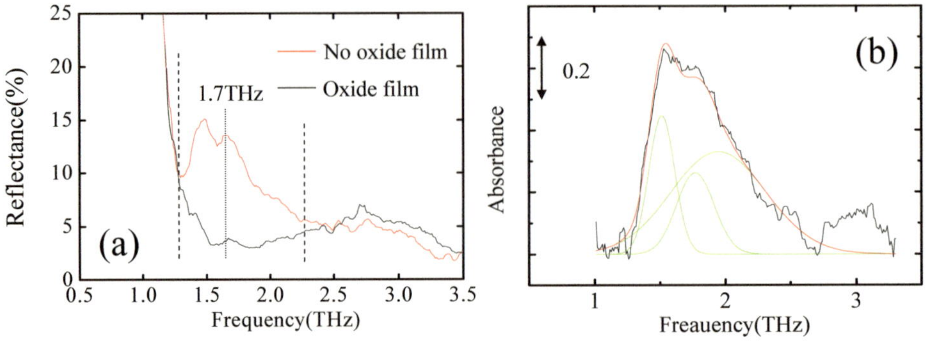

Fig. 14. THz spectra of copper surface covered with oxide and without oxides at room temperature.

5. Conclusion

Frequency-tunable terahertz (THz) wave pulse is generated by exciting a phonon-polariton mode in a GaP crystal, which are based upon non-collinear phase-matched different frequency generation (DFG) of nonlinear optical effect. We have developed a compact THz-wave generator using two small Cr:Forsterite lasers with single Q-switched Nd:YAG laser pumping. A Cr:Forsterite laser was constructed with diffraction gratings, by which the pulse duration and delay time of the Cr:Forsterite laser depend on Cr:Forsterite laser energy and the cavity length. The Cr:Forsterite laser energy was tuned using the optical alignment and pumping energy. Temporal overlap of two Cr:Forsterite laser pulses was realized at the GaP crystal. A single-frequency THz wave was generated at energy of 4.7 pJ around 2.95 THz using a 30-cm-long Cr:Forsterite laser system. The tunable range was approximately 0.3–7.5 THz.

The THz waves are located between radio waves and light waves. Thus, a THz wave has unique properties with high transparency for non-polarized materials such as with radio waves and easy handling as in the case of light. THz waves are expected to be a promising frequency for the non-destructive diagnosis of the interior of non-polarized materials. THz non-destructive diagnosis are possible for tensile strain in deformed polyethylene and metal conductor surfaces covered with insulating polyethylene.

6. References

[1] J. Nishizawa, "Semiconductor Maser", Japanese Patent 273217, April 1957.

[2] K. Suto and J. Nishizawa, "Low-threshold semiconductor Raman laser", *IEEE J. Quantum Electron.*, vol. 19, pp. 1251-1254 (1983).

[3] K. Suto, T. Kimura, J. Nishizawa, "Fabrication and characteristics of tapered waveguide semiconductor Raman laser", *IEE Proc.-Optoelectron*, vol. 143, pp. 113-118 (1996).

[4] T. Tanabe, K. Suto, J. Nishizawa, T. Kimura and K. Saito, "Frequency-tunable high-power terahertz wave generation from GaP", *J. Appl. Phys.*, vol. 93, pp. 4610-4615 (2003).

[5] T. Tanabe, K. Suto, J. Nishizawa, K. Saito, T. Kimura, "Tunable Terahertz Wave Generation in the 3- to 7-THz region from GaP", *Appl. Phys. Lett.*, vol. 83, pp. 237-239 (2003).
[6] T. Tanabe, K. Suto, J. Nishizawa, K. Saito, T. Kimura, "Frequency-tunable terahertz wave generation via excitation of phonon-polaritons in GaP", *J. Phys. D: Appl. Phys.*, vol. 36, pp. 953-957 (2003).
[7] K. Suto, T. Sasaki, T. Tanabe, K. Saito, J. Nishizawa, M. Ito, "GaP THz wave generator and THz spectrometer using Cr:Forsterite lasers", *Rev. Sci. Instrum.*, vol. 76, pp. 123109 1-3 (2005).
[8] J. Nishizawa, T. Tanabe, K. Suto, Y. Watanabe, T. Sasaki, Y. Oyama, "Continuous-wave frequency-tunable terahertz-wave generation from GaP", *IEEE Photo. Tech. Lett.*, vol. 18, pp. 2008-2010 (2006).
[9] B. B. Hu, M. C. Nuss, "Imaging with terahertz waves" *Opt. Lett.* vol. 20, pp. 1716-1719 (1995).
[10] D. M. Mittleman, M. Gupta, R. Neelamani, R. G. Baraniuk, J. V. Rudd, M. Koch, "Recent advances in terahertz imaging" *Appl. Phys. B* vol. 68, pp. 1085-94 (1999).
[11] B. Ferguson, X. -C. Zhang "Materials for terahertz science and technology" *Nature Mater.* vol. 1, pp. 26-33 (2002).
[12] K. Kawase, Y. Ogawa, Y. Watanabe, H. Inoue "Non-destructive terahertz imaging of illicit drugs using spectral fingerprints" *Opt Express* vol. 11, pp. 2549-54 (2003).
[13] J. Nishizawa, "Development of THz wave oscillation and its application to molecular sciences" *Proc. Jpn. Acad. Ser. B* vol. 80, pp. 74-81 (2004).
[14] M. Tonouchi, "Cutting-edge terahertz technology" *Nature Photo.* vol. 1, pp. 97-105 (2007).
[15] Y. Oyama, L. Zhen, T. Tanabe, M. Kagaya, "Sub-Terahertz Imaging of Defects in Building Blocks" *NDT & E Int.* vol. 42, pp. 28-33 (2009).
[16] J. Nishizawa, T. Yamada, T. Sasaki, T. Tanabe, T. Wadayama, T. Tanno, K. Suto, "Terahertz dichroism of MBBA liquid crystal on rubbed substrate" *Appl. Surf. Sci.* vol. 252, pp. 4226-29 (2006).
[17] T. Tanabe, K. Watanabe, Y. Oyama, K. Seo, "Polarization sensitive THz absorption spectroscopy for the evaluation of uniaxially deformed ultra-high molecular weight polyethylene" *NDT & E Int.* vol. 43, pp. 329-333 (2010).
[18] J. Nishizawa, T. Sasaki, T. Tanabe, N. Hozumi, Y. Oyama, Ken Suto "Single-frequency coherent terahertz-wave generation using two Cr:forsterite lasers pumped using one Nd:YAG laser" *Rev. Sci. Instrum.* vol. 79, p. 036101 (2008).
[19] A. Mooradian and G. B. Wright, "First order Raman effect in III–V compounds", *Solid State Commun.* vol. 4, pp. 431-434 (1966).
[20] V. G. Dmitriev, G. G. Gurzadyan, D. N. Nikogosyan, Handbook of Nonlinear Optical Crystals (Berlin: Springer) pp. 68-240 (1997).
[21] M. Tasumi, T. Shimanouchi, "Crystal Vibrations and Intermolecular Forces of Polymethylene Crystals" *J. Chem. Phys.* vol. 43, pp. 1245-58 (1965).
[22] M. Tasumi, S. Krimm "Crystal Vibrations of Polyethylene" *J. Chem. Phys.* vol. 46, pp. 755-766 (1967).
[23] K. Yamamoto, M. Yamaguchi, M. Tani, M. Hangyo, S. Teramura, T. Isu, N. Tomita, "Degradation diagnosis of ultrahigh-molecular weight polyethylene with terahertz time-domain spectroscopy" *Appl Phys Lett* vol. 85, pp. 5194-5196 (2004).

[24] R. P. Wool, R. S. Bretzlaff, B. Y. Li, C. H. Wang, R. H. Boyd, "Infrared and raman spectroscopy of stressed polyethylene " *J. Polym. Sci. B.* vol. 24, pp. 1039-1066 (1986).

Part 2

Laser Beam Manipulation

7

Laser Pulse Contrast Ratio Cleaning in 100 TW Scale Ti: Sapphire Laser Systems

Sylvain Fourmaux, Stéphane Payeur, Philippe Lassonde, Jean-Claude Kieffer and François Martin
Institut National de la Recherche Scientifique, Énergie, Matériaux et Télécommunications - Université du Québec
Canada

1. Introduction

Extreme laser peak intensities can be produced with current laser technology, using both the chirped pulse amplification (CPA) technique (Strickland & Mourou, 1985) and Ti:Sapphire amplification crystals (Le Blanc et al., 1993). Laser systems using this technology are commercially providing instantaneous power in excess of 100 TW with a laser pulse duration $\sim$30 fs and energy per pulse of several Joules. By focusing these pulses to a few μm spot size, high intensity laser matter interaction studies are now routinely performed at peak intensities of $10^{18} - 10^{20}$ W/cm^2.

For fundamental physics research, increased laser intensities enhances current interaction processes and can lead to new and more efficient interaction regimes. A peak intensity above 10^{22} W/cm^2 has already been reported by Yanovski et al. (Yanovsky et al., 2008) and a facility such as the Extreme Light Infrastructure (ELI) (Gerstner, 2007) envisions peak intensities in the range of 10^{23} W/cm^2 which are needed for experiments on radiation reaction effects (Zhidkov et al., 2002).

For applications development, recent progress of laser systems combining high intensity and high repetition rate have attracted considerable interest for the production of solid target based secondary sources where high mean brightness is required. In high field science, this includes bright x-ray sources (Chen et al., 2004; Schnürer et al., 2000; Teubner et al., 2003; Thaury et al., 2007), high energy particle acceleration (Fritzler et al., 2003; Steinke et al., 2010; Zeil et al., 2010) and nuclear activation (Grillon et al., 2002; Magill et al., 2003).

To illustrate this interest for high peak intensities, recently published scaling laws for laser based proton acceleration on thin film solid targets (Fuchs et al., 2006) have shown that an important increase of the on target laser intensity is necessary to reach the expected energy required for biomedical application in the proton therapy field (60 - 250 MeV). Moreover, intensities greater than 10^{20} W/cm^2 will allow access to the non collisional shock acceleration regime where >100 MeV maximum energy protons could be produced (Silva et al., 2004).

For currently available peak intensities ($10^{18} - 10^{20}$ W/cm^2 range), where the field strength is sufficient to accelerate particle to relativistic energies, the laser pulse contrast ratio (LPCR) is a crucial parameter to take into consideration. Considering the laser pulse intensity temporal profile, the LPCR is the ratio between its maximum (peak intensity) and any fixed delay before it. A low contrast ratio can greatly modify the dynamics of energy coupling between the laser pulse and the initial target by producing a pre-plasma that can change the interaction mechanism. This issue will be even more important for future laser systems with higher intensities. Thus, LPCR characterization and improvement are of great importance for such laser systems.

In this chapter, the second section presents the basic concepts: Laser Pulse Contrast Ratio, ionization and damage thresholds definitions. Section 3 is devoted to the presentation of the typical structure and performances of a high power laser system based on CPA Ti:Sapphire technology which is currently the most popular way to reach high intensity at a high repetition rate. Section 4 presents the measurement techniques to characterize the pre-pulse, the Amplification of Spontaneous Emission (ASE), and the coherent contrast of a laser system. We can divide the pulse cleaning techniques that have been proposed to improve the LPCR into two categories: first those used before compression which mainly reduce pre-pulses and ASE; second those applied after compression which allow a coherent contrast enhancement. In the first category, we present in section 5 the technique based upon high energy injection through saturable absorbers before power amplification and the technique based on cross wave polarization (XPW). To improve the coherent contrast, only the cleaning techniques used after pulse compression are efficient (second category); we detail in section 6 the techniques of second harmonic generation and the technique using plasma mirrors. We conclude this chapter in section 7.

2. Basic concepts

2.1 Laser Pulse Contrast Ratio (LPCR) definition

We define the LPCR as the ratio R between the laser pulse peak intensity I_{peak} and any pre-pulse or pedestal intensity I_p, i.e. $R = I_{peak}/I_p$. A low contrast ratio laser pulse can generate a pre-plasma before the arrival of the laser pulse peak intensity and completely change the interaction mechanism between the laser pulse and the initial target (Workman et al., 1996; Zhidkov et al., 2000). For high intensity laser systems using the CPA technique, several overlapping time scales of the intensity temporal profile should be distinguished, each corresponding to a specific source of LPCR degradation (Konoplov, 2000). On the ns time scale, from a few ns up to approximately 20 ns, insufficient contrast of the amplified laser pulse relative to neighbouring pulses could occur in the oscillator or a regenerative amplifier cavity resulting in pre-pulses. Below several ns, amplified fluorescence produced by the pumped crystal during the amplification process may propagate down the amplifier system; this Amplification of Spontaneous Emission (ASE) produces a plateau shaped ns pedestal. On the ps time scale, below a few $10'$s of ps, imperfect pulse compression due to deficient laser spectrum manipulations results in a LPCR degradation close to the peak intensity. On this timescale, the LPCR is called the coherent contrast and corresponds to the rising edge of the laser pulse toward the peak intensity.

2.2 Target ionization and damage thresholds

When considering the effect of low LPCR, two target parameters are of importance: the ionization threshold and the damage threshold. The ionization threshold is intensity dependent. It corresponds to the ionization of the atomic species present in the target and production of a plasma; for example, the ionization threshold is close to 10^{12} W/cm^2 for metals and close to 10^{13} W/cm^2 for dielectrics targets. The presence of a pre-plasma near the target surface before the laser pulse peak intensity modifies the electron density gradient and the nature of the processes leading to laser energy absorption in the target. The damage

threshold is fluence dependent. It corresponds to the onset of a phase transition following energy deposition in the solid target and subsequent morphology change. This threshold is a function of the laser pulse duration and the nature of the target (bulk or thin film). For a low fluence, the value depends on the target material, the damage may affect only the first atomic layers such that the damage threshold may be difficult to characterize. A more precisely measured characteristic, used here, is the ablation threshold that corresponds to a strong solid to gas phase transition with ejection of matter. The ablation threshold has been covered in several papers; in metals, its value is close to a few J/cm^2 for laser pulse durations of approximately 1 ns (Cabalin et al., 1998; Corkum et al., 1988). Such a morphology change certainly affects the energy absorption process and may damage or destroy the target in some cases, such as thin foils or velvet targets (Kulcsár et al., 2000).
Pre-pulses incident on the target long before the peak intensity (several ns) can modify the target surface morphology or damage it. Thus, it is absolutely necessary to check the presence of any ns pre-pulses. This measurement has to be performed before every experimental measurement set as every alignment of the laser system can detune the Pockels cells or produce reflections that can increase laser pre-pulses. Usually the duration of the pre-pulses are in the sub-ns time range, and the ablation and ionization threshold are similar.
Even if the ns pedestal is below the ionization threshold, the energy deposited over a few ns can be above the ablation threshold and can change the target morphology or damage it. Thus, it is important to check the pedestal temporal profile several ns before the laser peak intensity. A situation where the energy contained in the pedestal over a few ns is a problem is illustrated in section 5.1.

3. High intensity laser systems based on Ti:Sapphire technology

The architecture of a typical laser system based on CPA Ti:Sapphire technology with its amplification stages is shown in figure 1. In this example, the laser consists of: an oscillator; a grating based stretcher which increases the laser pulse duration; a regenerative amplifier; three multi-pass amplification stages to reach the final maximum laser pulse energy; and a grating compressor to restore the short laser pulse duration.

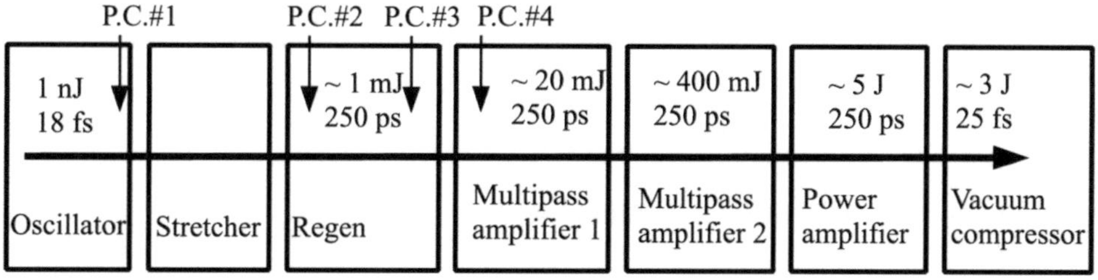

Fig. 1. Architecture of a high intensity laser system. P.C. indicates the position of the Pockels cells in the laser system. Typical output energy and pulse duration (FWHM) are indicated at each location.

The Ti:Sapphire oscillator operates at several 10's of MHz repetition rate and generates laser pulses with a central operating wavelength close to 800 nm. It typically produces an initial pulse duration of 18 fs with 80-100 nm spectral bandwidth. An acousto-optic programmable dispersive filter (AOPDF) is usually used to control the spectral phase distribution of the pulse and compensate for the gain narrowing that occurs in the different amplification stages. It is normally placed after the stretcher to control the spectral phase and amplitude distribution of the laser pulse. The grating based stretcher increases the laser pulse duration to $\sim$ 250 ps in order to reduce the peak intensity on the optical components during the energy amplification process. A 55 nm bandwidth FWHM can typically be achieved after compression. In a typical high power laser system, the minimum pulse duration routinely produced after compression is currently 25-30 fs. Note that the grating based compressor has to be located inside a vacuum chamber as the laser pulse instantaneous power after compression on the last grating is high enough that self-focusing could occur during propagation in air leading to pulse distortion and potential damage to the optics.

The regenerative cavity provides a high amplification factor of $10^5 - 10^6$. Note that some laser systems do not use a regenerative amplifier but only multi-pass amplifiers (Pittman et al., 2002). In our example, the output energy of the first two multi-pass amplifiers is approximately 20 mJ and 400 mJ, respectively. The final amplification stage increases the laser pulse energy up to several Joules. This amplifier usually includes a large number of YAG pump laser, each typically providing close to 1 J in energy. The amplification crystals are usually water cooled except for the final stage which requires cryogenic cooling to avoid thermal lensing. As an example, a 30 fs laser pulse with an output energy of 3 J after compression corresponds to an instantaneous power of 100 TW.

Pockels cells are used through the laser system to isolate the main pulse and increase the LPCR: the first, the pulse picker reduces the laser repetition rate from the MHz range down to the Hz range; the second, at the entrance of the regenerative cavity, serves to seed it; the third, inside the regenerative cavity, allows the amplified pulse to exit and the fourth, between the regenerative cavity and the first multi-pass amplifier, reduces the pre-pulse level. These Pockels cells rotate the polarization and effectively isolate the amplified laser pulse with a typical switching time of approximately 4 ns. For a laser system with a power greater than 10 TW up to a few 100's of TW, the laser repetition rate is typically 10 Hz.

We can illustrate the architecture of a high intensity laser with the performance of a commercially available system based on CPA Ti:Sapphire technology. The technical specifications of the 10 TW laser system of the Advanced Laser Light Source (ALLS) facility located in Varennes (Canada) are the following: the minimum pulse duration is 32 fs, limited by the gain narrowing occurring during the amplification; only two multi-pass amplifiers are used, providing 400 mJ before compression; the repetition rate is 10 Hz and the central wavelength is 800 nm.

4. Laser Pulse Contrast Ratio measurements techniques

4.1 Pre-pulse measurement by a fast photodiode

The pre-pulses in the nanosecond time range are usually detected at the nominal laser energy using the leakage through a high reflectivity mirror. A fast photodiode (typically a high-speed silicon photodiode with 1 ns rise time) coupled to neutral density (ND) filters calibrated at the laser wavelength are used to measure the ns pre-pulses. As the photodiode rise time is greater than the duration of any typical pre-pulse, the measured value corresponds to the integrated pre-pulse energy; it is important to measure the pre-pulse duration separately to determine its intensity and associated LPCR. Normally, the time delay between the pre-pulses and the peak intensity is fixed by the round trip time of the oscillator and the regenerative amplifier which is typically from a few ns up to 10's of ns.

Again using the ALLS facility 10 TW laser system as example, the pre-pulses are generated by the regenerative amplifier cavity at 8 ns before the peak intensity of the laser pulse. The time delay between these pre-pulses and the main amplified pulse is a multiple of a round trip inside the regenerative cavity, the first pre-pulse exits the cavity before the main pulse with one less round trip. The typical energy ratio between the amplified pulse maximum energy and the first pre-pulse is 3×10^4. The first pre-pulse duration is 500 fs corresponding to a LPCR of 4.7×10^5. The use of an additional Pockels cell after the regenerative cavity increases the LPCR of the pre-pulses by a factor 60 but has no effect on shorter timescales because of the limited rise time of the Pockels cells.

4.2 ASE and coherent contrast measurement using a high dynamic range third-order cross-correlator

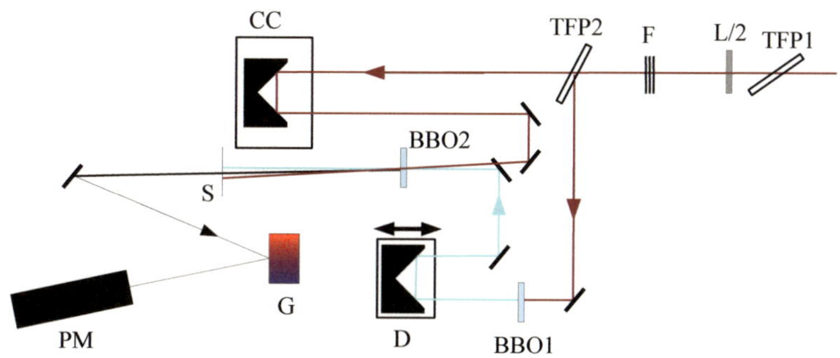

Fig. 2. Experimental set-up for LPCR measurement of the laser system pulse pedestal and coherent contrast using a high dynamic range third order cross-correlator. The arrows illustrate the laser beam propagation path, the 800 nm beam is indicated in red, the 400 nm (second harmonic) in blue, and the 266 nm (third harmonic) in black. TFP1 and TFP2 are thin film polarizer, F is a set of calibrated filters, L/2 is a wave-plate; BBO1 and BBO2 are BBO crystals, D is a motorized delay line and CC is a corner cube, S is a slit to select the third harmonic beam, G a grating and PM the photo-multiplier detector.

The LPCR for the time range of a few ns is characterized using a high dynamic range third-order cross-correlator. For this measurement, part of the laser beam is usually sampled while the laser system is working at nominal energy or close to it. During the measurement, the last amplification stage can only be partly pumped, this does not significantly affect the final result as long as most of the amplification has been obtained (Kiriyama et al., 2010). The experimental set-up is described on figure 2. The sampled beam is divided using a polarizer beamsplitter (TFP2). The reflected part is frequency doubled using a 200 μm thick type I BBO crystal and reflected by a motorized delay line (D). The transmitted part at the fundamental frequency is directed toward a corner cube (CC). Both beams are then combined in a non-collinear geometry into a 100 μm thick type II BBO crystal to generate the third harmonic. The 266 nm third harmonic is separated from the other wavelengths using a grating and recorded by a photomultiplier.

The use of calibrated neutral density filters allows the cross-correlator to cover a dynamic range typically of 12 decades. The range of the motorized delay line (D) is limited to 600 ps in the standard commercial cross-correlator. In routine use, usually only the pedestal

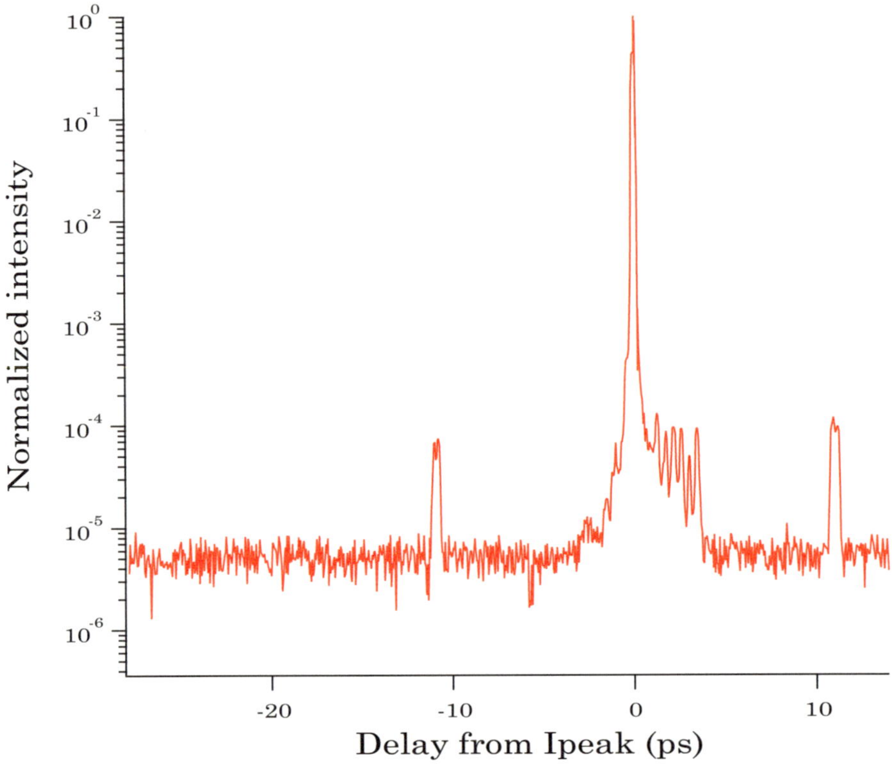

Fig. 3. Normalized intensity (inverse of LPCR) for the time interval ranging before the laser pulse peak intensity from -28 ps up to +14 ps. Courtesy of A. Laramée and F. Poitras of the ALLS facility technical team.

level several 10's of ps before the peak amplified laser pulse and the rising edge of the laser pulse are checked. The corner cube (CC) can be set onto a second manual delay line to allow successive scans with different offset values to cover delays greater than 600 ps between the fundamental and the second harmonic. The pedestal has been studied in this way in the ns range (Fourmaux et al., 2011; Hong et al., 2005; Itatani et al., 1998). The background level limits the dynamic range over which the measurement can be achieved. This is determined by blocking the fundamental frequency on the transmitted part of the beam. It should also be mentioned that the cross-correlator measurement increases the pulse duration up to $\sim$ 100 fs due to pulse dispersion by the vacuum window, filters and BBO crystals used in the cross-correlator and to sample the beam. Several pre-pulses are usually observed before the main pulse (see in figure 3 the pre-pulse at -10 ps). These correspond to replica of post pulses due to the presence of windows or optics in the laser system and the cross-correlator diagnostic (the 266 nm third harmonic results from a convolution of the 800 nm fundamental wavelength by the 400 nm second harmonic).

Using the ALLS facility 10 TW laser system as example, figure 3 shows the normalized intensity for a time interval before the laser peak intensity of -28 ps up to 14 ps. The LPCR due to the pedestal level is 2×10^5. Assuming a solid metallic target, a focused intensity of 5×10^{17} W/cm^2 is then high enough to ionize the target. This clearly demonstrates that standard high power laser systems need to improve the LPCR in order to avoid pre-plasma production.

5. Pulse cleaning techniques for ultrafast laser system

Several techniques have been proposed in order to enhance the LPCR in ultrafast CPA laser system: saturable absorbers (Hong et al., 2005; Itatani et al., 1998), double CPA laser system (Kalashnikov et al., 2005), non linear birefringence (Jullien et al., 2004), cross polarized wave generation (XPW) (Petrov et al., 2007), plasma mirror (Lévy et al., 2007) or second harmonic generation (Toth et al., 2007). Details of these techniques can be found in the cited references. Few of these pulse cleaning techniques have been implemented in high repetition rate 100 TW scale laser systems and usually, the LPCR have only been characterized to a few 100 ps before the laser pulse peak intensity. This section covers the cleaning techniques used before compression, these mainly reduce pre-pulses and ASE on high intensity Ti:Sapphire laser system. The following section will cover the cleaning techniques used after compression.

5.1 Saturable absorber cleaning and high energy injection before power amplification

This cleaning technique is based on removing the pedestal through saturable absorber transmission and high energy injection before power amplification. A saturable absorber is an optical component with a high optical absorption which is reduced at high intensities. As the intensity of the pedestal and any pre-pulses is not intense enough to change the absorption of the saturable absorber, they are not transmitted. But the high intensity portion of the laser pulse will trigger a change in absorption on its rising edge, enhancing the laser pulse temporal contrast ratio for the transmitted laser pulse. This technique has been applied previously to low power laser systems (Hong et al., 2005; Itatani et al., 1998) and is now applied with 100 TW class CPA laser systems , producing an LPCR of 10^8-10^9.

The Canadian ALLS 200 TW laser system uses such a saturable absorber to improve the LPCR before injection into the regenerative amplifier. This laser system produces 25 fs pulses with an energy of 5.4 J after compression at a10 Hz repetition rate. The maximum intensity obtained with a f/3 off axis parabola is 3×10^{20} W/cm^2 (Fourmaux et al., 2008). The cleaning stage, located between the oscillator and the stretcher, consists of a 14 pass ring amplifier pumped with 8 mJ of energy, followed by a 2 mm thick RG 850 saturable absorber used to increase the LPCR at this stage of amplification. The laser pulse energy after the cleaning stage is 10 μJ, the gain amplification factor is 10^4 and the saturable absorber transmission is 45%. The fluence in the saturable absorber is 10 mJ/cm^2. This allows the injection of a pre-amplified laser pulse with a high LPCR, reducing the required subsequent amplification and hence decreasing ASE production.

The normalized intensity of the laser system is shown in figure 4 for a time delay before I_{peak} ranging from -5.4 ns up to +100 ps. This measurement is achieved, using the third order cross-correlator, by setting the corner cube (CC) onto a manual delay line and moving it several times in order to achieve a time delay up to 5.4 ns before I_{peak}. The LPCR exhibits a minimum 55 ps before I_{peak} at 10^9, increasing for longer delays; for example, the LPCR is 2.5×10^8 at -500 ps. The inset in figure 4 shows the normalized intensity for a time delay before I_{peak} ranging from -500 up to +100 ps; this is basically the coherent contrast. Small bumps with a 500-600 ps periodicity can be observed in figure 4. These correspond to a reproducible instrument artifact from an angular shift that occurs at the centre of the motorized translation's travel (delay line D). This does not affect the aspect of the ASE signal.

There is a general deterioration of the LPCR for increasing delays before I_{peak}, especially in the interval between 250 ps and 2.1 ns with a value close to 2.5×10^8. We attribute this trend to the ASE generated inside the regenerative amplification. Indeed, during this delay range,

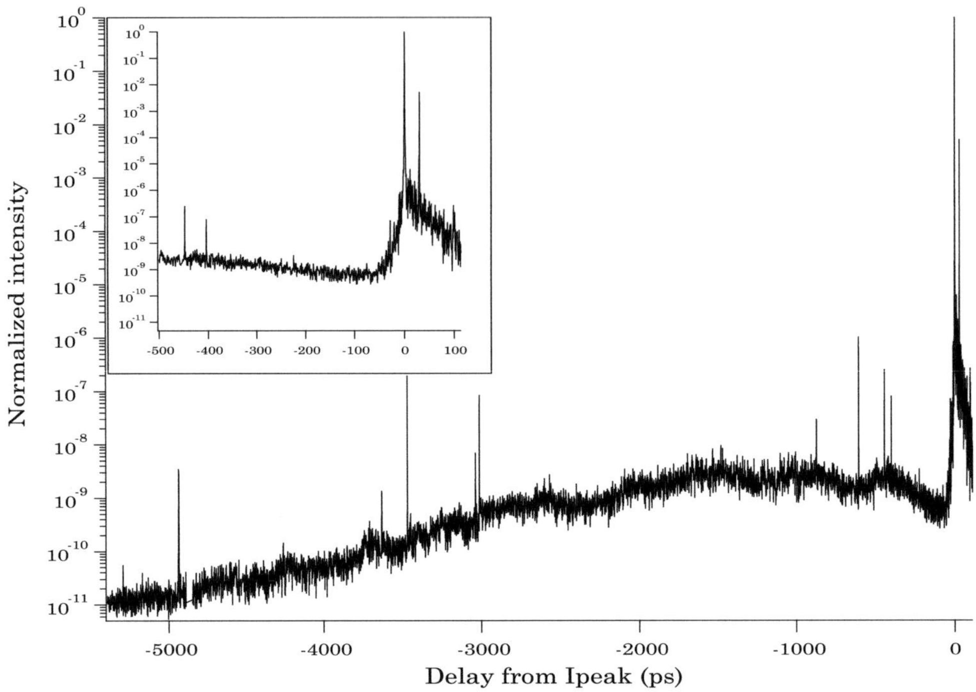

Fig. 4. Normalized intensity (inverse of LPCR) for large delays before the laser pulse peak intensity up to 5.4 ns. The inset shows a magnification of the time interval ranging from -500 ps up to +100 ps.

the amplification is sufficiently high to produce an important amount of ASE. During the amplification process, injecting the chirped laser pulse temporally decreases the ASE. For an amplifier in the saturation regime, this occurs because of the competition between the ASE and the stretched pulse to deplete the population of excited states. As the number of photons is greater in the chirped pulse compared to ASE, this depletion occurs to the detriment of the ASE. This explains why the LPCR remains high for delays close to the amplified laser pulse. Considering this LPCR, a maximum peak intensity of 2.5×10^{20} W/cm^2 is acceptable for the pedestal intensity to remain below the ionization threshold 10^{12} W/cm^2. Taking into account the correction factor due to the pulse broadening increases this intensity to 2×10^{21} W/cm^2; this is greater than the maximum peak intensity that can be reached by such a laser system with a f/3 off-axis parabola. Thus with these focusing conditions the LPCR is high enough to avoid pre-plasma production on target before the rising edge of the laser pulse. However, the energy deposited before the peak intensity must also remain below the ablation threshold (1 J/cm^2). Due to the LPCR degradation a few ns before the peak laser intensity, this limits the peak intensity to 1.7×10^{18} W/cm^2, well below the maximum intensity that can be obtained with this laser system. Thus, even though the LPCR is 10^9 a few10's of ps before the peak laser intensity, the ASE of this laser system is an issue. A simple method to reduce the ASE in this system has been proposed: it consists of adding an additional saturable absorber after the regenerative cavity in order to reduce the pre-pulses and the pedestal (Fourmaux et al., 2011).

5.2 XPW technique

XPW generation is a third order non-linear process that rotates the polarization of the laser pulse; the intensity of the rotated pulse is cubic with the input intensity. Thus XPW generation is well adapted for enhancing the LPCR of a femtosecond laser pulse. When used between crossed-polarizers, XPW generation rejects the low-intensity parts of the pulse to improve the LPCR. A barium fluoride (BaF2) nonlinear crystal is usually used for XPW generation. Details of the XPW technique and research related to the prospects of its implementation on high power laser system can be found in several publications (Cotel et al., 2006; Jullien et al., 2004; Jullien, Albert et al.;K; Ramirez et al., 2011).

Techniques using XPW generation have been implemented in a double CPA geometry on a few 10 TW class laser system: "salle jaune" (10 Hz, 1.5 J, 30 fs) at LOA (Flacco et al., 2010) and LOASIS laser system (10 Hz, 500 mJ, 40 fs) at LBNL (Plateau et al., 2009), but except for measurements in the ps time range, no detailed characterization of the LPCR is available.

The XPW technique has also been implemented on the HERCULES laser at CUOS without the use of an additional CPA. To achieve this, a higher input energy (1 μJ) is used for XPW generation in two BaF2 crystals mounted in series with an AOPDF to compensate for the dispersion in the optics. The laser system produces after 1.5 J of energy in a 30 fs pulse after compression at a 0.1 Hz repetition rate (Chvykov et al., 2006).

Their measurement indicate an LPCR improvement of 3 orders of magnitude for the pedestal level at a few hundred ps before the peak intensity. Chvykov et al. have demonstrated an LPCR of 10^{11} using XPW in this single CPA geometry. No detailed characterization on the ns range is presented in their work and the low repetition rate allows only 0.15 W of average power, insufficient for high brightness laser based applications. According to the authors, this LPCR level is high enough to work with a laser pulse focused to an on target intensity of 10^{22} W/cm^2 and avoid pre-plasma production.

6. Coherent contrast improving cleaning techniques

The previously described techniques are used to clean the laser pulse before compression. Thus, they do not improve the coherent contrast which is limited by the imperfect pulse compression associated with deficient manipulations of the laser spectrum. Steepening the rising edge of the laser pulse is important for laser generated processes such as proton acceleration or harmonics generation on solid targets, to properly control the laser plasma interaction (Grismayer & Mora, 2006; Teubner et al., 2003). Using a cleaning technique after compression is challenging for high power laser systems as an energy level of several Joule precludes the use of standard transmission optical components.

6.1 Second harmonic generation

This technique is commonly used on high power laser systems with longer pulse durations, from 100's of fs to ns (using Nd:glass amplification crystals), because high conversion efficiencies are easily attained using non-linear crystals such as KDP (potassium dihydrogen phosphate), KD*P (potassium dideuterium phosphate), BBO (β-barium borate) or LBO (lithium triborate). For example, close to 70 % second harmonics conversion efficiency can be obtained using a 1054 nm laser pulse with ps pulse duration into a KDP crystal of a few mm length at a laser intensity of 100 GW/cm^2 (Amiranoff et al., 1999).

Second harmonic generation is more difficult for Ti:Sapphire high power laser systems with pulse durations between 25 - 60 fs and an energy of several Joule (Begishev et al., 2004; Marcinkevičius et al., 2003). A very short pulse duration requires that a thin doubling crystal be used because of the group velocity mismatch between the fundamental and the second harmonic wavelength at 800 and 400 nm respectively. For example, group velocity mismatch limits the optimum KDP crystal thickness to 600 μm for a 50 fs laser pulse duration, reducing the length over which efficient second harmonic generation can occur.

Usually, when considering high power laser system with large beams operated under vacuum after compression, it is easier to directly propagate the beam in the doubling crystal instead of using an afocal system, with optical components in reflection, to reduce the beam diameter. Another reason to keep the beam diameter large is to keep a sufficiently low laser intensity to avoid self phase modulation effects that decreases the conversion efficiency and distorts the beam wavefront. Using a laser system with 3 J of energy, 90 mm beam diameter at full aperture is typical. Today, the only crystal that can be found with such large dimensions is KDP.

Moreover, the second harmonics crystal must not distort the wavefront to avoid degraded beam focusing, thus requiring good optical quality surfaces. Unfortunately, the larger diameter crystal require a greater thickness to maintain a good optical quality: for example, 1 mm thickness for a 100 mm full aperture diameter KDP crystal. Another problem with using a very short pulse duration is the frequency bandwidth; a typical bandwidth of 80 nm is necessary to obtain a 25 fs laser pulse duration. The second harmonic generation will remain efficient as long as the phase matching conditions are satisfied. This occurs only over a limited range of wavelength, thus reducing the effective bandwidth.

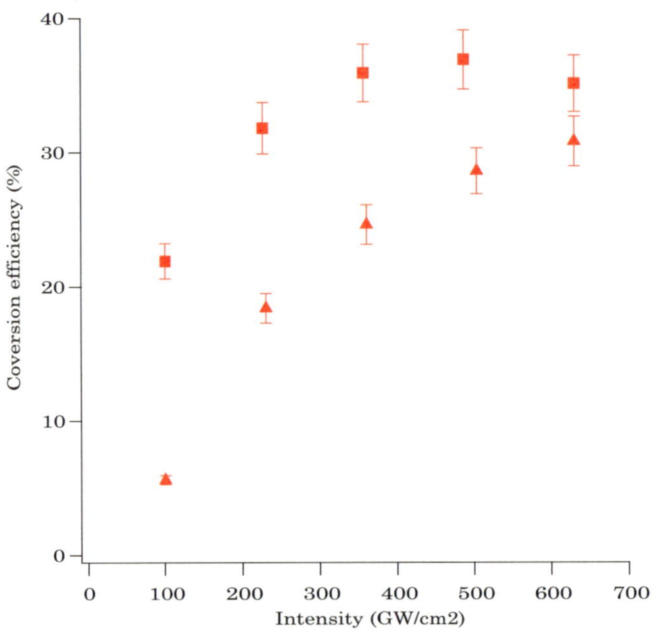

Fig. 5. Second harmonic conversion efficiency as a function of the intensity on the KDP doubling crystal using the 10 TW ALLS laser system. The results have been obtained with two crystals thickness: 1mm (square) and 600 μm (triangle).

The optimum conversion efficiency has been measured as a function of the incident energy using a KDP doubling crystal with the 10 TW ALLS laser system having a 32 fs laser pulse duration. The result is shown on figure 5 for two crystal dimensions: 1 mm thick $\times$ 60

mm diameter (square) and 600 μm thick $\times$ 40 mm diameter (triangle). Note that the pulse duration has been optimized for each crystal thickness in order to obtain the maximum conversion efficiency (the pulse duration was set by changing the optical distance between the compression gratings). The maximum energy used in this measurement for both KDP crystals thickness is 180 mJ after compression.
This cleaning technique has been used with the INRS 10 TW laser system in order to limit the hydrodynamic plasma expansion when laser pulses where focused on solid target to produce Kα x-ray line emission; the laser system was producing 60 fs pulses of 600 mJ energy per pulse at a 10 Hz repetition rate. In this particular case, the contrast ratio at the fundamental frequency was close to 10^5, for both the picosecond and the nanosecond time range. In these experiments, the beam was frequency doubled with a KDP crystal to achieve a high contrast ratio exceeding 10^9. A good beam quality was maintained at 400 nm in the near and far field. A conversion efficiency of 40% was achieved using a 1 mm thick KDP crystal. This reduces the available energy before focusing down to 200 mJ but maintains the repetition rate without any limitation in the number of pulses (Toth et al., 2005; 2007).
When a sufficiently high conversion efficiency is achieved, the advantages of this techniques, are that it maintains the nominal repetition rate of the laser system and improves the LPCR (square of the initial value) while requiring only a relatively simple experimental setup (direct propagation inside the doubling crystal is enough). It should be pointed out that, when frequency doubling the laser pulse, a large fraction of the energy remains a the fundamental frequency. To take fully advantage of the LPCR improvement of the second harmonic, the 400 nm wavelength has to be discriminated from the 800 nm wavelength as this last one arrives in advance in time compare to the 400 nm wavelength (usually this is realized by using several dichroic mirrors). The drawbacks of this technique are the low conversion efficiency because of unadapted crystal (for lack of availability) and the difficulty in obtaining the optimum phase matching conditions.

6.2 Plasma mirror

The principle of the plasma mirror is shown in figure 6: a glass plate with an anti-reflection coating is inserted at 45° into the converging laser beam, after the the focusing optics (typically an off-axis parabola) and a few mm before the laser focus. The laser fluence on the plasma mirror is approximately 100 J/cm^2, corresponding to a peak intensity of 10^{15}-10^{16} W/cm^2

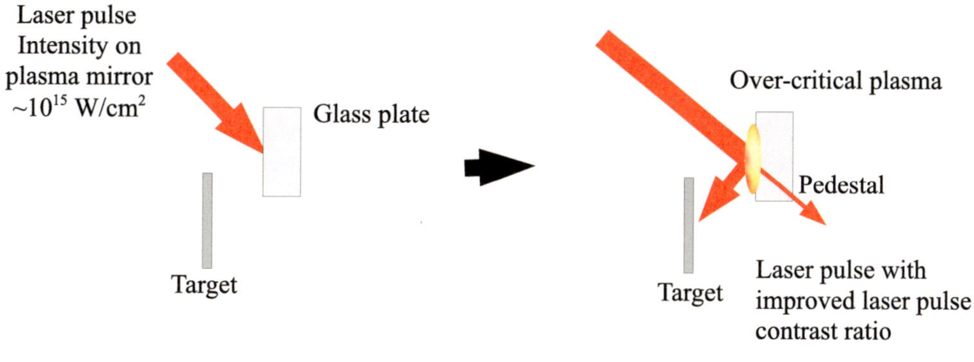

Fig. 6. Principle of the plasma mirror in the case of a simple geometry close to the target.

on the plate (considering a 30 fs pulse duration). Such an intensity generates a solid-density plasma on the glass plate surface, reflecting the laser pulse at its critical density. Since the laser pedestal and any pre-pulses are not intense enough to produce dielectric breakdown, these are not reflected and propagate through the plate. There is thus an enhancement of the laser pulse temporal contrast ratio (LPCR) for the reflected laser pulse. The plasma mirror is laterally translated after each shot to present a fresh surface for the next shot. The advantage of such a technique is to reduce both the pre-pulse, ASE and rising font of the laser pulse.
Several group have developed plasma mirror systems (Doumy et al., 2004; Dromey et al., 2004; Hörlein et al., 2008; Nomura et al., 2007; Wittmann et al., 2006).
A double plasma mirror geometry has been implemented on a 10 TW laser system at CEA-Saclay. This laser nominally produces 60 fs pulses with 700 mJ energy at a 10 Hz repetition rate. Implementing the plasma mirror reduces the repetition rate to 1 Hz with a total reflectivity of 50 % and a total shot number limited to 2000 (Thaury et al., 2007).

7. Conclusion

Characterization and improvement of the LPCR is a crucial problem for laser plasma interaction on solid targets as ultra-short laser systems delivering increasing energy and peak intensity become available. Several techniques are available to control the laser-target interaction to avoid damaging the target or producing a pre-plasma, but none of these can compensate all the different causes of LPCR degradation (pre-pulses, ASE and rising edge of the laser front) with good efficiency while maintaining the laser repetition rate. For the study of basic physics processes, the plasma mirror has proven to be an important tool but its implementation is non trivial and the limited number of shots available prevents its applications outside the field of basic research. The two more widespread techniques used today on commercial high intensity laser system, XPW and high energy injection after a saturable absorber, do not improve the coherent contrast. Moreover, the pre-pulses or the ASE level have to be checked carefully so they remain low enough not to affect the interaction conditions.
Improvement of these techniques is still an active area of research both for the use of high intensity laser system in the field of fundamental physics studies and for the development of technological applications. Until now, LPCR reduction techniques are clearly advantaged for laser based process and applications using gas targets since the requirements are important but not critical as for solid targets.

8. Acknowledgment

We thank the ALLS technical team for their support, especially F. Poitras and A. Laramée for their help on the ALLS 10 TW laser system characterization. The ALLS facility was funded by the Canadian Foundation for Innovation (CFI). This work was funded by NSERC, the Canada Research Chair program and Ministère de l'Éducation du Québec.

9. References

F. Amiranoff, V. Malka, M. Salvati, E. L. Clark, A. E. Dangor, K. Krushelnick, A. Modena, Z. Najmudin, M. I. Santala, M. Tatarakis *et al.*, (1999). Electron acceleration in a pre-formed plasma at 527 nm : an experiment performed with funding from the TMR large-scale facilities access programme. RAL *Technical report*, RAL-TR-99-008

I. A. Begishev, M. Kalashnikov, V. Karpov, P. Nickles, H. Schonnagel, I. A. Kulagin & T. Usmanov (2004). Limitation of second-harmonic generation of femtosecond Ti:sapphire laser pulses *J. Opt. Soc. Am. B*, Vol. 21, pp. 318-322

L.M. Cabalin & J. J. Laserna, (1998). Experimental determination of laser induced breakdown thresholds of metals under nanosecond Q-switched laser operation *Spectrochimica Acta Part B*, Vol. 53, pp. 723-730

L. M. Chen, P. Forget, S. Fourmaux *et al.*, (2004). Study of hard x-ray emission from intense femtosecond Ti:Sapphire laser-solid target interactions *Phys. of Plasmas*, Vol. 11, pp. 4439

V. Chvykov, P. Rousseau, S. Reed, G. Kalinchenko & V. Yanovsky, (2006). Generation of 1011 contrast 50 TW laser pulses *Opt. Lett.*, Vol. 31, pp. 1456-1458

P. B. Corkum, F. Brunel, N. K. Sherman & T. Srinivasan-Rao, (1998). Thermal Response of Metals to Ultrashort-Pulse Laser Excitation *Phys. Rev. Lett.*, Vol. 61, pp. 2886-2889

A. Cotel, A. Jullien, N. Forget, O. Albert, G. Chériaux & C. Le Blanc, (2006). Nonlinear temporal pulse cleaning of a 1-μm optical parametric chirped-pulse amplification system *Appl. Phys. B*, Vol. 83, pp. 7

G. Doumy, F. Quéré, O. Gobert, M. Perdrix, Ph. Martin, P. Audebert, J. C. Gauthier, J.-P. Geindre & T. Wittmann (2004). Complete characterization of a plasma mirror for the production of high-contrast ultraintense laser pulses *Phys. Rev. E*, Vol. 69, pp. 026402-12

B. Dromey, S. Kar, M. Zepf & P. Foster (2004). The plasma mirror: subpicosecond optical switch for ultrahigh power lasers *Rev. Sci. Instrum.*, Vol. 75, pp. 645-649

A. Flacco, T. Ceccotti, H. George, P. Monot, Ph. Martin, F. Réau, O. Tcherbakoff, P. d'Oliveira, F. Sylla, M. Veltcheva, F. Burgy, A. Tafzi, V. Malka & D. Batani, (2010). Comparative study of laser ion acceleration with different contrast enhancement techniques *Nuclear Instruments and Methods in Physics Research Section A*, Vol. 620, pp. 18-22

S. Fourmaux, S. Payeur, A. Alexandrov, C. Serbanescu, F. Martin, T. Ozaki, A. Kudryashov & J. C. Kieffer, (2008). Laser beam wavefront correction for ultra high intensities with 100 TW laser system at the Advanced Laser Light Source *Opt. Express*, Vol. 16, pp. 11987

S. Fourmaux, S. Payeur, S. Buffechoux, P. Lassonde, C. St-Pierre, F. Martin & J. C. Kieffer, (2011). Pedestal cleaning for high laser pulse contrast ratio with a 100 TW class laser system *Opt. Express* Vol. 19, pp. 8486-8497

S. Fritzler, V. Malka, G. Grillon, J. P. Rousseau, F. Burgy, E. Lefebvre, E. d'Humières, P. McKenna & K. W. D. Ledingham, (2003). Proton beams generated with high-intensity lasers: Applications to medical isotope production *Appl. Phys. Lett.*, Vol. 83, pp. 3039

J. Fuchs, P. Antici, E. d'Humières *et al.*, (2006). Laser-driven proton scaling laws and new paths towards energy increase *Nature Physics*, Vol. 2, pp. 48

E. Gerstner, (2007). Laser physics: Extreme light *Nature*, Vol. 446, pp. 16 - 18

G. Grillon, Ph. Balcou, J.- P. Chambaret, D. Hulin, J. Martino, S. Moustaizis, L. Notebaert, M. Pittman, Th. Pussieux, A. Rousse, J- Ph. Rousseau, S. Sebban, O. Sublemontier & M. Schmidt (2002). Deuterium-Deuterium Fusion Dynamics in Low-Density Molecular-Cluster Jets Irradiated by Intense Ultrafast Laser Pulses *Phys. Rev. Lett.*, Vol. 89, pp. 065005

T. Grismayer & P. Mora, (2006). Influence of a finite initial ion density gradient on plasma expansion into a vacuum *Physics of Plasmas*, Vol. 13, pp. 032103-7

K.-H. Hong, B. Hou, J.A. Nees, E. Power & G.A. Mourou, (2005). Generation and measurement of >108 intensity contrast ratio in a relativistic kHz chirped-pulse amplified laser *Appl. Phys. B* Vol. 81, pp. 447-457

R. Hörlein, B. Dromey, D. Adams, Y. Nomura, S. Kar, K. Markey, P. Foster, D. Neely, F. Krausz, G. D. Tsakiris & M. Zepf, (2008). High contrast plasma mirror: spatial filtering and second harmonic generation at 10^{19} W cm^2 *New Journal of Physics*, Vol. 10, pp. 083002-12

J. Itatani, J. Faure, M. Nantel, G. Mourou & S. Watanabe, (1998). Suppression of the amplified spontaneous emission in chirped-pulse-amplification lasers by clean high-energy seed-pulse injection *Optics Communications* Vol. 148, pp. 70-74

A. Jullien, F. Augé-Rochereau, G. Chériaux, J.-P. Chambaret, P. d'Oliveira, T. Auguste & F. Falcoz, (2004). High-efficiency, simple setup for pulse cleaning at the millijoule level by nonlinear induced birefringence *Opt. Lett.*, Vol. 29, pp. 2184-2186

A. Jullien, O. Albert, G. Chériaux, J. Etchepare, S. Kourtev, N. Minkovski & S. M. Saltiel, (2006). Two crystal arrangement to fight efficiency saturation in cross-polarized wave generation *Opt. Express*, Vol. 14, pp. 2760

A. Jullien, S. Kourtev, O. Albert, G. Chériaux, J. Etchepare, N. Minkovski & S. M. Saltiel, (2006). Highly efficient temporal cleaner for femtosecond pulses based on cross-polarized wave generation in a dual crystal scheme *Appl. Phys. B*, Vol. 84, pp. 409-414

M. P. Kalashnikov, E. Risse, H. Schonnagel & W. Sandner, (2005). Double chirped-pulse-amplification laser: a way to clean pulses temporally *Optics Lett.*, Vol. 30, pp. 923-925

H. Kiriyama, M. Mori, Y. Nakai, T. Shimomura, H. Sasao, M. Tanoue, S. Kanazawa *et al.*, (2010). High temporal and spatial quality petawatt-class Ti:sapphire chirped-pulse amplification laser system *Opt. Lett.*, Vol. 35, pp. 1497-1499

O. A. Konoplov, (2000). Temporal high-contrast structure of ultrashort pulses: a plea for standardization of nomenclature and conventions *Proc. of SPIE*, Vol. 3934, pp. 102-112

G. Kulcsár, D. Al Mawlawi, F. W. Budnik, P. R. Herman, M. Moskovits, L. Zhao & R. S. Marjoribanks, (2000). Intense Picosecond X-Ray Pulses from Laser Plasmas by Use of Nanostructured 'Velvet' Targets *Phys. Rev. Lett.*, Vol. 84, pp. 5149-5152

C. Le Blanc, G. Grillon, J. P. Chambaret, A. Migus & A. Antonetti, (1993). Compact and efficient multipass Ti:sapphire system for femtosecond chirped-pulse amplification at the terawatt level *Opt. Lett.*, Vol. 18, pp. 140-142

A. Lévy, T. Ceccotti, P. d'Oliveira, F. Réau, M. Perdrix, F. Quéré, P. Monot, M. Bougeard, H. Lagadec, P. Martin, J.-P. Geindre & P. Audebert, (2007). Double plasma mirror for ultrahigh temporal contrast ultraintense laser pulses *Opt. Lett.*, Vol. 32, pp. 310-312

J. Magill, H. Schwoerer, F. Ewald *et al.*, (2003). Laser transmutation of iodine-129 *Appl. Phys. B*, Vol. 77, pp. 387

A. Marcinkevičius, R. Tommasini, G.D. Tsakiris, K.J. Witte, E. Gaižauskas & U. Teubner (2004). Frequency doubling of multi-terawatt femtosecond pulses *Applied Physics B*, Vol. 79, pp. 547-554

Y. Nomura, L. Veisz, K. Schmid, T. Wittmann, J. Wild & F. Krausz, (2007). Time-resolved reflectivity measurements on a plasma mirror with few-cycle laser pulses *New Journal of Physics*, Vol. 9, pp. 9

G. I. Petrov, O. Albert, J. Etchepare & S. M. Saltiel, (2001). Cross-polarized wave generation by effective cubic nonlinear optical interaction *Opt. Lett.*, Vol. 26, pp. 355-357

M. Pittman, S. Ferré, J. P. Rousseau, L. Notebaert, J. P. Chambaret, & G. Chériaux, (2002). Design and characterization of a near-diffraction-limited femtosecond 100-TW 10-Hz high-intensity laser system *Appl. Phys. B*, Vol. 74, pp. 529

G. R. Plateau, N. H. Matlis, O. Albert, C. Tóth, C. G. R. Geddes, C. B. Schroeder, J. van Tilborg, E. Esarey & W. P. Leemans, (2009). Optimization of THz Radiation Generation from a Laser Wakefield Accelerator *Proceedings of Advanced Accelerator Concepts: 13th Workshop*, CP1086, edited by C. B. Schroeder, W. Leemans and E. Esarey, AIP, pp. 707-712

L. P. Ramirez, D. N. Papadopoulos, A. Pellegrina, P. Georges, F. Druon, P. Monot, A. Ricci, A. Jullien, X. Chen, J. P. Rousseau & R. Lopez-Martens, (2011). Efficient cross polarized wave generation for compact, energy-scalable, ultrashort laser sources *Optics Express*, Vol. 19, pp.93-98

M. Schnürer, R. Nolte, A. Rousse *et al.*, (2000), Dosimetric measurements of electron and photon yields from solid targets irradiated with 30 fs pulses from a 14 TW laser *Phys. Rev. E*, Vol. 61, pp. 4394

L. O. Silva, M. Marti, J. R. Davies *et al.*, (2004). Proton Shock Acceleration in Laser-Plasma Interactions *Phys. Rev. Lett.*, Vol. 92, pp. 015002

S. Steinke, A. Henig, M. Schnurer, T. Sokollik, P.V. NicklesI, D. Jung, D. Kieffer, R. Horlein, J. Schreiber, T. Tajima, X.Q. Yan, M. Hegelich, J. Meyer-Ter-Vehn, W. Sandner & D. Habs (2010). Efficient ion acceleration by collective laser-driven electron dynamics with ultra-thin foil targets *Laser and Particle Beams*, Vol. 28, pp. 215-221

D. Strickland & G. Mourou, (1985). Compression of amplified chirped optical pulses *Optics Communications*, Vol. 55, pp. 447-449

U. Teubner, G. Pretzler, Th. Schlegel, K. Eidmann, E. Forster & K. Witte, (2003). Anomalies in high-order harmonic generation at relativistic intensities *Phys. Rev. A*, Vol. 67, pp. 0138161-11

C. Thaury, F. Quéré, J.-P. Geindre, A. Lévy, T. Cecotti, P. Monot, M. Bougeard, F. Réau, P. D'Oliveira, P. Audebert, R. Marjoribanks & P. Martin, (2007). Plasma mirrors for ultrahigh-intensity optics *Nature Physics*, Vol. 3, pp. 424-429

R. Toth, J. C. Kieffer, S. Fourmaux, T. Ozaki & A. Krol, (2005). In-line phase-contrast imaging with a laser-based hard x-ray source *Rev. Sci. Instrum.*, Vol. 76, pp. 083701

R. Toth, S. Fourmaux, T. Ozaki, M. Servol, J. C. Kieffer, R. E. Kincaid & A. Krol, (2007). Evaluation of ultrafast laser-based hard x-ray sources for phase-contrast imaging *Phys. Plasmas*, Vol. 14, pp. 053506-8

T. Wittmann, J. P. Geindre, P. Audebert, R. S. Marjoribanks, J. P. Rousseau, F. Burgy, D. Douillet, T. Lefrou, K. Ta Phuoc & J. P. Chambaret (2006). Towards ultrahigh-contrast ultraintense laser pulsesÑcomplete characterization of a double plasma-mirror pulse cleaner *Rev. Sci. Instrum.*, Vol. 77, pp. 083109-6

J. Workman, A. Maksimchuk, X. Liu, U. Ellenberger, J. S. Coe, C.-Y. Chien & D. Umstadter, (1996). Picosecond soft-x-ray source from subpicosecond laser-produced plasmas *JOSA B*, Vol. 13, pp. 125-131

V. Yanovsky, V. Chvykov, G. Kalinchenko *et al.*, (2008). Ultra-high intensity-300 TW laser at 0.1 Hz repetition rate *Opt. Express*, Vol. 16, pp. 2110

K. Zeil, S. D. Kraft, S. Bock, M. Bussmann, T. E. Cowan, T. Kluge, J. Metzkes, T. Richter, R. Sauerbrey & U. Schramm (2010). The scaling of proton energies in ultrashort pulse laser plasma acceleration *New Journal of Physics*, Vol. 12, pp. 045015 1-16

A. Zhidkov, A. Sasaki, T. Utsumi, I. Fukumoto, T. Tajima, F. Saito, Y. Hironaka, K. G. Nakamura, K-I Kondo & M. Yoshida, (2000). Prepulse effects on the interaction of intense femtosecond laser pulses with high-Z solids *Phys. Rev. E* Vol. 62, pp. 7232 -7240

A. Zhidkov, J. Koga, A. Sasaki & M. Uesaka, (2002). Radiation Damping Effects on the Interaction of Ultraintense Laser Pulses with an Overdense Plasma *Phys. Rev. Lett.*, Vol. 88, pp. 185002

8

Controlling the Carrier-Envelope Phase of Few-Cycle Laser Beams in Dispersive Media

Carlos J. Zapata-Rodríguez[1] and Juan J. Miret[2]
[1]*Departament of Optics, University of Valencia, Burjassot*
[2]*Departament of Optics, Pharmacology and Anatomy, University of Alicante, Alicante*
Spain

1. Introduction

During the last decade it has been practicable to achieve a full control of the temporal evolution of the wave field of ultrashort mode-locked laser beams (1). Advances in femtosecond laser technology and nonlinear optics have made possible to tailor the phase and magnitude of the electric field leading to a wide range of new applications in science. Many physical phenomena are dependent directly on the electric field rather than the pulse envelope such as electron emission from ionized atoms (2) and metal surfaces (3), or carrier-wave Rabi-flopping (4). Moreover, attosecond physics is for all practical purposes accessible by using femtosecond pulses with controlled carrier-envelope (CE) phase conducting to coherent light generation in the XUV spectral regions (5). Additional applications in the frequency domain includes optical metrology where the laser spectrum is employed (6).

In this chapter we apply fundamental concepts of three-dimensional wave packets to illustrate not only transverse but what is more fascinating on-axis effects on the propagation of few-cycle laser pulses (7). The frequency-dependent nature of diffraction behaves as a sort of dispersion that makes changes in the pulse front surface, its group velocity, the envelope form, and the carrier frequency. The procedure lays on pulsed Gaussian beams, in which these changes are straightforwardly quantified. In particular, the carrier phase at any point of space near the beam axis is evaluated. Anomalous pulse front behavior including superluminality in pulsed Gaussian beams is also found. Finally the CE phase is computed in the focal volume and in the far field.

Generally focused pulses manifest a strong phase dispersion in the neighborhood of the geometrical focus, so that enhanced spatial resolution is achieved in CE phase-dependent phenomena. In some circumstances, however, increased depth of focus may be of convenience so that a stationary CE phase should be required in the near field. It is noteworthy that Gouy wave modes (8) show some control over on-axis phases demonstrating undistorted pulse focalization even in dispersive media. Practical realizations may be driven by angular dispersion engineering of ultrashort laser beams. In this concern we introduced the concept of dispersive imaging (9) as a tool for controlling the dispersive nature of broadband wave fields. Achromatic (10) and apochromatic (11) corrections of the angular spectrum of diffracted wave fields may be achieved with the use of highly-dispersive lenses such as kinoform-type zone plates. This procedure has been employed previously to compensate the longitudinal

chromatic aberration (12; 13) and also the diffraction-induced chromatic mismatching of Fraunhofer patterns (14; 15).
We exploit dispersive imaging assisted by zone plates to gain control over the waveforms of focused laser beams drawing near to Gouy waves modes. The theoretical analysis is addressed only to pulsed Gaussian beams for simplicity. We carry to term a lens system design applying the $ABCD$ matrix formalism. Fundamental attributes of a dispersive beam expander are provided in order to adjust conveniently the spatial dispersion of the collimated input beam. Numerical simulations evidence CE-phase stationarity of few-cycle focused pulses in dispersive media along the optical axis near the focus.

2. Carrier-envelope phase

Let us review in this section the basic grounds on linear wave propagation of few-cycle plane waves in dispersive media. In particular we consider the effect of the CE phase on the spectrum of a pulse train as it is launched by a mode-locked laser. Besides this sequel, nonlinear effects also give rise to a relative phase shift between carrier and envelope which are briefly discussed.

2.1 Pulsed plane waves

We start by considering a plane wave propagating in the z direction. Then we write the electric field amplitude at $z = 0$ as (16)

$$E(0,t) = \mathrm{Re} \int_{-\infty}^{\infty} S(\omega) \exp(-i\omega t) d\omega \tag{1}$$

The effect of propagation over a distance z is to multiply each Fourier component in the time domain by a factor $\exp(ikz)$. Note that the wavenumber varies upon the frequency as $k(\omega) = \omega n(\omega)/c$, where n is the index of refraction of the medium. Therefore we may write

$$E(z,t) = \mathrm{Re}\, E_0(z,t) \exp\left(ik_0 z - i\omega_0 t + i\varphi_0\right) \tag{2}$$

We are assuming a carrier wave with frequency and wavenumber ω_0 and $k_0 = k(\omega_0)$, respectively. Additionally, there is an envelope of the wave field that we denote by

$$E_0(z,t) = \int_{-\infty}^{\infty} S(\omega) \exp\left\{i[k(\omega) - k_0]z - i(\omega - \omega_0)t - i\varphi_0\right\} d\omega \tag{3}$$

The absolute phase $\varphi_0 = \arg \int S(\omega) d\omega$ represents the argument of the complex wave field at $z = 0$ and $t = 0$, leading to a real valued envelope E_0 at the origin.
The Taylor expansion of $k(\omega)$ around the carrier frequency,

$$k(\omega_0 + \Omega) = k_0 + \left(\frac{\partial k}{\partial \omega}\right)_{\omega_0} \Omega + \frac{1}{2}\left(\frac{\partial^2 k}{\partial \omega^2}\right)_{\omega_0} \Omega^2 + \ldots \tag{4}$$

where $\Omega = \omega - \omega_0$, gives

$$\frac{\partial E_0}{\partial z} + \dot{k}_0 \frac{\partial E_0}{\partial t} + \frac{i\ddot{k}_0}{2}\frac{\partial^2 E_0}{\partial t^2} + \ldots = 0. \tag{5}$$

From here on out we assume that a dot over a parameter stands for a derivative with respect to ω, and a subscript 0 denotes its evaluation at the specific frequency $\omega = \omega_0$. Therefore the inverse velocity $\dot{k}_0$ represents $\partial_\omega k$ computed at $\Omega = 0$. This is also consistent with the fact that the carrier wavenumber k_0 symbolizes $k(\omega_0)$. It oftentimes comes to happen that $\ddot{k}_0 \partial_t^2 E_0$ and all the higher-derivative terms in (5) are negligible compared with the first-derivative terms. As a consequence Eq. (5) approaches

$$\frac{\partial E_0}{\partial z} + \frac{1}{v_g} \frac{\partial E_0}{\partial t} = 0. \tag{6}$$

The group velocity

$$v_g = \frac{1}{\dot{k}_0} = \frac{c}{n_0 + \omega_0 \dot{n}_0}, \tag{7}$$

where n_0 represents the real part of the refractive index, and c denotes the speed of light in vacuum. We point out that a negligible absorption at ω_0 is taken for granted, concluding that $\dot{k}_0$ may be set as purely real. In this approximation, the evolution of the the electric field is simply

$$E(z,t) = \text{Re } E_0(0, t - z/v_g) \exp\left[-i\omega_0(t - z/v_p) + i\varphi_0\right] \tag{8}$$

Accordingly the pulse envelope propagates without change of shape or amplitude at the group velocity v_g. However, the phase of the field evolves with a velocity $v_p = \omega_0/k_0 = c/n_0$. In dispersive media, the group velocity and the phase velocity are clearly dissimilar. If otherwise we neglect dispersion by setting $\dot{n}_0 = 0$, it is finally inferred from (7) that both velocities coindice.

2.2 Few-cycle wave fields

For a short pulse of a duration much longer than an optical cycle, its envelope E_0 provides efficiently the time evolution of the wave field that is necessary to include in the vast majority of time-resolved electromagnetic phenomena. If, however, the pulse length is of the order of a single cycle, the phase variation of the wave becomes relevant upon the appropriate description of the electric field. The issue of the absolute phase of few-cycle light pulses was first addressed by Xu et al. (17). Their experiments revealed that the position of the carrier relative to the envelope is generally rapidly varying in the pulse train emitted from a mode-locked laser oscillator.

To understand the origin of this carrier phase shift, it is convenient to introduce a coordinate system that is moving with the pulse at the group velocity. Thus we follow the pulse evolution in this system. For that purpose we perform the coordinate transformation $t' = t - z/v_g$ and $z' = z$. Equation (8) in this moving frame of reference can then be rewritten as

$$E(z', t') = \text{Re } E_0(0, t') \exp\left[-i\omega_0 t' + i\varphi(z')\right] \tag{9}$$

where the phase

$$\varphi(z') = \varphi_0 + \omega_0 \left(\frac{1}{v_p} - \frac{1}{v_g}\right) z' = \varphi_0 - \frac{\omega_0^2 \dot{n}_0}{c} z'. \tag{10}$$

Note that $\varphi(z')$ determines the position of the carrier relative to the envelope.

In Fig. 1 we represent the pulse evolution in the time domain at different planes z' of an ultrashort wave field propagating in sapphire. The numerical simulations make use of a bandlimited signal $S(\Omega') = 1 - 2\Omega'^2 + \Omega'^4$ for $|\Omega'| < 1$, where $\Omega' = (\omega - \omega_0)/\Delta_0$. The mean frequency is $\omega_0 = 3.14$ fs^{-1} and the width of the spectral window is $0.8\omega_0$, i.e. $\Delta_0 = 0.4\omega_0$, providing a 4.8 fs (FWHM) transform-limited optical pulse. By inspecting $\varphi(z')$

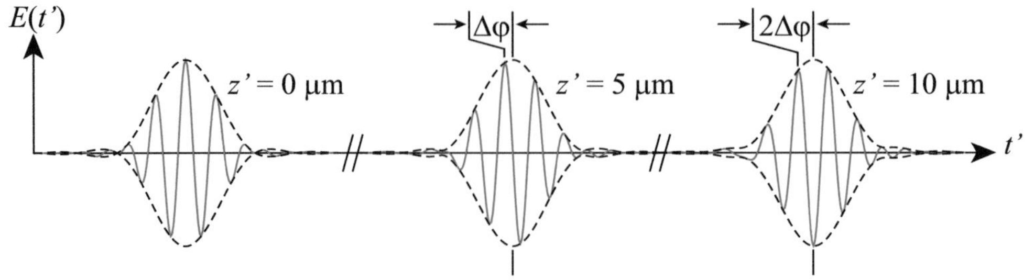

Fig. 1. Plane wave propagation in sapphire. The velocities $v_p = 0.566\,c$ and $v_g = 0.556\,c$ are estimated for a wavelength $\lambda_0 = 600$ nm in vacuum. The envelope is shown in dashed line.

we can identify the difference between the phase delay z'/v_p and the group delay z'/v_g as the reason that the carrier slides under the envelope as the pulse propagates in the dispersive medium. With regard to this matter it is common to introduce the dephasing length L_d as the propagation length over which the carrier is offset by a phase of π with respect to the envelope. Since we express the change in the carrier phase shift upon passage through a medium of length L as $\Delta\varphi = -\left(\omega_0^2 \dot{n}_0/c\right) L$, we finally obtain

$$L_d = \frac{\pi c/\omega_0^2}{|\dot{n}_0|} = \frac{1}{2}\left|\frac{\partial n}{\partial \lambda}\right|_{\lambda_0}^{-1}. \tag{11}$$

In Eq. (11), λ_0 represents the carrier wavelength as measured in vacuum. The dephasing length can be as short as 10 μm in sapphire, as illustrated in Fig. 1. Now it becomes obvious that the difference between phase and group delay in transparent optical materials originates from the wavelength dependence of the (real) refractive index.

2.3 Pulse train spectrum in a mode-locked laser

Considering a single pulse, for instance at $z = 0$, it will have an amplitude spectrum $S(\omega)$ that is centered at the optical frequency ω_0 of its carrier. From a mode-lock laser, however, we would obtain a train of pulses separated by a fixed interval $T = L_c/v_g$,

$$F(t) = \sum_m E(mL_c, t), \tag{12}$$

where L_c is the round-trip length of the laser cavity. For this case, the spectrum can easily be obtained by a Fourier series expansion,

$$\tilde{F}(\omega) = S(\omega)\sum_m \exp\left[im\Delta\varphi(L_c) - im\omega T\right] = S(\omega)\frac{2\pi}{T}\sum_m \delta\left(\omega - m\omega_r - \omega_{CE}\right). \tag{13}$$

This gives a comb of regularly spaced frequencies, where the comb spacing $\omega_r = 2\pi/T$ is inversely proportional to the time between pulses (18). Additionally we have considered that the CE phase is evolving with space, such that from pulse to pulse emitted by a mode-locked laser there is a phase increment of $\Delta\varphi(L_c) = \varphi(L_c) - \varphi_0$. Therefore, the carrier-wave is different for successive pulses and repeats itself with the frequency

$$\omega_{CE} = \frac{\Delta\varphi(L_c)}{2\pi}\omega_r, \tag{14}$$

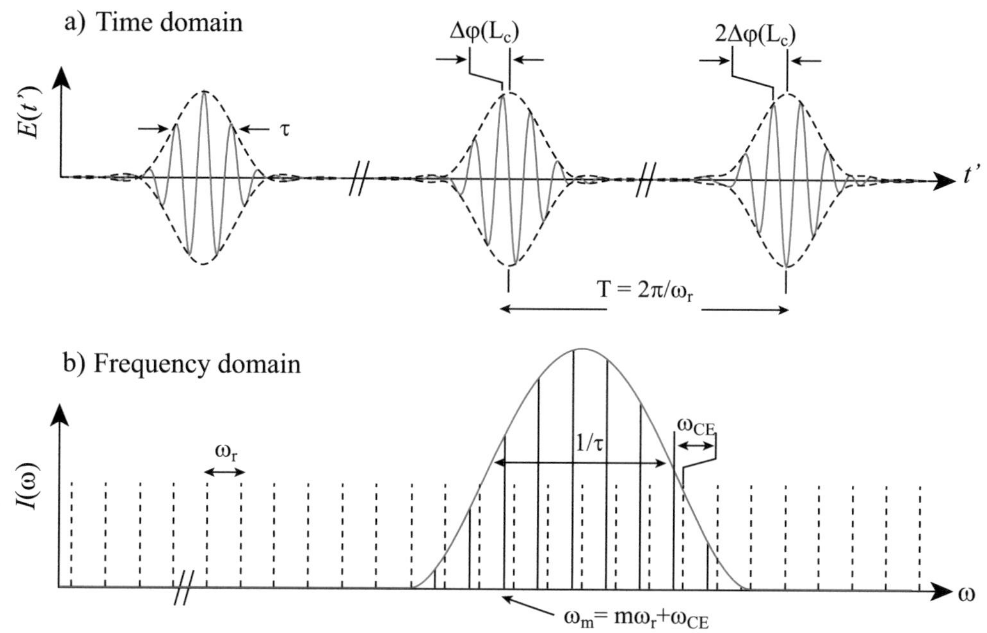

Fig. 2. Illustration of the time-frequency correspondence for a pulse train with evolving carrier-envelope phase.

that is called the carrier-envelope frequency (19). Then in the spectral domain, a rigid shift $\omega_{CE} = \Delta\varphi(L_c)/T$ will occur for the frequencies of the comb lines, as shown in Figure 2. Changing the linear or nonlinear contributions to the round-trip phase shift $\Delta\varphi(L_c)$ inside the laser cavity changes the CE frequency.

2.4 Nonlinear effects

In mode-locked lasers, nonlinear effects also give rise to a relative phase shift between the carrier and the envelope. This is not surprising as there is a nonlinear contribution to the phase shift of the intracavity pulse as it passes through the gain crystal. The most remarkable effect in femtosecond lasers is the third order Kerr nonlinearity responsible for self-phase modulation, which produces a significant spectral broadening. The Kerr effect leads to a self-phase shift of the carrier, similar to the soliton self-phase shift in fiber optics (20). The Kerr effect also induces a distortion of the envelope called self-steepening causing a group delay of the envelope with respect to the underlying carrier.

One of the first experiments to measure carrier-envelope phase evolution observed its intensity dependence (17). Further experiments showed that for shorter pulses, $\Delta\varphi$ was much less sensitive to changes in the pulse intensity (21). In order to give a response to this effect we may consider that the group velocity also depends on intensity. In fact the group velocity changes twice as fast with intensity as does the phase velocity (20). Novel theoretical treatment considered also the fact that dispersion and nonlinearity in the laser are not constant as a function of position in the cavity (22). At this stage, a number of phenomena have been identified in theory and experiment, although some aspects of the connection between them remain unclear.

We point out that the intensity dependence of the CE phase provides a parameter by which the carrier-envelope phase can be controlled. However, it means that the amplitude noise will be converted to phase noise.

3. CE phase of Gaussian laser beams

Since many lasers produce output beams in which the transverse intensity distribution is approximately Gaussian, there has been an appreciable interest in studying the properties of Gaussian beams. The exact Gaussian-like solution to the paraxial wave equation is introduced in this section. Thereafter we obtain analytical expressions of the Gouy phase shift and the CE phase shift of Gaussian beams. Finally we demonstrate that CE phase stationarity, i.e. conservation of the CE phase along with the z coordinate, requires a given spatial dispersion of the wave field to balance dispersion of the dielectric host medium.

3.1 The Gouy phase shift

The so-called Gouy phase is the subject of continuous investigation (23–26) since in 1890 Louis G. Gouy published a celebrated paper (27) on the longitudinal phase delay of spherical beams. Its relevance is a great deal more than purely academic. For instance, the superluminal phase velocity found in the focal region of paraxial Gaussian beams may be understood in terms of this phase anomaly (7). More recently, direct observation of Gouy phases has attracted the interest in the framework of single-cycle focused beams (28–32). In this context, the Gouy phase shift results of great importance for a wide variety of phenomena and applications involving high field physics and extreme nonlinear optics.

It is shown that an exact solution to the paraxial wave equation

$$\frac{\partial^2 \psi}{\partial x^2} + \frac{\partial^2 \psi}{\partial y^2} + 2ik(\omega)\frac{\partial \psi}{\partial z} = 0, \tag{15}$$

for a monochromatic beam of Gaussian cross section traveling in the $+z$ direction and centered on the z axis is given by (33)

$$\psi(\vec{R},\omega) = S(\omega)\frac{z_R}{iq(z,\omega)} \exp\left[ik(\omega)\frac{x^2+y^2}{2q(z,\omega)}\right] \tag{16}$$

where $\vec{R} = (x,y,z)$, (x,y being Cartesian coordinates in a plane perpendicular to the beam axis), $S(\omega) = \psi(\vec{0},\omega)$ is the in-focus time-domain spectrum of the field, and $q(z,\omega) = z - iz_R(\omega)$ stands for the complex radius of curvature. The parameter $z_R(\omega) = k(\omega)s^2/2$ is called the Rayleigh range, which depends on the spot size s at the beam waist. In fact, this term is commonly used to describe the distance that a collimated beam propagates from its waist ($z = 0$) before it begins to diverge significantly.

The monochromatic wave field of a Gaussian beam is completely described by including the phase-only term $\exp(ikz - i\omega t)$. In the vicinity of the beam axis, $(x,y) = (0,0)$, the phase front is approximately flat and the Gaussian beam behaves essentially like a plane wave. However, the evolution of the phase front along with the propagation distance z ceases to be linear. In particular, the phase shift of the complex wave field stored at different transverse planes from the beam waist is $\phi = kz - \phi_G$, where

$$\phi_G(z) = \arctan\left(\frac{z}{z_R}\right) \tag{17}$$

is the Gouy phase. Therefore, the wave accumulates a whole phase of π rads derived from the term ϕ_G, which is produced for the most part near the beam waist.
Let us provide an intuitive explanation of the physical origin of the Gouy phase shift, which is based on the transverse spatial confinement of the Gaussian laser beam. Through the uncertainty principle, localization of the wave field introduces a spread in the transverse momenta and hence a shift in the expectation value of the axial propagation constant (25). At this stage we consider a smooth variation of ϕ_G along the optical axis, which allows us to substitute ϕ by its first-order series expansion $[k - \partial_z\phi_G(z_0)]\,z - \phi_G(z_0) + z_0\partial_z\phi_G(z_0)$ in the neighborhood of a given point z_0. Within a short interval around z_0, the wave front propagates with a spatial frequency $k - \partial_z\phi_G(z_0)$. Accordingly we conveniently introduce the local wavenumber $k_z(z,\omega) = \partial_z\phi$. For a Gaussian beam, the local wavenumber is simply

$$k_z = k - \frac{z_R}{z^2 + z_R^2}. \tag{18}$$

Note that k_z approaches the wavenumber k associated with a plane wave in the limit $z \to \pm\infty$. In the near field, however, k_z is lower than k and it reaches a minimum value at the beam waist.

3.2 The CE phase shift

We will show that the Gouy phase dispersion is determinant in the waveform of broadband optical pulses (7). This is commonly parametrized by the CE phase. For that purpose, let us consider a pulsed Gaussian beam once again. As mentioned above, the electromagnetic field near the beam axis evolves like a plane wave whose wave number k_z given in (18) will change locally. In this context, the temporal evolution of the wave field is fundamentally given by Eq. (9). This is a particularly accurate statement within the Rayleigh range of highly-confined Gaussian beams, which is based on the short-path propagation in the region of interest.
For non-uniform beams, the on-axis pulse propagation evolves at a local phase velocity $v_p(z) = \omega_0/k_{z0}$ and group velocity $v_g(z) = \dot{k}_{z0}^{-1}$. Particularly, the phase velocity of the Gaussian beam in terms of the normalized axial coordinate $\zeta = z/z_{R0}$ is

$$v_p = c\left[n_{p0} - \frac{1}{\mathcal{L}_0\,(1+\zeta^2)}\right]^{-1}, \tag{19}$$

where z_{R0} is the Rayleigh range at the carrier frequency ω_0, and the Gaussian length $\mathcal{L} = kz_R/n$. Note that $\mathcal{L}_0 = k_0^2 s_0^2/2n_{p0}$ also gives the area (which is conveniently normalized) of the beam waist at ω_0. On the other hand, the group velocity yields

$$v_g = c\left[n_{g0} + \frac{\mathcal{F}_0\,(1-\zeta^2)}{\mathcal{L}_0\,(1+\zeta^2)^2}\right]^{-1}, \tag{20}$$

where $n_g = c\dot{k}$ is the group index, and $\mathcal{F} = \omega\dot{z}_R/z_R$. In order to understand the significance of $\mathcal{F}$, let us conceive a Gaussian beam exhibiting an invariant $\mathcal{F}$ within a given spectral band around ω_0. This case would consider a dispersive Rayleigh range of the form $z_R = z_{R0}\,(\omega/\omega_0)^{\mathcal{F}}$, a model employed elsewhere (32). Therefore, $\mathcal{F}$ parametrizes the longitudinal dispersion of the Gaussian beam. This can be deduced also from the relationship $\mathcal{F} = \omega\dot{\mathcal{L}}/\mathcal{L} - 1$.
In Fig. 3(a) we compare graphically the phase velocity and the group velocity of pulsed Gaussian beams with $\mathcal{L}_0 = 34$ and different values of the parameter $\mathcal{F}_0$, which are

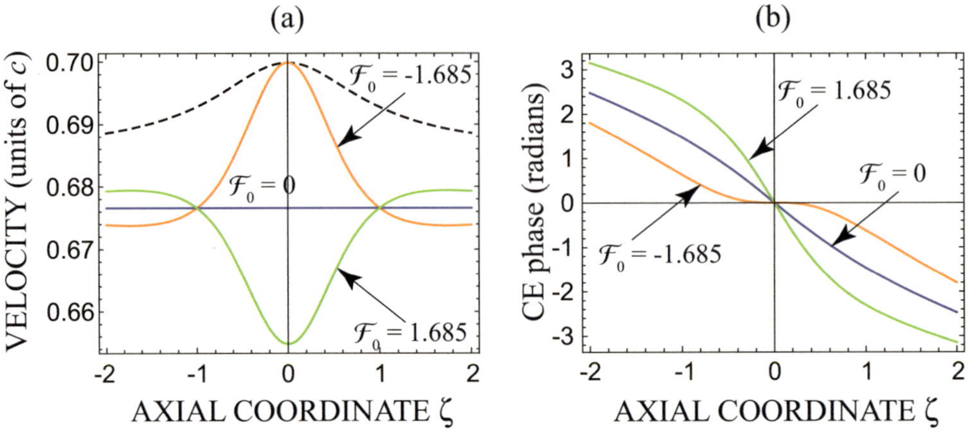

Fig. 3. Phase velocity (dashed line) and group velocity (solid lines) represented in (a) for pulsed Gaussian beams of different $\mathcal{F}_0$ ($\mathcal{L}_0 = 34$) propagating in fused silica at $\omega_0 = 3.14$ fs^{-1}. In (b) we also show the CE phase evolution along the optical axis.

propagating in fused silica. At the mean frequency $\omega_0 = 3.14$ fs^{-1} we have $n_{p0} = 1.458$ and $n_{g0} = 1.478$. As expected, isodiffracting Gaussian beams evidencing an invariant Rayleigh range ($\mathcal{F} = 0$) exhibit a constant group velocity along the optical axis. Of particular importance is the case $\mathcal{F}_0 = -1.685$ for which the phase velocity and the group velocity of the pulsed Gaussian beam matches at the focal point. At the boundaries of the focal volume, $|\zeta| = 1$, a significant mismatch of velocities is clear, which is quantitatively equivalent for different values of $\mathcal{F}_0$.

In order to estimate the CE phase we will not employ Eq. (10) straightforwardly. For convenience we start by considering the fact that

$$\partial_z \varphi(z) = \omega_0 \left(\frac{1}{v_p} - \frac{1}{v_g} \right), \tag{21}$$

which is a magnitude that is accurate at least locally. Therefore, the CE phase of the Gaussian laser beam may be computed as

$$\varphi(\zeta) = \mathcal{L}_0 \int_0^{\zeta} \left(\frac{c}{v_p} - \frac{c}{v_g} \right) d\zeta = -\arctan(\zeta) - \mathcal{L}_0 \Delta n_0 \zeta - \mathcal{F}_0 \frac{\zeta}{\zeta^2 + 1}, \tag{22}$$

assuming that $\varphi = 0$ at the beam waist. Here $\Delta n = n_g - n$ is the difference of group index and refractive index in the dispersive medium. The CE phase will accumulate the difference shown by the phase shift of the carrier and the phase delay of the wave packet as the pulsed Gaussian beam propagates along the focal region. As a result, the CE phase displays the mismatch growth of the phase and group velocities, as shown in Fig. 3(b). Note that, within the Rayleigh range, the absolute value of the CE phase reaches a minimum for $\mathcal{F}_0 = -1.685$.

3.3 Stationarity of the CE phase

Sensitivity to CE phase is observed for instance in photoionization of atoms (34) and in photoelectron acceleration at metal surfaces mediated by surface plasmon polaritons (35; 36). For those experiments, it may result of convenience to neutralize the CE phase shift of the pulsed laser beam as it propagates in the near field. Stationarity of the CE phase, i.e.

conservation of φ along with the z coordinate, entails a spatiotemporal evolution of the wave field (9) expressed in terms of an spatially-invariant waveform (irrespective of its amplitude). In free space, a rigorous stationary CE phase has been reported solely for X-waves (37). However, finite-energy focal pulses demonstrate a variation of φ upon z due to material dispersion and, importantly, the presence of the Gouy phase. At most, a stationary CE phase may be achieved in a small region around a point of interest, which in our case represents the centre of the waist plane. Thus we reach the stationarity condition $\partial_z \varphi = 0$ at $z = 0$.
From the discussion given above we derive that CE phase stationarity is attained when the phase velocity matches the group velocity in the region of interest. In this case, the spatial-temporal dynamics of the electric field given in (9) in the vicinity of the beam waist may be given in terms of a single variable of the form $t - z/v$. For a pulsed Gaussian beam, the condition of velocity matching at focus, that is $v_p = v_g$ at $\zeta = 0$, is satisfied if

$$\mathcal{F}_0 = -\left(1 + \mathcal{L}_0 \Delta n_0\right). \tag{23}$$

Equivalently Eq. (23) reads $\dot{\mathcal{L}}_0 = -\mathcal{L}_0^2 \Delta n_0 / \omega_0$.
We conclude that CE phase stationarity requires a given spatial dispersion of the wavefield, represented by the parameter $\mathcal{F}$, in order to balance dispersion of the dielectric material Δn_0. In particular, $\Delta n = 0$ in vacuum and therefore $\mathcal{F} = -1$, independently of the beam length. From the numerical simulations shown in Fig. 3 we may also employ the values $n_0 = 1.458$ and $n_{g0} = 1.478$ for fused silica at the frequency $\omega_0 = 3.14\ \text{fs}^{-1}$, and $\mathcal{L}_0 = 34$ for the pulsed Gaussian beam. In this case Eq. (23) gives $\mathcal{F}_0 = -1.685$. In the vicinity of the beam waist Eq. (22) proves stationarity features at the origin, that is $\partial_\zeta \varphi = 0$, as shown in Fig. 3(b). Moreover, Eq. (23) leads to ultraflattened curves of the CE phase evolution in the focal region since, in fact, $\zeta = 0$ is a saddle point where $\partial_\zeta^2 \phi_0 = 0$.

4. Managing the CE offset

Provided that propagation of ultrashort laser beams is produced with a CE phase commonly running within the Rayleigh range, in this section we analyze a procedure to induce a controlled spatial dispersion leading to keep the CE phase stationary. In particular we exploit dispersive imaging assisted by zone plates to gain control over the waveforms of Gaussian laser beams. Using the *ABCD* matrix formalism, we disclose fundamental attributes of a dispersive beam expander capable of adjusting conveniently the spatial dispersion of the collimated input beam. Some optical arrangements composed of hybrid diffractive-refractive lenses are proposed.

4.1 Focusing pulses with stationary CE phase

We consider a collimated Gaussian beam propagating in vacuum, which has an input Rayleigh range $z_{Rin} = \omega s_{in}^2/2c$. This pulsed laser beam impinges over an objective lens in order to produce the required wave field embedded in a dispersive medium of refractive index n_p. The Rayleigh range of the focused field is denoted by z_R. For convenience we assume a nondispersive infinity-corrected microscope objective of focal length f. We also ignore beam truncation. Under the Debye approximation, the width of the Gaussian beam at the back focal plane is (38)

$$s = \frac{2f}{ks_{in}}, \tag{24}$$

provided that $z_R \ll f$. In this model, the focused laser beam is free of focal shifts induced by either a low Fresnel number or longitudinal chromatic aberrations. As a consequence,

the Gaussian lengths of the input ($\mathcal{L}_{in}$) and focused ($\mathcal{L}$) beams satisfies $\mathcal{L}_{in}\mathcal{L} = \omega^2 f^2/c^2 n_p$. Moreover, on-axis dispersion of the fields obeys $\mathcal{F}_{in} + \mathcal{F} = -\Delta n/n_p$.
Equation (23) provides the condition that the focused pulse must fulfil in order to keep its CE phase stationary. This equation leads to a new constraint for the input Gaussian beam. Therefore we recast Eq. (23) showing explicitly the specifications for the input pulse, that yields

$$\mathcal{F}_{in0} = 1 + \frac{\Delta n_0}{n_{p0}} \left(\frac{\omega_0^2 f^2}{c^2 \mathcal{L}_{in0}} - 1 \right). \tag{25}$$

As a consequence, dispersive tailoring of the laser beam reaching $\mathcal{F}_{in0}$ of Eq. (25) leads to CE-phase stationarity near the beam waist of the converging field. Note that if focusing is performed in vacuum, that is assuming $\Delta n_0 = 0$, then $\mathcal{F}_{in} = 1$. In this case the input pulsed field will have a Gaussian width that is independent upon the frequency. In fact, this is a well-established assumption in numerous studies (33).
When focusing is carried out in dispersive bulk media, apparently, Eq. (25) is not generally satisfied by the input laser beam. In principle, modification of $\mathcal{L}_{in0}$ and f by using beam expanders and different microscope objectives, respectively, may result of practical convenience. However, these laser-beam tunings are produced at the cost of resizing the beam spot at the region of interest. This is evident if we express Eq. (25) as $\mathcal{F}_{in0} = 1 + \Delta n_0 \left(\mathcal{L}_0 - n_{p0}^{-1} \right)$. In a majority of applications this is undesirable.
We might conserve typical lengths of inputs and focused beams at ω_0 if, alternatively, spatial dispersion of the laser pulse is altered by modifying $\mathcal{F}_{in0}$. In this case we switch the group velocity of the focused pulse at its waist in order to match a given phase velocity. Next we propose an optical arrangement specifically designed to convert a given collimated pulsed beam, which has a Gaussian length $\tilde{\mathcal{L}}_{in0}$ and a dispersion parameter $\tilde{\mathcal{F}}_{in0}$ that violates Eq. (25), into an ultrashort collimated laser beam of the same length $\mathcal{L}_{in0} = \tilde{\mathcal{L}}_{in0}$ but a different parameter $\mathcal{F}_{in0}$ such that Eq. (25) is satisfied. Thus it conforms a previous step to the focusing action, which will be performed by the microscope objective.
First we may give general features of the required system using the *ABCD* matrix formalism. A perfect replica (image) of the Gaussian beam is generated if we impose $B = 0$ and $C = 0$, simultaneously. This afocal system provides a lateral magnification $A = D^{-1}$ of the image. Therefore the relationship $z_{Rin} = A^2 \tilde{z}_{Rin}$ is inferred. Moreover, since $\tilde{\mathcal{L}}_{in0} = \mathcal{L}_{in0}$ we derive that $A_0^2 = 1$. In other words, the beam expansion will be unitary at ω_0. Also it may be obtained that

$$\mathcal{F}_{in0} = \tilde{\mathcal{F}}_{in0} + \frac{2\omega_0 \dot{A}_0}{A_0}. \tag{26}$$

As a consequence, dispersion of the matrix element A is required in order to change the spatial dispersion properties of the Gaussian beam, thus switching $\tilde{\mathcal{F}}_{in0}$ for $\mathcal{F}_{in0}$.
The dispersive nature of A also has certain implications with regard to the temporal response of the *ABCD* system. In particular, the wave field spectrum of the input laser beam will be altered at the output plane of the optical system, since it will be multiplied by the spectral modifier A^{-1}. Under strong corrections carried out at this preprocessing stage, the spectrum of the collimated laser beam might be greatly distorted (39). Ultimately, a spectral shift of the mean (carrier) frequency would be induced, thus urging to recalculate Eq. (25). However, this effect will be neglected in this chapter.

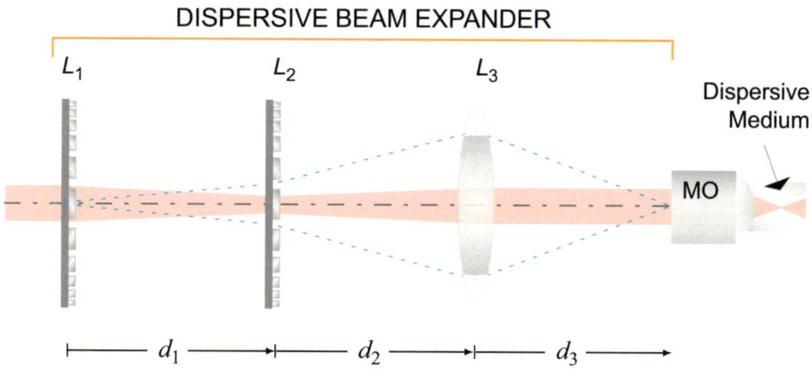

Fig. 4. Schematic diagram of the focusing setup: L_1 and L_2 are components of the diffractive doublet. Also L_3 represents a refractive lens, and MO is the microscope objective.

4.2 Diffraction-induced manipulation of CE dephasing

We propose the thin-lens triplet depicted in Fig. 4 that will perform the necessary spatio-spectral processing of the pulsed Gaussian beam in order to keep stationary its CE shift near the beam waist. This is inspired in the setup shown in (9). Firstly we introduce a doublet composed of kinoform-type zone plates, which are named L_1 and L_2 in Fig. 4, whose dispersive focal lengths are $f_1 = f_{10}\omega/\omega_0$ and $f_2 = f_{20}\omega/\omega_0$. The diffractive doublet is followed by a nondispersive refractive lens (L_3) of focal length $f_3 = f_{30}$. Note that phase-only diffractive lenses may have an optical efficiency of nearly 100% at the carrier frequency ω_0. Neglecting losses induced for instance by material absorption and optical reflections, the kinoform-lens efficiency may be estimated as (40)

$$\eta(\omega) = \text{sinc}^2\left(\frac{\omega}{\omega_0} - 1\right), \tag{27}$$

where $\text{sinc}(x) = \sin(\pi x)/(\pi x)$. We point out that Eq. (27) ignores chromatic dispersion in the lens material to simplify our discussion. This assumption may be given in practice using for instance a diffractive mirror (41). Therefore, the strength of the response η decreases severely at frequencies different of ω_0 due to the appearance of undesirable diffraction orders. This is a relevant effect for subcycle pulses, however it may be negligible for ultrashort beams of sufficiently narrow power spectrum as considered below. Interestingly, diffractive optical elements with high-efficiency responses, which are significantly flatter than η given in Eq. (27) and spanning the visible wavelength range, have been reported in (42) using twisted nematic liquid crystals.

We point out that the $ABCD$ elements of the optical triplet are necessarily dispersive in virtue of the ω-dependent character of f_1 and f_2. The length of the diffractive doublet is d_1, and the coupling distance from L_2 to L_3 is d_2. Furthermore, we assume that L_1 is placed at the waist plane of the input Gaussian beam, whose Rayleigh range is $\tilde{z}_{Rin}$. After propagating through the triplet, the output field is examined at a distance

$$d_3 = \frac{[d_1 d_2 - (d_1 + d_2) f_2] f_3}{d_1(d_2 - f_2 - f_3) + f_2(f_3 - d_2)} \tag{28}$$

from L_3 where the waist image is found ($B = 0$). Therefore the Rayleigh range turns to be $z_{Rin} = -\text{Im}\{q_{in}\}$. For completeness, let us remind the well-known matrix equation for

Gaussian beams

$$q_{in} = \frac{A\tilde{q}_{in} + B}{C\tilde{q}_{in} + D}, \tag{29}$$

being $\tilde{q}_{in} = -i\tilde{z}_{Rin}$ the complex radius of the input laser beam at the plane of L_1.
Distance d_3 given in Eq. (28) depends upon ω. As a consequence, we have control over the waist location of the imaged Gaussian beam for a single frequency. This is of concern when placing the microscope objective. Let us impose d_3 at $\omega = \omega_0$ to be the distance from L_3 to the objective lens (see Fig. 4). The longitudinal chromatic dispersion of the waist plane leads to a dispersive spherical wavefront at the entrance plane of the objective. Ultimately this fact yields a longitudinal chromatic aberration of the broadband focused field. However it may be neglected upon evaluation of the Gaussian Fresnel number $N_G = \mathrm{Re}\{q_{in}\}/z_{Rin}$ (38) of the beam impinging onto the objective lens. Specifically $|N_G| \ll 1$ should be satisfied.
Furthermore, a parametric solution satisfying $B = 0$ and $C = 0$, simultaneously, cannot be found. A sufficient condition may be established by imposing $C_0 = 0$ and $\dot{C}_0 = 0$ so that C vanishes in the vicinity of the carrier frequency. Such a tradeoff is given when

$$f_3 = d_2 + \frac{(d_1 - f_{10})^2}{d_1}, \tag{30}$$

$$f_{20} = -\frac{(d_1 - f_{10})^2}{f_{10}}. \tag{31}$$

This approach suggests that afocality of the the imaging setup is not rigorous but stationary around $\omega = \omega_0$.
Finally, an axial distance $d_2 = f_{10} - d_1$ yields an unitary magnification $A_0 = 1$. Alternatively we may consider that $d_2 = 3f_{10} - d_1 - 2f_{10}^2/d_1$, which gives also a unitary magnification, $A_0 = -1$. In both cases, if additionally

$$f_{10} = d_1\left(1 + \frac{2}{\Delta\mathcal{F}_{in0}}\right), \tag{32}$$

where the mismatching $\Delta\mathcal{F}_{in0} = \mathcal{F}_{in0} - \tilde{\mathcal{F}}_{in0}$, then it is found that $\dot{A}_0 = A_0\Delta\mathcal{F}_{in0}/2\omega_0$ as requested in (26). We conclude that the focal length of each lens composing the hybrid diffractive-refractive triplet is determined by the parameter $\Delta\mathcal{F}_{in0}$ and the positive axial distance d_1.
Note that $d_2 = f_{10} - d_1$ is positive if $\Delta\mathcal{F}_{in0} > 0$. In the same way $d_2 = 3f_{10} - d_1 - 2f_{10}^2/d_1$ yields a nonnegative real value in the case $\Delta\mathcal{F}_{in0} \leq -4$. As a consequence, our proposal cannot be applied if $-4 < \Delta\mathcal{F}_{in0} \leq 0$ since a negative value of the axial distance d_2 is demanded. In these cases we might consider a different arrangement: a triplet with L_1 being a nondispersive thin lens and L_3 a zone plate instead. Following the analysis given above, it can be proved that a satisfactory dispersive processing is also provided using positive values of d_1 and d_2 in the interval $\Delta\mathcal{F}_{in0} < 0$.
Let us illustrate the validity of our approach by means of a numerical simulation. We consider an isodiffracting Gaussian beam that is focused by a microscope objective lens of $f = 4$ mm. In the image space, the pulsed wave field propagates in fused silica. Note that, from a practical point of view, oil-immersion objectives are suitable to get rid of longitudinal chromatic aberration and focal shifts. An input width $s_{in} = 0.8$ mm ($\mathcal{L}_{in0} = 3.5\ 10^7$) allows that aperturing might be neglected in immersion objectives of NA > 0.75. Numerical computations are performed with the Fresnel-Kirchhoff diffraction formula. At focus we

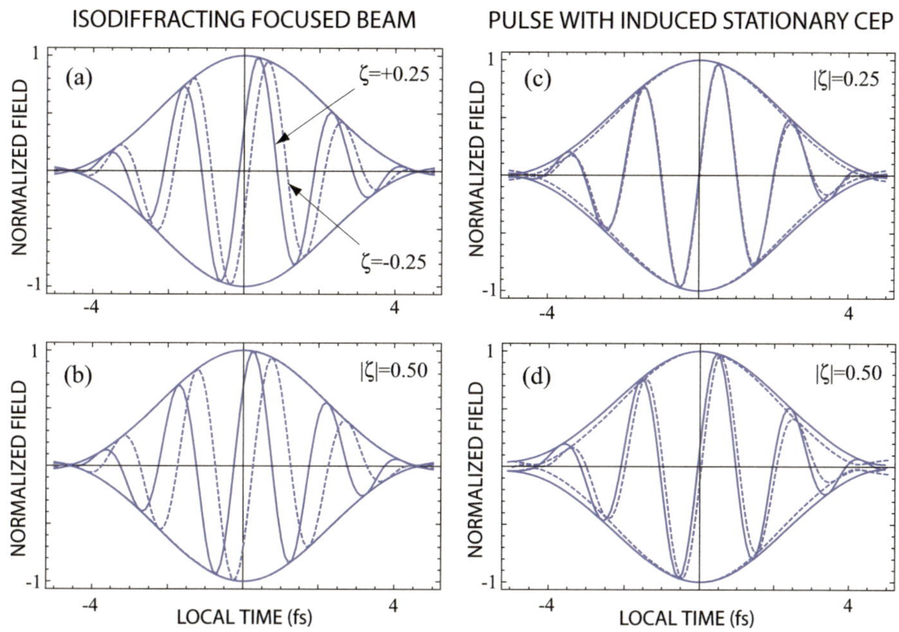

Fig. 5. Evolution of the wave field and envelope at different on-axis points ζ of the focal volume. Negative values of ζ are drawn in dashed lines and positive values in solid lines. Subfigures (a)–(b) correspond to isodiffracting Gaussian beams focused onto bulk fused silica. Time domain is given in terms of the local time t'. In (c)–(d) we employ a procedure to achieve stationarity of the CE phase.

employ a bandlimited signal ε of normalized amplitude spectrum $1 - 2\Omega'^2 + \Omega'^4$ for $|\Omega'| < 1$. Once again, the mean frequency is $\omega_0 = 3.14$ fs^{-1} and the width of the spectral window is $0.8\omega_0$. Figures 5(a) and 5(b) show on-axis waveforms near the focal point. The field envelopes are unaltered in spite of material dispersion. However, the carrier shifts inside the envelope as a reply to the velocity mismatching that it is found at focus, where $v_p = 0.700c$ and $v_g = 0.677c$.

Insertion of the dispersive beam expander of Fig. 4 provides focal waveforms shown in Figs. 5(c) and 5(d). We observe that CE phase is significantly stabilized around the focus, where $\mathcal{L}_0 = 34$ and $\mathcal{F}_0 = -1.685$. In order to minimize the Fresnel number of the pulsed beam to be focused, we have selected $d_1 = 200$ mm leading to a unitary maginfication $A_0 = 1$ and giving $|N_G| < 0.4$ in our spectral window. Equations (30)–(32) yield values for the focal distances of the triplet, $f_{10} = 439$ mm, $f_{20} = -130$ mm, and $f_3 = 525$ mm. We also derive the on-axis distance $d_2 = f_{10} - d_1 = 239$ mm. Importantly, evaluation of Eq. (28) provides a negative axial distance d_3 at the carrier ω_0. This inconvenience is relieved by placing the microscope objective immediately behind L_3 ($d_3 = 0$), as considered in the numerical simulations. This procedure is robust since such an adjustment still maintains the value of v_g switched up to $0.700c$. Inevitably a minor asymmetric pulse broadening is observable. We point out that a relay system might be employed in cases where pulse distortions are dramatic. Placed between the dispersive beam expander and the microscope objective, the relay system may translate the appropriately-corrected virtual pattern of the broadband Gaussian pulse to the entrance plane of the focusing element.

In the numerical simulations, the contribution due to unwanted higher orders of a given kinoform lens is neglected. Considering an input signal S like that observed at focus, the mean efficiency of an ideal diffractive lens may be estimated as

$$\bar{\eta} = \frac{\int_{-1}^{1} |S|^2 \eta \, d\Omega'}{\int_{-1}^{1} |S|^2 d\Omega'}, \tag{33}$$

which theoretically reaches a value $\bar{\eta} = 0.954$. This fact reveals that higher-order foci carry only 4.6% of the input intensity. Correspondingly, the spectral amplitude shaping the transmitted beam and induced by the thin lenses may be disregarded. Finally, the spectral modifier $\sqrt{\eta}$ altering the input signal S through a kinoform zone plate would also induce a small pulse stretching. As a consequence, the FWHM is estimated to increase from 4.8 fs up to 4.9 fs. Again this phenomenon may be ignored.

5. Conclusions

In summary, we investigated the problem of focusing a few-cycle optical field of Gaussian cross section in order to exhibit a stationary time-domain regime within its depth of focus. For that purpose we established some necessary requirements to be hold by an optical setup to drive pulsed beams with adjustable spatial dispersion. This configuration allows the group velocity of the pulse to be tuned ad libitum keeping the phase velocity unaltered. Therefore dispersive imaging opens the door to fine-tune the electric field of ultrashort laser beams (43; 44). Ultimately the group velocity is regulated at the focal point in order to match the prescribed phase velocity.
We also set forth a dispersive beam expander consisting of a hybrid diffractive-refractive triplet, which is capable of preparing the spatiotemporal response of ultrashort Gaussian pulses to maintain a stationary CE phase along with the optical axis when it is focused within a dispersive medium. Robustness of the optical arrangement is demonstrated upon its coupling with immersion microscope objectives. Finally, the theoretical approach shown in this chapter results promising but still unaccomplished since it must be ratified by experimental evidence.

6. Acknowledgments

This research was funded by the Spanish Ministry of Science and Innovation under the project TEC2009-11635.

7. References

[1] D. J. Jones, S. A. Diddams, J. K. Ranka, A. Stentz, R. S.Windeler, J. L. Hall, and S. T. Cundiff. Carrier-envelope phase control of femtosecond mode-locked lasers and direct optical frequency synthesis. *Science*, 288:635–639, 2000.

[2] G. G. Paulus, F. Lindner, H. Walther, A. Baltuska, E. Goulielmakis, M. Lezius, and F. Krausz. Measurement of the phase of few-cycle laser pulses. *Phys. Rev. Lett.*, 91:253004, 2003.

[3] A. Apolonski, P. Dombi, G.G. Paulus, M. Kakehata, R. Holzwarth, Th. Udem, Ch. Lemell, K. Torizuka, J. Burgdörfer, T.W. Hänsch, and F. Krausz. Observation of light-phase-sensitive photoemission from a metal. *Phys. Rev. Lett.*, 92:073902, 2004.

[4] O. D. Mücke, T. Tritschler, M.Wegener, U. Morgner, F. X. Kärtner, G. Khitrova, and H. M. Gibbs. Carrier-wave rabi flopping: role of the carrier-envelope phase. *Opt. Lett.*, 29:2160–2162, 2004.
[5] A. Baltuska, T. Udem, M. Uiberacker, M. Hentschel, E. Goulielmakis, C. Gohle, R. Holzwarth, V. S. Yakovlev, A. Scrinzi, T.W. Hänsch, and F. Krausz. Attosecond control of electronic processes by intense light fields. *Nature*, 421:611, 2003.
[6] P. Cancio Pastor, G. Giusfredi, P. De Natale, G. Hagel, C. de Mauro, and M. Inguscio. Absolute frequency measurements of the $2^3S_1 \rightarrow 2^3P_{0;1;2}$ atomic helium transitions around 1083 nm. *Phys. Rev. Lett.*, 92:023001, 2004.
[7] M. A. Porras. Diffraction effects in few-cycle optical pulses. *Phys. Rev. E*, 65:026606, 2002.
[8] M. A. Porras, C. J. Zapata-Rodríguez, and I. Gonzalo. Gouy wave modes: undistorted pulse focalization in a dispersive medium. *Opt. Lett.*, 32:3287–3289, 2007.
[9] C. J. Zapata-Rodríguez and M. T. Caballero. Ultrafast beam shaping with high-numerical-aperture microscope objectives. *Opt. Express*, 15:15308–15313, 2007.
[10] C. J. Zapata-Rodríguez and M. T. Caballero. Isotropic compensation of diffraction-driven angular dispersion. *Opt. Lett.*, 32:2472–2474, 2007.
[11] C. J. Zapata-Rodríguez, M. T. Caballero, and J. J. Miret. Angular spectrum of diffracted wave fields with apochromatic correction. *Opt. Lett.*, 33:1753–1755, 2008.
[12] T. E. Sharp and P. J. Wisoff. Analysis of lens and zone plate combinations for achromatic focusing of ultrashort laser pulses. *Appl. Opt.*, 31:2765–2769, 1992.
[13] E. Ibragimov. Focusing of ultrashort laser pulses by the combination of diffractive and refractive elements. *Appl. Opt.*, 34:7280–7285, 1995.
[14] G. M. Morris. Diffraction theory for an achromatic fourier transformation. *Appl. Opt.*, 20:2017–2025, 1981.
[15] G. Mínguez-Vega, E. Tajahuerce, M. Fernández-Alonso, V. Climent, J. Lancis, J. Caraquitena, and P. Andrés. Dispersion-compensated beam-splitting of femtosecond light pulses: Wave optics analysis. *Opt. Express*, 15:278–288, 2007.
[16] P. W. Milonni. *Fast light, slow light and left-handed light*. Institute of Physics Publishing, Bristol, 2005.
[17] L. Xu, Ch. Spielmann, A. Poppe, T. Brabec, F. Krausz, and T. W. Hänsch. Route to phase control of ultrashort light pulses. *Opt. Lett.*, 21:2008–2010, 1996.
[18] Steven T. Cundiff. Phase stabilization of ultrashort optical pulses. *J. Phys. D: Appl. Phys.*, 35:R43–R59, 2002.
[19] R. Ell, J. R. Birge, M. Araghchini, and F. X. Kärtner. Carrier-envelope phase control by a composite plate. *Opt. Express*, 14:5829–5837, 2006.
[20] H. A. Haus and E. P. Ippen. Group velocity of solitons. *Opt. Lett.*, 26:1654–1656, 2001.
[21] Kevin W. Holman, R. Jason Jones, Adela Marian, Steven T. Cundiff, and Jun Ye. Intensity-related dynamics of femtosecond frequency combs. *Opt. Lett.*, 28:851–853, 2003.
[22] Mark J. Ablowitz, Boaz Ilan, and Steven T. Cundiff. Carrier-envelope phase slip of ultrashort dispersion-managed solitons. *Opt. Lett.*, 29:1808–1810, 2004.
[23] A. Rubinowicz. On the anomalous propagation of phase in the focus. *J. Opt. Soc. Am.*, 54:931–936, 1938.
[24] R. W. Boyd. Intuitive explanation of the phase anomaly of focused light beams. *J. Opt. Soc. Am.*, 70:877–880, 1980.
[25] S. Feng and H. G. Winful. Physical origin of the Gouy phase shift. *Opt. Lett.*, 26:485–487, 2001.

[26] Carlos J. Zapata-Rodríguez, David Pastor, and Juan J. Miret. Gouy phase shift in Airy beams. *Opt. Lett.*, (submitted).

[27] L. G. Gouy. Sur une propriété nouvelle des ondes lumineuses. *Compt. Rendue Acad. Sci. (Paris)*, 110:1251–1253, 1890.

[28] Simin Feng, Herbert G. Winful, and Robert W. Hellwarth. Gouy shift and temporal reshaping of focused single-cycle electromagnetic pulses. *Opt. Lett.*, 23:385–387, 1998.

[29] Z. L. Horváth and Zs. Bor. Reshaping of femtosecond pulses by the Gouy phase shift. *Phys. Rev. E*, 60:2337–2346, 1999.

[30] A. B. Ruffin, J. V. Rudd, J. F. Whitaker, S. Feng, and H. G. Winful. Direct observation of the Gouy phase shift with single-cycle Terahertz pulses. *Phys. Rev. Lett.*, 83:3410–3413, 1999.

[31] F. Lindner, G. G. Paulus, H. Walther, A. Baltuška, E. Goulielmakis, M. Lezius, and F. Krausz. Gouy phase shift for few-cycle laser pulses. *Phys. Rev. Lett.*, 92:113001, 2004.

[32] T. Tritschler, K. D. Hof, M. W. Klein, and M. Wegener. Variation of the carrier-envelope phase of few-cycle laser pulses owing to the Gouy phase: a solid-state-based measurement. *Opt. Lett.*, 30:753–755, 2005.

[33] A. E. Siegman. *Lasers*. University Science Books, Mill Valley, 1986.

[34] E. Goulielmakis, M. Uiberacker, R. Kienberger, A. Baltuska, V. Yakovlev, A. Scrinzi, Th. Westerwalbesloh, U. Kleineberg, U. Heinzmann, M. Drescher, and F. Krausz. Direct measurement of light waves. *Science*, 305:1267–1269, 2004.

[35] S. E. Irvine, P. Dombi, Gy. Farkas, and A.Y. Elezzabi. Influence of the carrier-envelope phase of few-cycle pulses on ponderomotive surface-plasmon electron acceleration. *Phys. Rev. Lett.*, 97:146801, 2006.

[36] P. Dombi and P. Rácz. Ultrafast monoenergetic electron source by optical waveform control of surface plasmons. *Opt. Express*, 16:2887–2893, 2008.

[37] M. A. Porras, G. Valiulis, and P. Di Trapani. Unified description of Bessel X waves with cone dispersion and tilted pulses. *Phys. Rev. E*, 68:016613, 2003.

[38] C. J. Zapata-Rodríguez. Debye representation of dispersive focused waves. *J. Opt. Soc. Am. A*, 24:675–686, 2007.

[39] C. J. Zapata-Rodríguez. Focal waveforms with tunable carrier frequency using dispersive aperturing. *Opt. Commun.*, 281:4840–4843, 2008.

[40] V. Moreno, J. F. Román, and J. R. Salgueiro. High efficiency diffractive lenses: Deduction of kinoform profile. *Am. J. Phys.*, 65:556–562, 1997.

[41] J. M. Bendickson, E. N. Glytsis, and T. K. Gaylord. Metallic surface-relief on-axis and off-axis focusing diffractive cylindrical mirrors. *J. Opt. Soc. Am. A*, 16:113–130, 1999.

[42] C. Oh and M. J. Escuti. Achromatic diffraction from polarization gratings with high efficiency. *Opt. Lett.*, 33:2287–2289, 2008.

[43] C. J. Zapata-Rodríguez. Ultrafast diffraction of tightly focused waves with spatiotemporal stabilization. *J. Opt. Soc. Am. B*, 25:1449–1457, 2008.

[44] C. J. Zapata-Rodríguez and M. A. Porras. Controlling the carrier-envelope phase of few-cycle focused laser beams with a dispersive beam expander. *Opt. Express*, 16:22090–22098, 2008.

9

Laser Beam Shaping by Interference: Desirable Pattern

Liubov Kreminska
University of Nebraska-Kearney, Department of Physics and Physical Science USA

1. Introduction

Laser beam shaping is an active discipline in optics owing to its importance to both illumination and detection processes. The formation of single or multiple optical vortices in a laser beam has taken on recent interest in areas ranging from electron and atom optics to astronomy. Here we describe our efforts to create localized vortex cores using only the interference of several laser beams with Gaussian profile, a method that may be particularly suited to the application of vortex modes to intense femtosecond laser pulses.

For many years the study of optical vortex formation has been of interest as a problem in itself and one typical of many linear and nonlinear optical phenomena (Nye et al., 1988) such as the formation of speckles, the appearance of solitons, operation of laser and photorefractive oscillators in transverse modes, the self-action of the laser oscillation in nonlinear media (Swartzlander, Jr. & Law, 1992; Kreminskaya et al., 1995); creation of optical vortices in femtosecond pulses (I. Mariyenko et al., 2005); investigation of atomic vortex beams in focal regions (Helseth, 2004), etc. The investigation and explanation of the pattern formation by different optical systems was activated for recent years (Karman et al., 1997; Brambilla et al.,1991); efficient generation of optical vortices by a kinoform-type spiral phase plate (Moh et al., 2006; Kim et al., 1997).

These two apparently different problems of pattern formation and vortex formation are the manifestation of the same phenomenon, which we explain by interference of many plane waves (Angelski et al., 1997; Kreminskaya et al., 1999; Masajada & Dubik, 2001).

An optical vortex (or screw dislocation, phase defect or singularity) is defined as the locus of zero intensity accompanied by a jump of phase on $\pm 2\pi m$ radians, occurring during a round-trip (Nye & Berry, 1974; Basistiy et al., 1995; Abramochkin & Volostnikov; 1993). The integer m is the topological charge, the sign corresponds to the direction of the phase growth: "+" to counterclockwise and "-"to clockwise. In the transversal cross-section, the optical vortex reveals itself as a point, in the 3-D space it exists along the line. A doughnut mode of laser beam is the example of optical vortex. Other member of optical dislocations family is the edge dislocation. The phase changes here by π radians. The shape of the edge dislocation is a line in the transversal cross-section and a plane in 3-D space. Interference pattern of the Young's experiment is the example of the edge dislocation. For the complex amplitude of the light, the condition of the zero intensity means simultaneous zero of the real and imaginary parts of the complex amplitude.

The experimental detection of vortices is accomplished by interference with a reference beam that results in m new fringes appearing in the off-axis interference scheme, etc. In our experiment, the detection of the phase was performed by the Electronic Speckle Pattern Interferometry (ESPI) system (Khizhnyak & Markov, 2007; Malacara, 2005).
In this paper the process of optical vortex arising from the interference of plane waves was studied both theoretically and experimentally as a continuation of the fundamental prediction of (Rozanov, 1993). On the base of this study the simple and convenient method of the vortex creation is proposed. An explanation of an appearance of phase singularities and hexagonal pattern in an initially smooth laser beam is given.

2. Theoretical approach and numerical modelling

The plane wave is the solution of the wave equation and could be described in Cartesian coordinates as (Born&Wolf, 1980):

$$U = A_0 \exp(i\varphi) \tag{1}$$

$$\varphi = \vec{k}\vec{r} + \varphi_0 = kx\sin\Theta\cos\psi + ky\sin\Theta\sin\psi + kz\cos\Theta + \varphi_0 , \tag{1.1}$$

where U is the complex field amplitude, A_0 is the amplitude, φ is the phase, $\vec{k}$ is the wave vector, $\vec{r}$ is the radius- vector, φ_0 is the initial phase shift, ψ and Θ are polar and azimuthal angles in the spherical system of coordinates. Wavefronts of the plane wave are equally spaced parallel planes perpendicular to the wave vector and have zero topological indices.
For our problem of search of singularities, the condition of zero intensity is equivalent to simultaneous zero of real and imaginary part of the complex field amplitude.
All calculations were performed using Wolfram Mathematica 6.

2.1 The interference of two plane waves

Let two plane waves interfere. The well-known periodical pattern of white and black fringes appears in any cross-section of the intensity distribution, Fig.1,a:

$$|U_{12}|^2 = |U_1 + U_2|^2 = A_1^2 + A_2^2 + 2A_1A_2\cos(\varphi_1 - \varphi_2) , \tag{2}$$

The period of the pattern depends on the angle α between interfering waves and is equal to

$$\Lambda = \lambda / (2\sin(\alpha / 2)) \tag{2.1}$$

where λ is the wavelength. The phase undergoes shifts by π in crossing every black fringe, producing edge dislocation, Fig.1,b. If the amplitudes of waves are equal (A, for example), the range of intensity (Eq.2) goes from maximum value of $4A^2$ to exact zero. The contrast of this pattern equals to one.
Further, according to the principle of independent propagation, two interfering waves move forward and interfere again and again, producing patterns with the same period Λ in subsequent cross-sections. Sets of planes of equal amplitude and/or phase are shifted on π

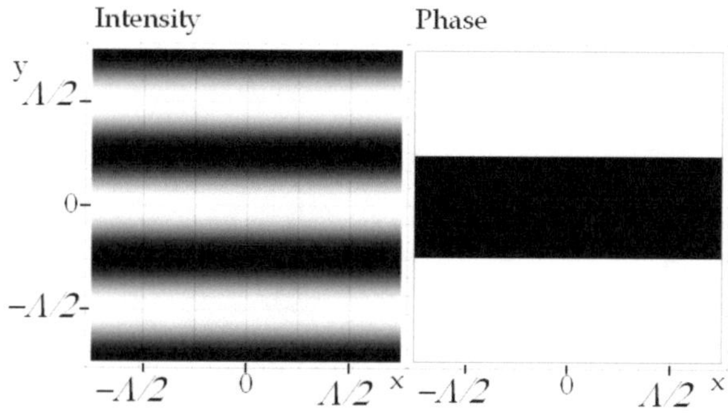

Fig. 1. (a) Intensity and (b) phase distribution of the interference of two plane waves of equal amplitudes in transversal cross-section. Set of alternating bright and black fringes is typical for the intensity. Jump of phase by π radians occurs along each black fringe forming a line of edge dislocations. Result of numerical odelling.

with respect to each other. For our case of equal amplitudes A, the condition of the black fringe (edge dislocation and zero intensity) is

$$A^2 + A^2 + 2A\ A\ \cos(\varphi_1 - \varphi_2) = 0 \tag{2.2}$$

Resulting in

$$\cos(\varphi_1 - \varphi_2) = -1 \tag{2.3}$$

and

$$\Delta\varphi = (\varphi_1 - \varphi_2) = (2n+1)\pi \tag{2.4}$$

where n is an integer number. We apply Eq.1.1 to describe the the planes of zero intensity in 3-D space as following:

$$\Delta\varphi = k(\sin\Theta_1\cos\psi_1 - \sin\Theta_2\cos\psi_2)x + k(\sin\Theta_1\sin\psi_1 - \sin\Theta_2\sin\psi_2)y + k(\cos\Theta_1 - \cos\Theta_2)z + \varphi_{12} = (2n+1)\pi \tag{2.5}$$

$$(\sin\Theta_1\cos\psi_1 - \sin\Theta_2\cos\psi_2)x + (\sin\Theta_1\sin\psi_1 - \sin\Theta_2\sin\psi_2)y + (\cos\Theta_1 - \cos\Theta_2)z + \varphi_{12}\lambda/(2\pi) = (n+1/2)\lambda \tag{2.6}$$

This equation defines the plane of edge dislocation in 3-D space.
The same way we defined the planes of the maximum intensity as

$$(\sin\Theta_1\cos\psi_1 - \sin\Theta_2\cos\psi_2)x + (\sin\Theta_1\sin\psi_1 - \sin\Theta_2\sin\psi_2)y + (\cos\Theta_1 - \cos\Theta_2)z + \varphi_{12}\lambda/(2\pi) = n\lambda \tag{2.7}$$

Theoretically the planes of edge dislocations are unlimited in the cross-section, in practice they exist in the *volume* of the crossing of the plane waves.

2.2 The interference of three plane waves

Let have three plane waves interfere, as shown in Fig. 2. For the sake of simplicity lets use the paraxial aproximation of small angles.

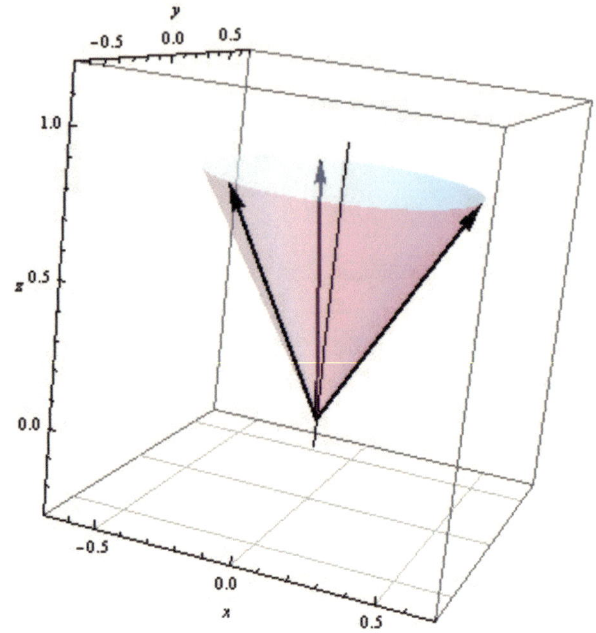

Fig. 2. The direction of wave vectors for three plane waves interfering at symmetric angles of $\psi=2\pi/3$ radians and azimuth $\theta=\pi/10$ radians.

The complex field of three-wave interference can be represented in the next form:

$$U_{\Sigma} = U_1 + U_2 + U_3 = (U_{12} + U_{13} + U_{23}) / 2 \tag{3}$$

which is the superposition of three systems of fringes of pairwize interference, where indices correspond to notation of Eq.2.
This definition looks artificial until the description of the intensity of this field:

$$|U_{\Sigma}|^2 = A_1^2 + A_2^2 + A_3^2 + 2A_1A_2\cos(\varphi_1 - \varphi_2) + 2A_1A_3\cos(\varphi_1 - \varphi_3) + 2A_2A_3\cos(\varphi_2 - \varphi_3) \tag{3.1}$$

when the addition of three fringe patterns (it is natural to call them partial) occurs, Fig.3.

2.2.1 Interference of three plane waves of equal amplitude

Let us simplify Eq. (3.1) by introducing the equal amplitudes of waves A_i=A.
By definition of the intensity of optical field, it has to be always positive and we have to analyze inequality

$$A^2(3 + 2\cos(\varphi_1 - \varphi_2) + 2\cos(\varphi_1 - \varphi_3) + 2\cos(\varphi_2 - \varphi_3)) \geq 0 \tag{3.2}$$

The maximum of intensity happens when all cosines are equal to 1. This corresponds to the intersection of three bright fringes (white lines, Fig.3, or schematically shown in Fig.4), producing the brightest spot of intensity of $9A^2$ in the point. According to the condition of Eq.3.2, the minimum of the function will be always zero, that coincide with the definition of singularities. For illustration purposes we had chosen polar angles ψ between wavevectors equal to $2\pi/3$ radians.

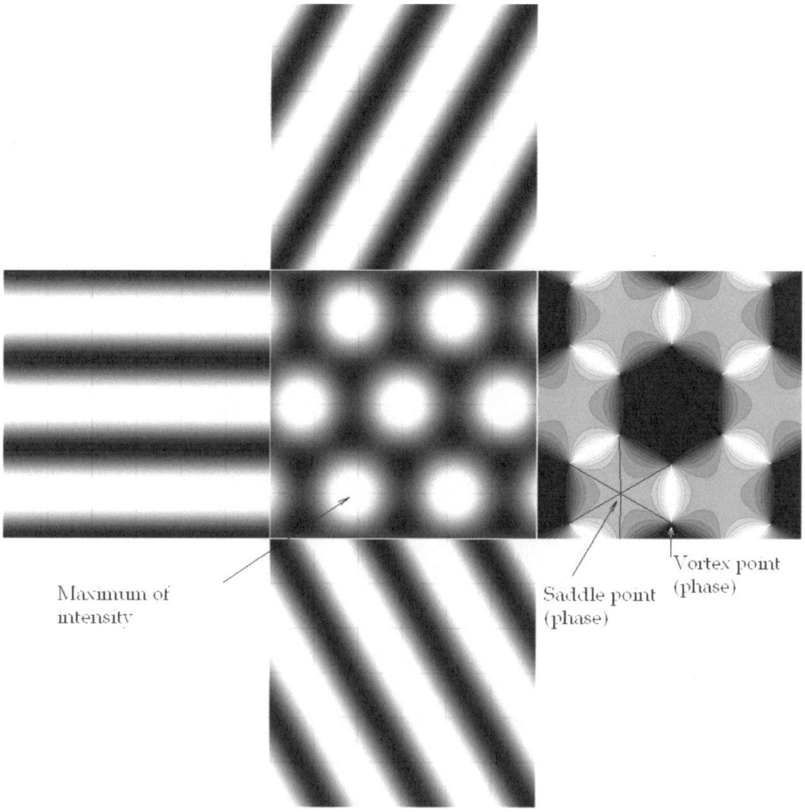

Fig. 3. Intensity distribution of three plane waves (center), pairs of plane waves (top, bottom, left) and the phase distribution (right). The phase distribution proves the existence of vortices of opposite signs in the apexes of black hexagons. The amplitudes of interfering waves are equal. The angles between the waves are equal. The crossing of izophase lines corresponds to a saddle point. The scale is the same as in Fig.1. Grey scale, black/white colour correspond to minimum/maximum values. Numerical calculations.

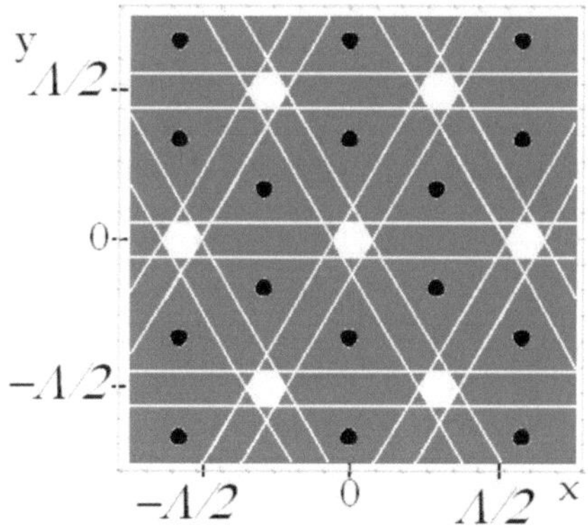

Fig. 4. Illustration of how three partial fringes of maximum intensity (in between of white lines) cross in one point, producing the bright spot of maximum intensity.

The important restrictions on the type of resulting pattern follows from the Eq.3.2, specifically, the partial black fringes never cross in a point (cosines can not be equal to -1 simultaneously, because the intensity could not be equal to a negative number -3).

2.2.2 Interference of three plane waves at different angles

We keep investigating interference of three plane waves of equal amplitude. Let's tilt one wave by 0.5 radians. The resulting pattern is shown in Fig.5.

(a) (b)

Fig. 5. The interference pattern of three plane waves: (a) intensity and (b) the phase distribution. The amplitudes of interfering waves are equal. The polar angle of one wave was changed by 0.5 radians. The scale is the same as in Fig.1.

2.2.3 Interference of three plane waves of different amplitudes

The pattern becomes complicated when waves are of different amplitudes.

The numerical results shown in Fig.3 and Fig.5 were produced under the assumption that the amplitude of each plane wave was equal and the contrast of the pattern was equal to one. During the lab procedure we can control relative contrast to certain accuracy. That is why it is important to know that optical vortices will persist under amplitude perturbation. To locate and confirm the existence of singularities at each of the six points around the bright spot, we search for points in the plane where the real and imaginary parts of the complex amplitude simultaneously go to zero.

For this part we had chosen three plane waves with equal angles between them (as shown in Fig.2). The intensities of two plane waves are equal, and the intensity of the third wave changes. The results of numerical modeling are presented in Fig.6 and Fig.7. In Fig.7 we plotted position of vortices for different values of A_3, inside of one cell of the pattern.

Line of $Re(U_\Sigma)=0$ coincides with line of $Im(U_\Sigma)=0$ for the edge dislocation in Fig.6,a for $A_3=0$. Fig.6,b shows the snake-like rolls with vortices in the wriggles, $A_3=0.2A$. Here we observe irregular hexagon of the six vortices. Vortices along the line x=0 move away from the fringes of case Fig.6a, while vortices along lines $\pm x_v$ move closer to each other. For $A_3=A$ in the Fig. 6c the crossing of three lines $Im(U_\Sigma)=0$ corresponds to a saddle point and regular hexagon of vortices is formed around. At $A_3=1.5A$ the vortices are still formed. Vortices along lines $\pm x_v$ move closer to each other. At $A_3=2A$ the vortices from the close neighbor pair annihilate at $\pm x_v$. We observe honey comb structure both for bright spots and vortices. At larger A_3 vortices are not created.

The vortices persist under such perturbations, and as the amplitude of A_3 varies up to $2A$.

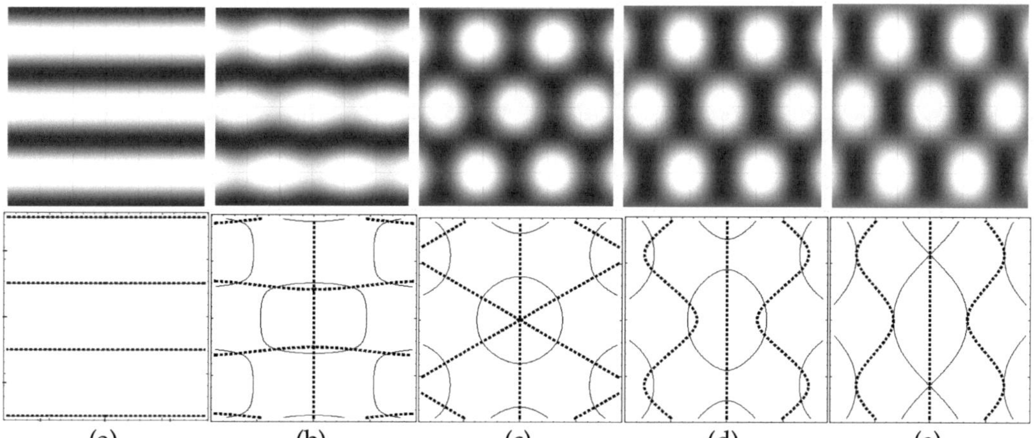

Fig. 6. Numerical calculations of the interference pattern of three plane waves (top) and plot of lines Re=0 (solid line) and Im=0 (dashed line) (bottom), here Re is a real part and Im is imaginary part of the complex beam amplitude. Both polar and azimuthal angles between wave vectors are equal. The amplitudes $A_1=A_2=A$ of two interfering waves are equal. Amplitude of the third wave was changed as a fraction of A: (a) $A_3=0$; (b) $A_3=0.2A$; (c) $A_3=A$; (d) $A_3=1.5A$; (e) $A_3=2A$. The scale is the same as in Fig.1.

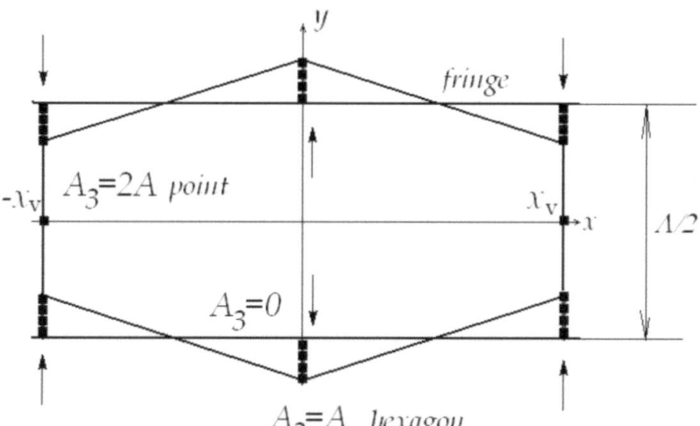

Fig. 7. The location of optical vortices within one cell of the pattern for different values of A_3. Note that every pair of vortices is displaced along verical line.

2.2.4 Topological analysis

Let us discuss the simplest case of interference of three plane waves of the same amplitude, with wavevectors oriented at the same angles relatively to each other (Fig.2, Fig.3). The sum of three partial fringes of the same period Λ results in a honey-comb pattern of bright spots (Fig.3 center) with the phase distribution as shown. The bright spots in , $\psi_2=2\pi/3$, $\psi_3=-2\pi/3$ and same $\Theta_1=\Theta_2=\Theta_3$ the condition of Eq.6 is satisfied for the matrix of Equations (4.1-4.3). The line of maximum intensity of the optical field exists in space with coordinates

$$a_x = 0\,,\ a_y = 0\,,\ a_z = \frac{3\sqrt{3}}{2}\sin^2\Theta\,.$$

Moreover, as each equation of fringe (4.1-4.3) is defined to within $\pm 2\pi n$, the set of parallel lines is observed. This means that pattern of bright spots of the stable transverse structure of the honey-comb cell, distribute in space along well determined lines.

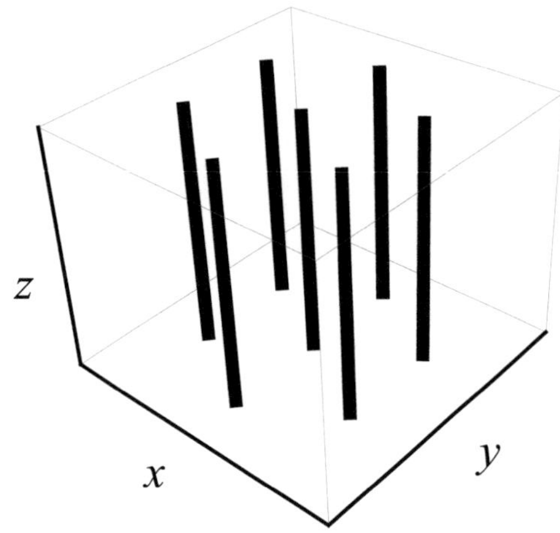

Fig. 8. Localization of one set of bright spots (one cell of the honey comb pattern) in 3-D space along the straight lines. All parameters of three plane waves are the same as in Fig.3.

2.2.5 Displacement of the whole pattern

By shifting the relative phases between three plane waves, it is possible to effectively shift the pattern so that the vortex is located at the center of a pattern, Fig.9. In the experiment this was done by using different optical path lengths. For our case of three plane waves with equal amplitudes and angles of $\psi_1=0$, $\psi_2=2\pi/3$, $\psi_3=-2\pi/3$ and same $\Theta_1=\Theta_2=\Theta_3$
To move the whole pattern in the transverse cross-section, we added initial phase shifts to each wave as following: $\varphi_{01}=0$, $\varphi_{02}=2\pi/3$, $\varphi_{03}=-2\pi/3$.

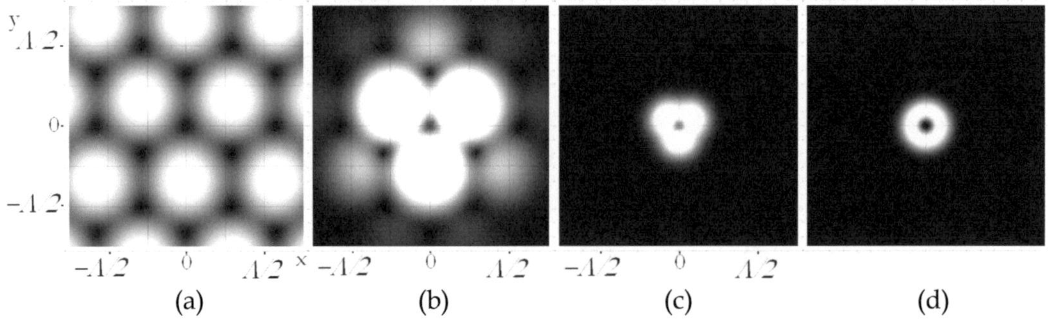

Fig. 9. The intensity distribution in transverse cross-section of (a) three plane waves interference pattern with the vortex in the center. This was achieved by shift of initial phases between interfering plane waves. The pattern repeats periodically; (b) Intensity distribution of interference pattern of three Gaussian beams. Here ω / Λ =2:1; (c) Intensity distribution of interference pattern of three Gaussian beams. Here ω / Λ =1:2; (d) Doughnut laser mode.

When compare the Fig 9a to the intensity distribution in Fig.3 center, one can see, that the vortex is located in point (0,0), as opposed to the bright spot with maximum intensity in the same point (0,0). The period of the pattern did not change.

2.2.6 Interference of three Gaussian beams

The real laser beam has a finite aperture, its amplitude depends on coordinates as

$$U=A_0 \exp^{-(x^2+y^2)/\omega^2} \tag{4}$$

with ω being the waist of the laser Gaussian beam. That is why we should not predict the periodic pattern to continue throughout all of space as a tessellation, both in transverse and longitudinal directions. Instead, the interference is confined locally. Mathematically, we achieve this finite spatial extent by modulating the amplitude of interfering plane waves with a Gaussian profile of Eq.4; this gives us a local area with interference pattern. This approach works well in the paraxial approximation. In the experiment, the small angles between interfering laser beams justify the use of this approximation. In Figs. 9.b,c we see that the pattern does not persist throughout the xy-plane, but is restricted to a small region of the transverse cross section. In this case the waist of the laser Gaussian beam ω was equal to the $2^*\Lambda$ (spatial period of the two-beam interference pattern). At this point, there exists a competitive interaction between two parameters: the period of the hexagonal pattern of vortices, which is controlled by the angle of interference, and the off axis attenuation, which is controlled by the waist of the laser beam. By increasing the attenuation, we can further decrease any off center intensity contributions and retain only the center black spot and its immediately surrounding bright ring, Fig.9,c. One would like to have this type of intensity profile to be as close as possible to the doughnut mode. Doughnut modes correspond to a first order Laguerre-Gaussian laser mode with a profile depicted in Fig.9d. This result is important for application like doughnut mode creation for the temporally focusing of electron pulses (Hilbert et al.,2009), (Helseth, 2004).

2.3 Interference of many plane waves

We performed modelling of interference of larger number of plane waves to explore the possibility of creation of different patterns and vortex tesselations. Various patterns can be created in the space, Fig.10. Between patterns of three-wave, five-wave and seven-wave interference the common features are as following: the bright central spot is formed, there is a radial symmetry, the number of vortices around the center is $2n$, where n is the number of interfering waves, in the phase distribution one can see the saddle point of the order (n-1). The black circles of edge dislocations and bright circles surround the central bright spot.

3. Experimental setup

The four-arm Mach-Zhender interferometer was assembled for our experimental investigation, Fig.11. Three arms were used to form the field under study and the fourth was used as the reference one. A He-Ne laser with Rayleigh range $z_R=\pi w^2/\lambda=1.5$ m was used as the light source. The reflectivity of the cube beamsplitters (BS) was 50/50 to obtain a set of interference patterns with equal high contrast. Optical attenuators were used to reach this condition precisely. All optical surfaces were covered with antireflection coatings to reduce reflection losses.

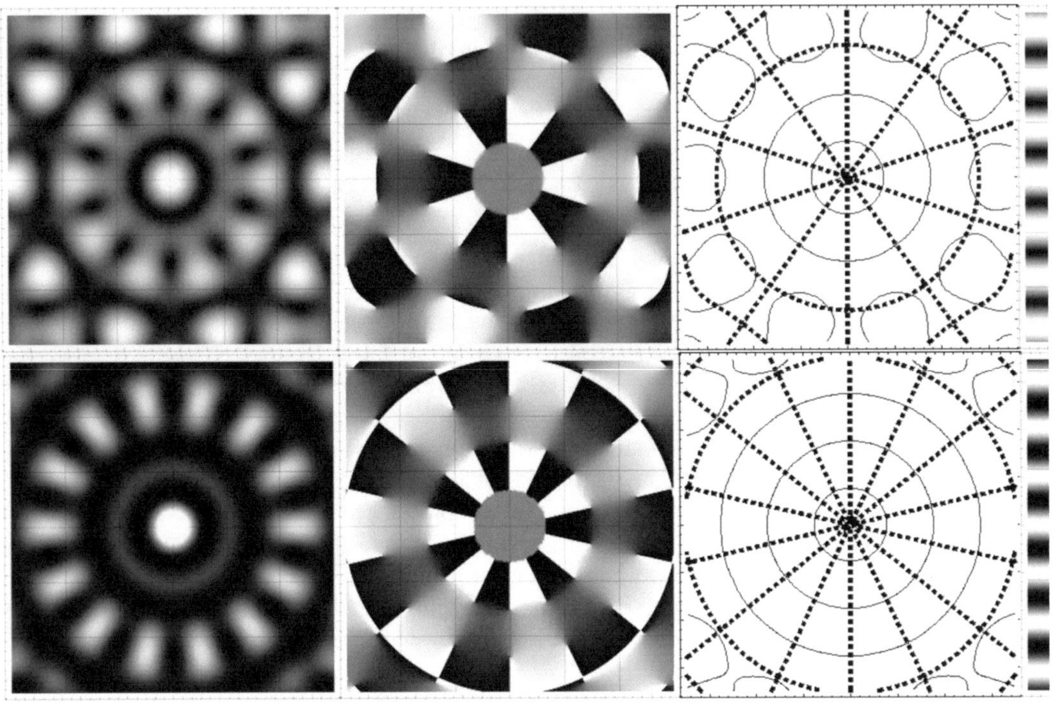

Fig. 10. Numerical calculations of the interference pattern of five plane waves (top) and of seven plane waves (bottom). Intensity distribution is shown in the first column, the phase distribution is shown in the second column and the plot of lines Re=0 (solid line) and Im=0 (dashed line) (third column). Angles between wave vectors are equal. The amplitudes of interfering waves are equal. The scale is related to the period of the two-waves interference as shown on the side.

The ESPI-system (Khizhnyak &Markov, 2007) was used for the phase reconstruction. To perform the phase reconstruction one of the totally reflective mirrors of the interferometer was placed to piezomotor, which allowed to move the mirror by fraction of the wavelength. We recorded pattern of interference of three beams under study and one reference beam (oriented at relatively large angle to obtain the frequent thin fringes to identify vortices as bifurcation of the fringe). Several snapshots were recorded by CD camera, stored in the memory of the computer and processed. As a result, we had a graph of the phase in the transversal cross-section.

With three-beam interference, different fields are formed depending on parameters of interfering waves, for example, the field of honey-comb structure of bright spots (Fig.12,a), for which optical vortices exist at the vertices of hexagons (Fig.3,b).

The frame of the reconstructed phase of the field proves the existence of vortices as its disruption on $\pm 2\pi$ radians inside one hexagonal cell. The neighbouring vortices are of opposite topological charge, i.e. the phase increases in opposite directions.

It was found experimentally that the location of vortex is the straight line along the axis of light propagation that supports theoretical predictions of the Chapter 2. In the transverse cross-section the pattern of vortices is the regular hexagonal for equal angles of plane waves, otherwise hexagons are irregular: the greater these angles the smaller the sizes of the hexagons.

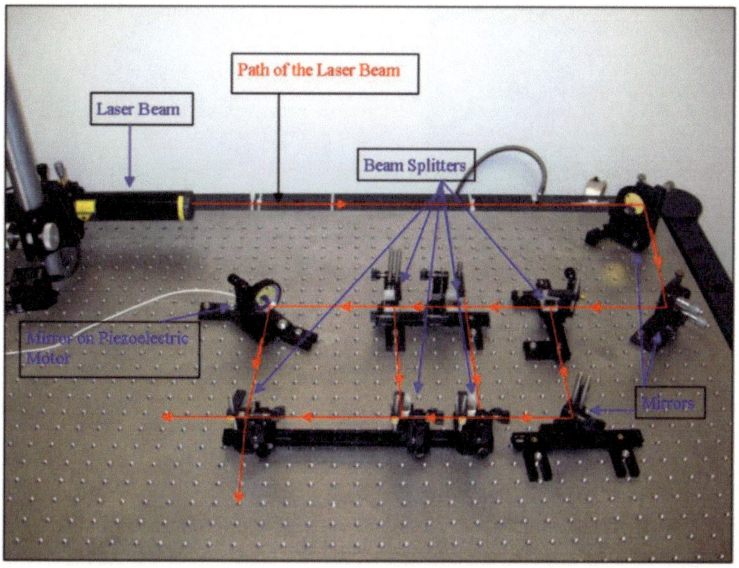

Fig. 11. A 633 nm linearly polarized HeNe laser at 5 mW sends a beam off one mirror. The beam entering a Mach-Zehnder configuration. The interferometer is a modified Mach-Zehnder which combines a total of four beams. In the fourth Mach-Zehnder arm the laser beam is redirected by a mirror, the position of which can be changed by a piezomotor stage. After recombination of the required number of beams the interference pattern is projected on a screen. A camera was used to capture the image. To help make finer adjustments to the pattern, we put a 3mm glass window in one of the beam paths. This allowed us to make minute changes to the path length of that beam by adjusting the angle of the window.

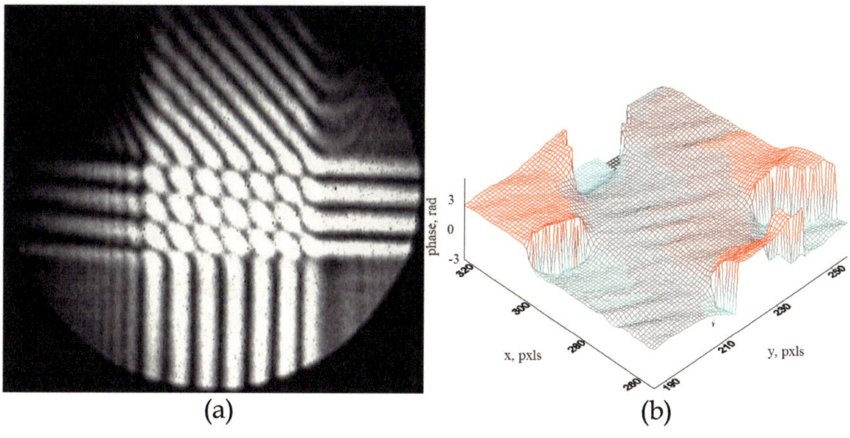

Fig. 12. (a) Experimental pattern of intensity distribution of the total field of three plane waves (center) and patterns of two plane waves (up, down, sides). The vortices reveal in vertices as black pots. To simplify the comparison with the numerical calculations of Fig.3, parts of each wave were closed. The diffraction effect is seen on the edges of screens. (b) the phase distribution in of one hexagonal cell of (a).

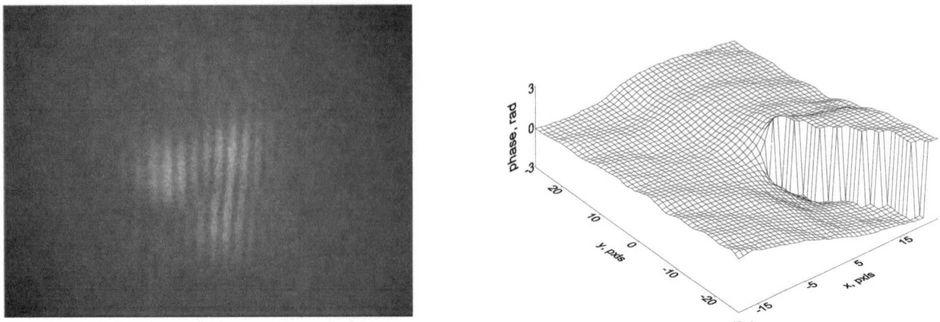

(a) (b)

Fig. 13. The experimental observation of one vortex formation: (a) with the reference beam, one can see the bifurcation of the single black fringe into two fringes, (b) the phase distribution in radians, transverse coordinates are in pixels.).

A single vortex was obtained when plane waves interfered at very small angles. This was seen as a black spot in the center of the light beam during three-beam interference (compare with Fig.9c). The phase distribution, reconstructed by ESPI-system shows the disruption on 2π radians during bypass around the vortex (Fig.13,b). The pattern is stable along the axis of light propagation in the volume of superposition of beams.

To change the sign of topological charge of vortex to the opposite it is sufficient to invert the angle of one of interfering beams.

Fig. 14 is a series of snapshots as an amplitude of one of interfering waves A_3 is incremented. In experiment this progression was achieved by placing an adjustable gray scale filter in one of the arms of the interferometer. As shown in Fig.6, the fringes are observed when $A_3=0$. For $A_3=0.2A$ the snake-like rolls reveal. The regular honey-comb cell of bright spots and the hexagon of black dots/vortices is formed for $A_3=A$.

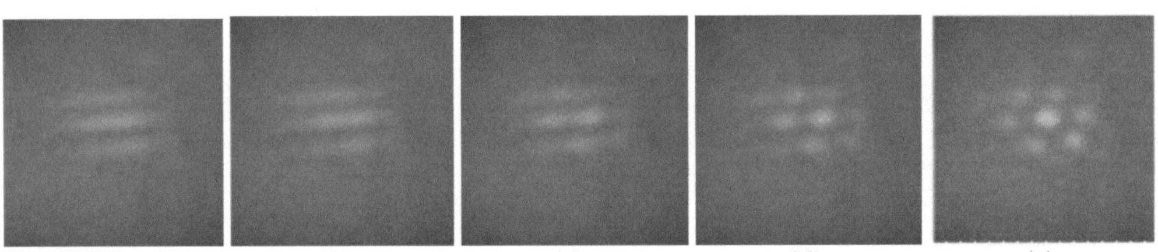

(a) (b) (c) (d) (e)

Fig. 14. Intensity patterns for interference of three laser beams with progressive change of amplitude of one of interfering waves. Here we go from (a) two beams $(A_3=0)$ to (e) complete three beams $(A_3=A)$ interference.

4. Discussion

The process of optical vortices and pattern formation resulting from the interference of many plane waves was studied theoretically and examined in experiment with the evident agreement.

The vortex nature of the regions of zero intensity inside the hexagonal patterns was proved both by the numerical analysis and ESPI reconstruction of the phase. The resulting pattern of vortices was transformed from regular hexagonal to irregular one by the change of parameters of interfering laser beams.

Such way of creation of vortices has several advantages over others:

- the vortices can be created at any desirable region in space.
- the vortices line up in a single straight line;
- the method is simple and precise in reproducing of desirable vortex configuration in experiment;
- the method used produces energy losses only from the beam-splitting.

5. Conclusions

The hexagonal structure of an optical field as a result of the interference of many plane waves is described. By interference of plane waves we have obtained a lattice of vortices of any desirable transverse structure. The law of conservation of the topological indices is fulfilled during this process. These vortices are stable objects that persist in space without annihilating.

6. Acknowledgment

Author acknowledges significant contribution of Dr. Anatoly Khizhnyak , help of Mr. Carl Corder and Mr. Jason Teten.

Author thanks the UNK Graduate Study program for the UNK's Undergraduate Research Fellowship, the NASA Nebraska Space Grant and Nebraska Research Initiative Grant for support that made this research possible.

7. References

Nye, J. F.; Berry, M. (1974). Dislocations in wave trains, *Proc. Roy. Soc. London*. Vol. A 336. pp. 165-190.

Rozanov, N. N. (1997). *Optical Bistability and Hysteresis in Distributed Nonlinear Systems*, Moscow: Nauka, Fizmatlit

Karman, G.P.; Beijersbergen, M.V.; Van Duijl, A.; Woerdman, J.P. (1997). Creation and annihilation of phase singularities in a focal field, *Optics Lett*. Vol. 22. pp.1503-1505.

Abramochkin, E. and Volostnikov, V. (1993). Spiral-type beams, *Optics Commun*. Vol.102. pp. 336-350.

Brambilla, M.; Lugiato, L.A.; Penna, V.; Prati, F.; Tamm, C.; Weiss, C.O. (1991). Transverse laser patterns. II. Variational principle for pattern selection, spatial multistability, and laser hydrodynamics, *Phys. Rev. A.*. Vol. 43. pp. 5114-5120.

Swartzlander Jr. , G.A. and Law, C.T. (1992). Optical vortex solitons observed in Kerr nonlinear media, *Phys. Rev. Lett*. Vol.69, pp.2503-2506.

Basistiy, I.V.; Soskin, M.S. and Vasnetsov, M.V.(1995). Optical wavefront dislocations and their properties, *Optics Commun*. Vol.119. pp. 604-612.

Nye, J.F.; Hajnal, J.V.; Hannay, J.H. (1988). Phase Saddles and Dislocations in Two-Dimensional Waves Such as tide, *Proc. Roy. Soc. Lond*. Vol. A 417. pp. 7-20.

Freund, I. (1999). Critical point explosions in two-dimensional wave fields, *Optics Commun*. Vol.159 .pp. 99-117.

Kreminskaya, L.; Soskin, M.; Khizhnyak, A.(1998). The gaussian lenses give birth to optical vortices in laser beams, *Optics Commun*. Vol. 145.pp. 377-384.

Born, M.; Wolf, E. (1980). *Principles of Optics*. Pergamon Press

Korn, G.; Korn, T. (1968). *Mathematical handbook*. McGraw-Hill Book Company. 832 p.

Khizhnyak, A.; Markov, V. (2007).Atmospheric turbulence profiling by detection the wave function of a test beam, *Proceedings of SPIE*, 6457, 64570L

Malacara, D., Serín, M., and Malacara, Z. (2005). *Interferogram Analysis For Optical Testing*, CRC Press. Second Edition

Hilbert, S.; Barwick, B.; Uiterwaal, K.; Batelaan, H.; Zewail, A. (2009). Temporal lenses for attosecond and femtosecond electron pulses, *Proceedings of the National Academy of Sciences*, Vol. 106, pp. 10558. , Available from www.physorg.com

Angelski, O.; Besaha, R .and. Mokhun, I.(1997).Appearance of wave front dislocations under interference among beams with simple wave fronts. *Optica Aplicata*, Vol.27. No.4, pp. 273-278

Kreminskaya, L.; Monroy, F.; Robles, W. (1999). Formación De Vórtices Ópticos En La Interferencia De Múltiples Haces. *Revista Colombiana de Física*, Vol.31, No.2, pp. 125–128

Masajada, J. and Dubik, B. (2001). Optical vortex generation by three plane wave interference, *Optics Communications*, Vol. 198, No 1-3

Mariyenko,I.; Strohaber, J. and Uiterwaal, C. (2005) Creation of optical vortices in femtosecond pulses, *Opt. Express*, Vol. 13, pp.7599-7608

Moh, K. J.; Yuan,X.-C.; Cheong,W. C.; Zhang,L. S.; Lin,J.; Ahluwalia,B. P. S.; and Wang H. (2006). High-power efficient multiple optical vortices in a single beam generated by a kinoform-type spiral phase plate, *Appl. Opt.* Vol. 45, pp. 1153-1161

Kim, G.-H.; Jeon,J.-H.; Ko,K.-H.; Moon,H.-J.; Lee J.-H., and Chang J.-S. (1997). Optical vortices produced with a nonspiral phase plate, *Applied Optics*, Vol. 36, No. 33, pp. 8614-8621

Shvedov,V. G.; Izdebskaya,Y. V.; Alekseev A. N., and Volyar A. V. (2002). The formation of optical vortices in the course of light diffraction on a dielectric wedge, *Tech. Phys. Lett.* ,Vol.28, pp.256-259

Helseth, L. E. (2004). Atomic vortex beams in focal regions, *Phys. Rev. A* , Vol.69, p. 015601

10

Nonlinear Pulse Reshaping in Optical Fibers

S. O. Iakushev[1], I. A. Sukhoivanov[2], O. V. Shulika[1],
J. A. Andrade-Lucio[2] and A.G. Perez[2]
[1]*Kharkov National University of Radio Electronics*
[2]*DICIS, University of Guanajuato*
[1]*Ukraine*
[2]*Mexico*

1. Introduction

The propagation of ultrashort laser pulses in the optical fibers is connected with a plenty of interesting and practically important phenomena. Unique dispersive and nonlinear properties of the optical fibers lead to various scenarios of the pulse evolution which are resulted in particular changes of the pulse shape, spectrum and chirp. The modern age of the optical fibers starts from the 1960s with the appearance of the first lasers. These fibers were extremely lossy but new suggestion on the geometry with the single mode operation (Kao et al., 1966) which was obtained by theoretical calculations based on the Maxwell's equations, and the development of a new manufacturing process (French et al., 1974) have led to the achievement of the theoretical minimum of the loss value 0.2 dB/m (Miya et al., 1979). The investigations of the nonlinear phenomena in the optical fibers have been continuously gained by the decreasing loss. Loss reduction in the fibers made possible the observation of such nonlinear processes which required longer propagation path length at the available power levels in the 1970s. Stimulated Raman Scattering (SRS) and Brillouin scattering (SBS) were studied first (Ippen et al., 1972). Optical Kerr-effect (Stolen et al., 1973), parametric four-wave mixing (FWM) (Stolen et al., 1974) and self-phase modulation (SPM) (Stolen et al., 1978) were observed later. The theoretical prediction of the optical solitons as an interplay of the fiber dispersion and the fiber nonlinearity was done as early as 1973 (Hasegawa et al., 1973) and the soliton propagation was demonstrated seven years later in a single mode optical fiber (Mollenauer et al., 1980).

Discovery of the optical solitons have revolutionized the field of the optical fiber communications. Nowadays solitons are used as the information carrying "bits" in optical fibers (Hasegawa et al., 2003). This is resulted from unique properties of the solitons. In general, the temporal and spectral shape of a short optical pulse changes during propagation in a medium due to the self-phase modulation and chromatic dispersion. This actually limits the transmission bit rate in optical fibers. Under certain circumstances, however, the SPM and dispersion can exactly cancel each other producing a self-localized waveform called the solitary wave. Due to the particle-like nature of these solitary waves during mutual interactions they were called solitons. Solitons are formed when GVD is anomalous, and also, the pulse energy and duration have to meet a particular condition in order to provide the formation of a stable soliton called the fundamental soliton. Fundamental solitons preserve their temporal and spectral shape even over long propagation distances. However, if the pulse energy is larger, e.g. the square of an integer number times the fundamental soliton energy, then a higher-order soliton is formed. Such pulses do not have a preserved shape, but their shape varies periodically.

Solitons possess a number of important features. An optical soliton is a solution of the nonlinear Schrödinger equation (NLSE) in the form of secant-shaped pulse (Zakharov et al., 1972). This soliton solution is very stable: even for substantial deviations of the initial pulse from the exact secant shape, the pulse tends to correct his shape towards the correct soliton form. It means that initial Gaussian or super-Gaussian pulse transforms to the soliton after some propagation distance in a fiber. Solitons are also very stable against changes of the properties of the medium, provided that these changes occur over distances which are long compared with a soliton period. It means that solitons can adapt their shape to slowly varying parameters of the medium.

Owing to the features mentioned above the solitons play important role not only in optical fiber communications but also in ultrashort lasers as well. Soliton mode locking is a frequently used technique for producing high-quality ultrashort pulses in the bulk and fiber lasers (Brabec et al., 1991; Duling III, 1991). Solitons have been the subject of intense theoretical and experimental studies in many different fields, including hydrodynamics, nonlinear optics, plasma physics, and biology (Gu, 1995; Akhmediev et al., 2008). Up to date many other kinds of solitons have been discovered depending on the dispersive and nonlinear properties of the fibers. In the context of the optical solitons, one can distinguish temporal or spatial solitons, depending on whether the confinement of light occurs in time or space. Temporal solitons have been mentioned above, they represent optical pulses which maintain their shape, whereas spatial solitons represent self-guided beams that remain confined in the transverse directions orthogonal to the direction of propagation. A spatial soliton arises when the self-focusing of an optical beam balances its natural diffraction-induced spreading. Spatiotemporal optical solitons can demonstrate both temporal and spatial localization of light (hence the term "light bullets"), which are nondiffracting and nondispersing wavepackets propagating in a nonlinear optical media. Other types of solitons include: dispersion-managed solitons, dissipative solitons, dark solitons, Bragg solitons, vector solitons, vortex solitons (Kivshar et al., 2003).

In spite of unique properties of the optical solitons the difficulties arise with generation of the high power pulses and their delivering through the optical fibers due to distortions and break-up effects. Particularly, the fundamental soliton exists at only one particular power level and propagation of higher-power pulses excites the higher-order solitons which are sensitive to perturbation and break-up through the soliton fission (Kodama et al., 1987; Dudley et al., 2006). Whereas in the normal dispersion regime the pulse propagation is subject to instability through the appearance of the optical wave breaking (OWB) (Tomlinson et al., 1985; Anderson et al., 1992). It was shown that optical intensity shock formation precedes the OWB appearance (Rothenberg, 1989). The shock occurs when the more intense parts of the pulse, owing to the nonlinearity, are travelling at the speeds which are different from those of the weaker parts. This reshaping evolves to a breaking singularity, when the top of the shock actually overtakes its bottom, resulting in a region of multivalued solutions.

Break-up effects in general restrict pulse propagation in an optical fiber, especially concerning the high-power pulses. However, in 1990s it was theoretically predicted that pulses with parabolic intensity profile and a linear frequency chirp can propagate without wave breaking in the normal dispersive optical fiber (Anderson et al., 1993). Then it was shown numerically that ultrashort pulses injected into a normal dispersion optical fiber amplifier evolve towards the parabolic pulse profile with amplification and, moreover, retained their parabolic shape even as they continued to be amplified to higher powers (Tamura et al., 1996). In 2000 an experimental observation of parabolic pulse generation in the fiber amplifier with normal dispersion applying advanced ultrashort pulse measurement technique of the frequency-resolved optical gating (FROG) was reported (Fermann et al., 2000).

This novel class of the optical pulses exhibits several fundamental features. Parabolic pulses propagate in a self-similar manner, holding certain relations (scaling) between the pulse power, width, and chirp parameter. The scaling of the amplitude and width of both the temporal amplitude and spectrum depends only on the amplifier parameters and the input pulse energy, and is completely independent of the input pulse shape. Moreover, the pulse chirp is completely independent of the propagation distance. A major result of the theoretical analysis becomes the demonstration that the self-similar parabolic pulse is an asymptotic solution to the NLSE with gain, representing a type of nonlinear "attractor" towards which any arbitrarily-shaped input pulse of the given energy would converge with sufficient distance (Fermann et al., 2000; Kruglov et al., 2002). Thus, initial arbitrary shaped pulse propagating in a normally dispersive fiber amplifier reshapes itself into a pulse having a parabolic intensity profile combined with a perfectly linear chirp. Theoretical and numerical results obtained later have shown that the self-similar parabolic pulses shape is structurally stable (Kruglov et al., 2003).

Self-similarity is a fundamental property of many physical systems and has been studied extensively in diverse areas of physics such as hydrodynamics, mechanics, solid-state physics and theory of elasticity (Barenblatt, 1996). Discovery of the self-similar pulse propagation in the optical fibers attracts much interest not only from theoretical standpoint but due to possible practical applications as well (Dudley et al., 2007). By analogy with the well-known stable dynamics of the solitary waves — solitons, these self-similar parabolic pulses have been termed as similaritons. One has to note that solitons themselves can also be interpreted as an example of self-similarity (Barenblatt, 1996). In contrast to the optical solitons however, the similaritons in optical fibers can tolerate strong nonlinearity without wave breaking. The normal GVD effectively linearizes the accumulated phase of the pulse allowing for the spectral bandwidth to increase without destabilizing the pulse. Solitons maintain their shape, width, and amplitude, whereas similaritons maintain their shape but not their width or amplitude; instead of that the certain relations between pulse power, width, and chirp parameter are hold.

Application of similaritons provided a significant progress in the field of pulse generation and amplification in fiber systems. The remarkable properties of the parabolic pulses have been applied to the development of a new generation of high-power optical fiber amplifier (Schreiber et al., 2006). Self-similar propagation regime allows avoiding the catastrophic pulse break up due to excessive nonlinearity. However, in contrast to the well-known chirped pulse amplification method, where the aim is to avoid nonlinearity by dispersive pre-stretching before amplification, a self-similar amplifier actively exploits the nonlinearity. This allows obtaining output pulses that are actually shorter than after recompression of the initial input pulse. Similariton approach has been applied also to the development of high-power fiber lasers ("similariton laser") allowing overcome existing power limitations of soliton modelocking (Ilday et al., 2004). Parabolic pulses are of great interest for a number of applications including, amongst others, the highly coherent continuum sources development and optical signal processing (Finot et al., 2009).

Parabolic pulses are generated usually in the active fiber systems such as amplifiers or lasers. However, last time it is also of interest to study alternative methods of generating parabolic pulses, especially in the context of non-amplification usage, such as optical telecommunications. Some approaches were proposed for the generation of parabolic pulses in the passive fiber systems, such as dispersion decreasing fibers (Hirooka et al., 2004) and fiber Bragg gratings (Parmigiani et al., 2006). Then it was found that nonlinearity and normal dispersion in the simple passive fiber can provide pulse reshaping towards the parabolic pulse at the propagation distance preceding the optical wave breaking (Finot et al., 2007). But with longer propagation distance the pulse shape doesn't remain parabolic. However, recently the possibility to obtain nonlinear-dispersive similaritons in the passive fibers at longer propagation distances was demonstrated (Zeytunyan et al., 2009; Zeytunyan et al., 2010). In this regime pulse propagates actually in some steady-state mode when pulse shape and spectrum does not change significantly with propagation distance and they repeat each other due to spectronic nature and moreover the chirp of such pulse is linear. Thus, there is some analogy with parabolic pulse formation in fiber with gain. However, the shape of the pulses obtained in passive fiber is different from the fully parabolic one. Under some special initial conditions even triangular pulses can be obtained (Wang et al., 2010).

Based on our recent results (Sukhoivanov et al., 2010; Yakushev et al., 2010) here we will discuss nonlinear pulse transformations in the normal dispersive regime of the passive fibers. Particularly in the weak nonlinear case (soliton order is less than 10) we will present numerical results regarding the pulse shape evolution towards parabolic profile depending on the initial pulse shape, soliton number and fiber length. We will show that some quasi-parabolic pulses with linear chirp can be obtained. The influence of the initial pulse parameters and fiber parameters on the shape of quasi-parabolic pulses will be investigated. We also reveal the role of the third-order dispersion.

In the strong nonlinear case (soliton order is larger than 10) optical wave breaking becomes significant. Previously, it was shown that oscillations induced by OWB are vanishing at longer distances and pulse transforms to the chirped trapezium-shaped pulse (Karlsson, 1994). However, there is no proper explanation of this phenomenon and further pulse evolution has not been investigated. We will present numerical results concerning to the pulse transformations at the much longer distance, until steady-stage regime is achieved. At this stage the pulse shape becomes nearly smooth with a linear chirp. We explain the OWB cancellation by the action of normal dispersion, which tends to flatten nonlinear chirp, induced by self-phase modulation during the pulse propagation in a fiber. It is possible to obtain the resulted parabolic pulse profile or a triangular one in the steady-state regime depending on the initial conditions.

2. Theoretical model

The evolution of an ultrashort pulse during its propagation in a normal-dispersion fiber with Kerr nonlinearity is well described by the nonlinear Schrödinger equation (NLSE) (Agrawal, 2007):

$$\frac{\partial A}{\partial z} = -\frac{\alpha}{2}A - \frac{i\beta_2}{2}\frac{\partial^2 A}{\partial T^2} + \frac{\beta_3}{6}\frac{\partial^3 A}{\partial T^3} + i\gamma|A|^2 A, \quad (1)$$

where $A(z,t)$ is the slowly varying complex envelope of the pulse; α is the loss; β_2 is the second order dispersion; β_3 is the third order dispersion; γ is the nonlinear coefficient; T is the time in a co-propagating time-frame; z is the propagation distance. The formation of the nearly parabolic pulses in the passive optical fibers is basically governed by the interplay of normal dispersion and nonlinearity. But here we extend standard NLSE by including the fiber loss and the third-order dispersion in order to investigate the impact of these factors on the parabolic pulse formation. Equation (1) is solved numerically using the split-step Fourier method (Agrawal, 2007). For the subsequent discussion it is convenient to use following notations for the dispersion length L_D, dispersion length associated with the third-order dispersion L'_D, nonlinear length L_{NL}, soliton order N and normalized length ξ:

$$L_D = T_0^2 / |\beta_2|,\ L'_D = T_0^3 / |\beta_3|,\ L_{NL} = 1/(\gamma P_0),\ N = \sqrt{L_D / L_{NL}},\ \xi = z / L_D. \quad (2)$$

here P_0 is the initial pulse peak power, T_0 is the initial pulse duration (half-width at 1/e-intensity level). At first, we investigate the weak nonlinear case when $1 < N < 10$. In this case the self-phase modulation (SPM) dominates over the group velocity dispersion (GVD) during the initial stages of the pulse evolution, then pulse evolution preliminary governed by GVD. Larger value of $N > 10$ leads to the stronger impact of the nonlinearity, leading to the optical wave breaking phenomenon.

In the following the quality of the pulse is estimated through the deviation of its temporal/spectral intensity profile $|A(T)|^2$ and a parabolic fit $|A_p(T)|^2$ of the same energy (Finot et al., 2006), which is expressed as the misfit parameter M:

$$M^2 = \frac{\int\left(|A|^2 - |A_p|^2\right)^2 d\tau}{\int|A|^4 d\tau}. \quad (3)$$

The expression for a parabolic pulse of energy $U_p = 4P_pT_p / 3\sqrt{2}$ is given by:

$$\begin{cases} A_p(T) = \sqrt{P_p}\sqrt{1 - 2T^2 / T_p^2}, & |T| \le T_p / \sqrt{2} \\ A_p(T) = 0, & |T| > T_p / \sqrt{2} \end{cases} \quad (4)$$

where P_p is the peak power of the parabolic pulse, T_p is the duration of the parabolic pulse (full-width at half-maximum). The misfit parameter M allows estimating the pulse shape imperfection as compared to the parabolic shape; the smaller value of M shows better fit to the parabolic waveform. Usually we can consider a pulse shape to be close enough to the parabolic one when $M < 0.04$.

3. Pulse reshaping in the weak nonlinear case

3.1 The influence of initial pulse shape and chirp on the resulted pulse profile

At first, we address pulse reshaping in the weak nonlinear case $1 < N < 10$ at the normal dispersive optical fiber. Basically pulse reshaping is governed by the interplay between SPM and GVD. One can separate two stages of the pulse evolution. In the first stage within the short propagation distance the SPM effect is strong. Owing to that the chirp becomes nonlinear and fast reshaping of the pulse temporal and spectral shapes occurs. It was shown that a parabolic pulse can be obtained from initial unparabolic pulse at this stage at the particular propagation distance (Finot et al., 2007). However, further pulse reshaping continues and pulse shape doesn't remain parabolic. Finally, SPM action goes down and GVD dominates over the longer propagation distance. At this stage pulse propagates in the nearly linear propagation regime. Here pulse chirp becomes linear and pulse temporal and spectral shape changes slowly and actually maintains some particular profile. However, the resulted pulse shape strongly differs from the initial pulse shape (Zeytunyan et al., 2009). Here, we investigate comprehensively pulse reshaping depending on the initial pulse parameters and fiber parameters.

Firstly, we investigate pulse reshaping in the case of initial Gaussian pulse shape. The M-map presented in Fig. 1 (a) demonstrates the dependence o f the misfit parameter M on the soliton order N and normalized length ξ. We also calculated the M-plot shown in Fig. 1 (b), which represents the dependence of the misfit parameter M versus ξ for a particular value N. Both the deviation of the temporal pulse shape and the deviation of the spectrum shape from the parabolic one were calculated. In the last case equations (3)-(4) are applied for the spectral pulse amplitude.

On Fig. 1(a) three specific areas are denoted by numbers. Number 1 indicates narrow and dip vertical area within the short propagation distance ($\xi < 1$). This area has been extensively investigated by Finot et al (Finot et al., 2007) and Boscolo et al (Boscolo et al., 2008). It is possible to achieve here fully parabolic pulse ($M \sim 0.04$). However, a suitable M is achieved only over the narrow range of propagation distances, then pulse shape changes. Moreover, from Fig. 1 (b) we can see that only temporal pulse profile achieves parabolic shape, whereas the spectral shape shows maximal deviation from the parabolic shape here. We should point out here also an adjacent wider vertical area (Number 2). Here the propagation distance is also small ($1 < \xi < 2$) and misfit parameter M is close to that one in the first area.

Now let us look at the pulse reshaping within the longer propagation distance. We are especially interested in this region because some steady-state regime is achieved here. This implies that pulse shape and spectrum do not change sufficiently with propagation distance any more. From Fig. 1(b) we can conclude that steady-state starts approximately from $\xi > 5$. Here both temporal and spectral curves are reduced very slowly and they are very close to each other. The value of N affects strongly on the misfit parameter M in the steady-state. From Fig. 1 (a) one can see that M sufficiently increases with increasing of N. It means that pulse shape in the steady-state will be different depending on the N. In Fig.1(a) we can point out also the horizontal valley $1.5 < N < 2.5$, where misfit parameter achieves smaller value as compared to the outside value of N (Number 3). In this area one can achieve $M \sim 0.07$.

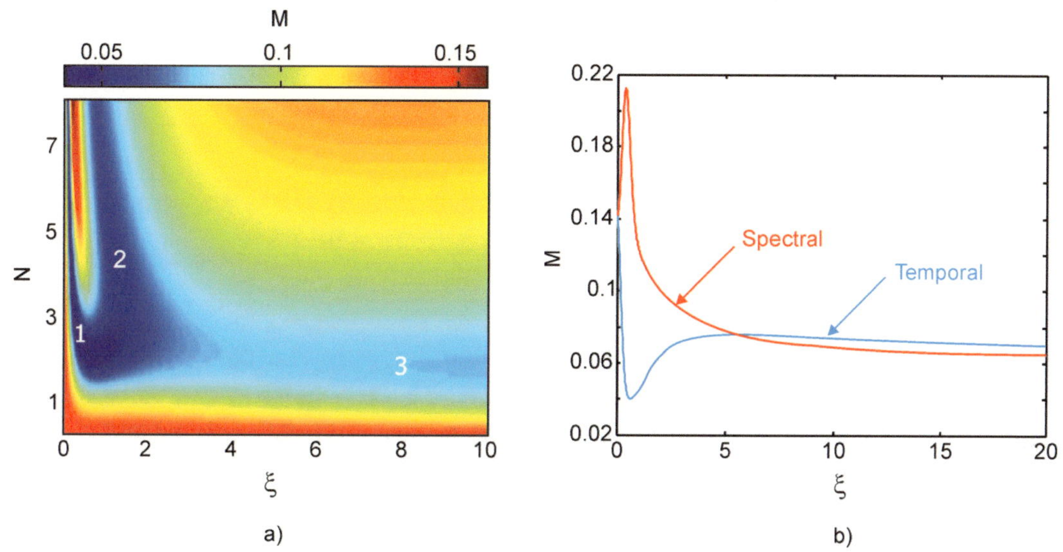

Fig. 1. a) Evolution of the misfit parameter M versus N and ξ in the case of initial Gaussian pulse shape. Numbers at the figure denote three specific areas described in the text. b) Evolution of the misfit parameter M versus ξ for the temporal and spectral pulse shapes ($N = 2$).

Now we investigate pulse reshaping in the case of initial secant pulse shape. The M-map for the secant pulse shape is shown in Fig. 2 (a) and appropriate M-plot is shown in Fig. 2 (b). From Fig. 2 we see that pulse reshaping process in the case of initial secant pulse is different from that one for Gaussian. Within the short propagation distance we find only one area where pulse achieves the parabolic shape. Moreover one can see that steady-state regime arises at much longer propagation distance as compared to the Gaussian. We find the same horizontal valley $1.5 < N < 2.5$, where misfit parameter achieves smaller value. However, here the resulted pulse shape in the steady-state differs stronger from the parabolic one as compared to the previous Gaussian pulse.

Now we investigate pulse reshaping in the case of initial super-Gaussian (2-nd order) pulse shape. The M-map for super-Gaussian pulse shape is shown in Fig. 3 (a) and corresponding M-plot is shown in Fig. 3 (b).

From Fig. 3 we can see that pulse reshaping process in the case of initial super-Gaussian pulse inevitably leads to the fully parabolic pulse for $N > 1$. Steady-state regime arises very fast as compared to the previously investigated initial pulse shapes. However, one should note that in the case of pulses with steeper leading and trailing edges (higher order super-Gaussian pulses) the resulted pulse profile in the steady-state regime differs stronger from the parabolic one.

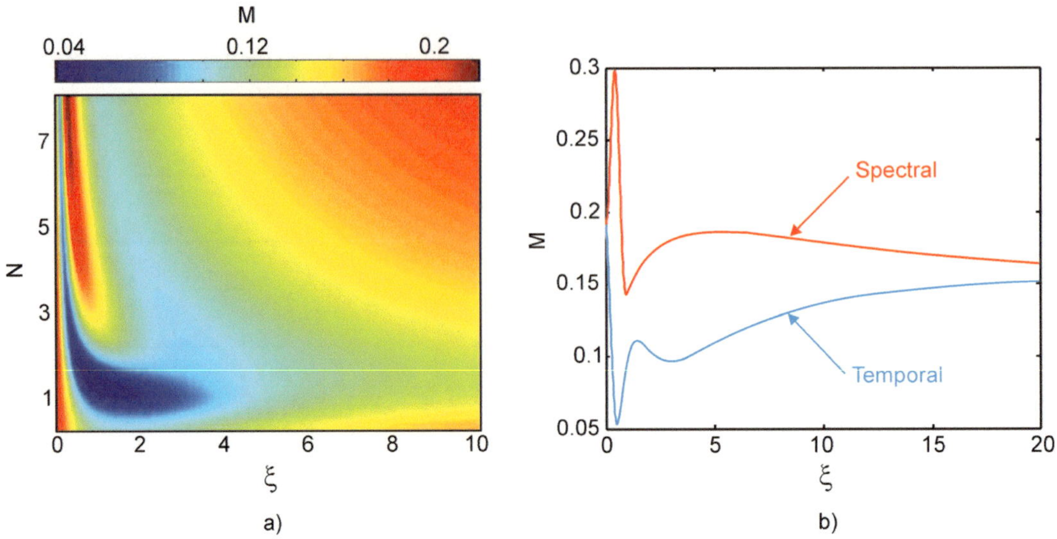

Fig. 2. a) Evolution of the misfit parameter M versus N and ξ in the case of initial secant pulse shape. b) Evolution of the misfit parameter M versus ξ for the temporal and spectral pulse shapes ($N=2$).

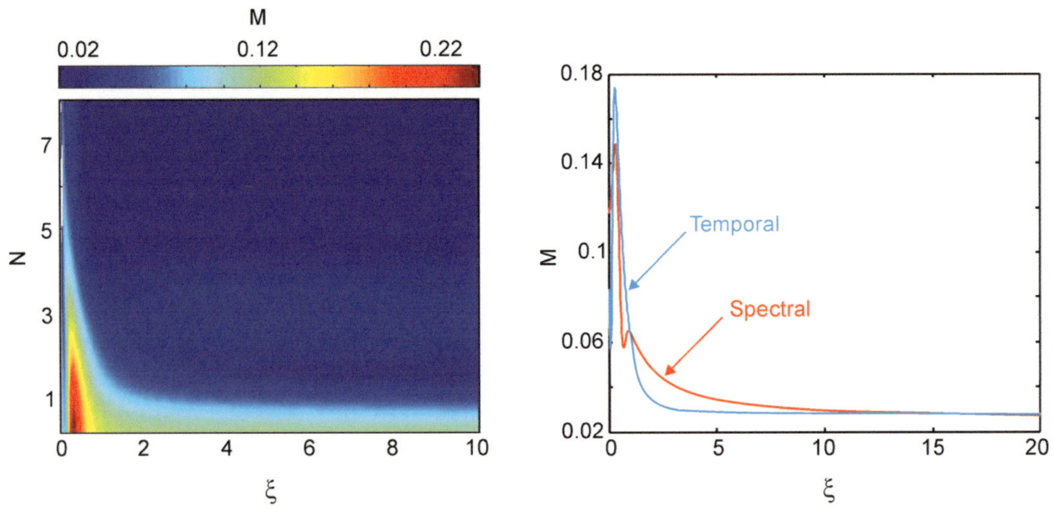

Fig. 3. a) Evolution of the misfit parameter M versus N and ξ in the case of initial super-Gaussian pulse shape. b) Evolution of the misfit parameter M versus ξ for the temporal and spectral pulse shapes ($N=2$).

Results presented above were obtained in the case of initially unchirped pulses, now we examine the influence of the initial pulse chirp on the resulted pulse shape in the steady-state regime. We include chirp in the initial pulse in the following way:

$$A_{chirp} = A_{unchirp} \exp\left(iC \frac{T^2}{2T_0^2} \right), \tag{5}$$

where $A_{unchirp}$ is the waveform of an unchirped initial pulse; C - chirp parameter.
This leads to the initial linear dependence of instantaneous frequency:

$$\omega = \omega_0 + C\frac{T}{T_0^2}, \tag{6}$$

where ω_0 is a central frequency of the pulse spectrum. The instantaneous frequency increases linearly from the leading to the trailing edge of the pulse for $C > 0$ (positive chirp) while the opposite process occurs for $C < 0$ (negative chirp). We compare pulse evolution for initial Gaussian pulse with positive chirp, negative chirp and unchirped ($C = 0$), Fig. 4.

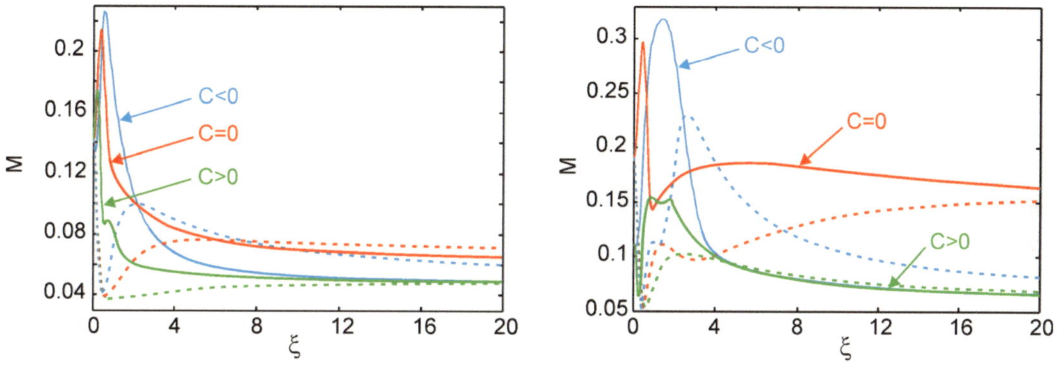

Fig. 4. Evolution of the misfit parameter M versus ξ for the temporal and spectral pulse shapes ($N = 2$) in the case of initial pulse with initial positive chirp, negative and unchirped. The absolute value of initial chirp parameter here is 0.5. a) - initial Gaussian pulse; b) - initial secant pulse. Dotted lines denote temporal curves; solid lines - spectral curves.

From Fig. 4 we can see that initial chirp is able to change significantly pulse evolution. We can see that the maxima of the spectral and temporal curves at the short propagation distances increase when initial chirp is negative and decrease when initial chirp is positive. In the steady-state regime positive chirp leads to the lower deviation from the parabolic shape both for temporal and spectral profiles and steady-state regime arises faster. In the case of negative chirp the spectral shape also becomes closer to the parabolic, but the temporal curve indicates stronger deviation. So, we can conclude that application of pulses with initial positive chirp is more preferable. For initial Gaussian pulse we can obtain misfit parameter $M \sim 0.05$ and for secant pulse $M \sim 0.08$ at $\xi = 8$.
In the case of initial super-Gaussian pulse the fully parabolic pulse is achieved already for unchirped pulse. Our calculations show that adding the initial chirp in this case does not change sufficiently pulse evolution presented in Fig. 3 a), there are only slight changes of the distance where steady-state regime is achieved. Increasing the magnitude of the positive chirp is able to provide actually a fully parabolic pulse from the initial Gaussian pulse. Fig. 5 shows the pulse shape, spectrum and chirp for $C = 1$ and $\xi = 4$. We see that both spectral and temporal profiles are parabolic ($M \sim 0.03$) and chirp is actually linear over the whole pulse.

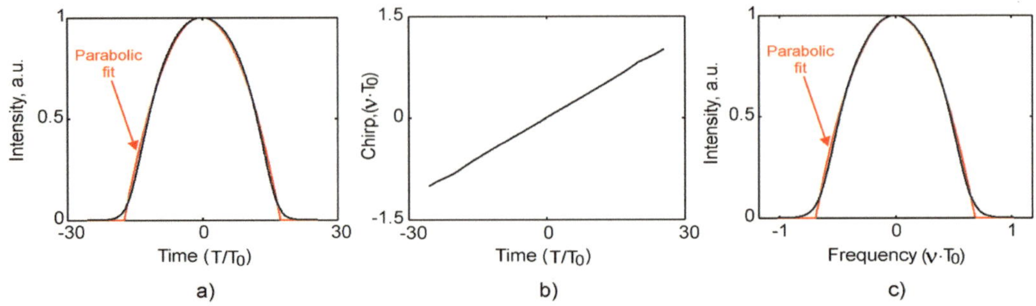

Fig. 5. Normalized pulse temporal intensity – (a), spectrum – (c) and chirp (b) in the steady-state regime produced from initial Gaussian pulse with positive chirp ($N=2$, $\xi=4$, $C=1$). Red curves show corresponding parabolic fits.

In the case of initial secant pulse a stronger initial positive chirp and longer propagation distance is required in order to achieve better quasi-parabolic pulse. For example, our calculations show that quasi-parabolic pulse with $M \sim 0.05$ for both spectral and temporal profile can be obtained when $N=2$, $\xi=8$, $C=2.5$.

3.2 Influence of the third-order dispersion, loss and gain

The self-phase modulation and the group velocity dispersion are main factors of pulse reshaping towards the parabolic shape in the passive optical fiber. However other factors are able to change the pulse reshaping. At first we investigate the impact of third-order dispersion (TOD) associated with the 3-rd order time derivative in the NLSE (1). In the linear regime ($\gamma=0$) TOD effects play a significant role only if $L_D' / L_D \leq 1$ (Agrawal, 2007). In this case TOD leads to the asymmetric broadening of the pulse with an oscillated tail near one of its edges. In the presence of nonlinearity the TOD influence becomes more complex. The impact of TOD in the steady-state regime of a passive fiber is quite similar to that one in the case of similariton pulse propagation (Latkin et al., 2007; Bale et al., 2010). From Fig. 6 we see that pulse temporal and spectrum profiles become asymmetric with the peak shifted towards one of the edges depending on the sign of the TOD; pulse chirp becomes nonlinear. Further increasing of TOD leads to the development of the optical shock-type instabilities. In the temporal profile we can observe fast and deep oscillations and a lateral satellite in the pulse spectrum arises.

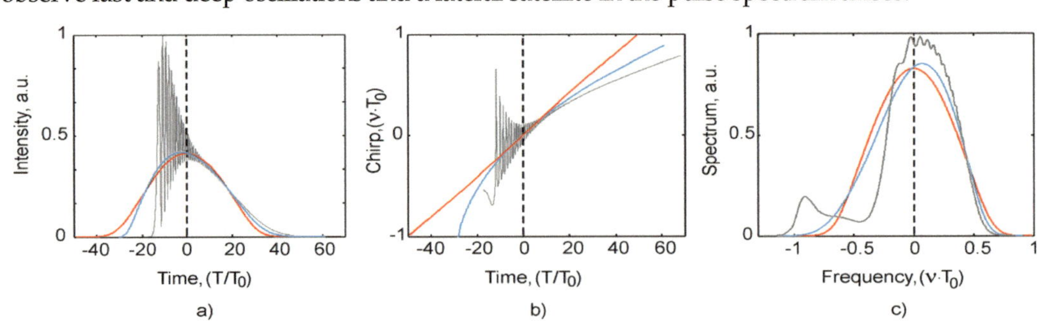

Fig. 6. Pulse evolution in the steady-state under the TOD impact ($N=2$, $\xi=8$). (a) - Pulse temporal profile. (b) - Chirp. (c) - Spectrum. Color indicates the relative magnitude of TOD: red curve – without TOD; blue curve - $L_D' / L_D = 7.3$; gray curve - $L_D' / L_D = 3.3$.

We estimate here distortion of the pulse shape due to the TOD impact applying (3) for perturbed pulse waveform and appropriate unperturbed pulse in the steady-state (instead of parabolic fit). Fig. 7 shows evolution of misfit parameter M versus L'_D / L_D for various initial pulse shapes.

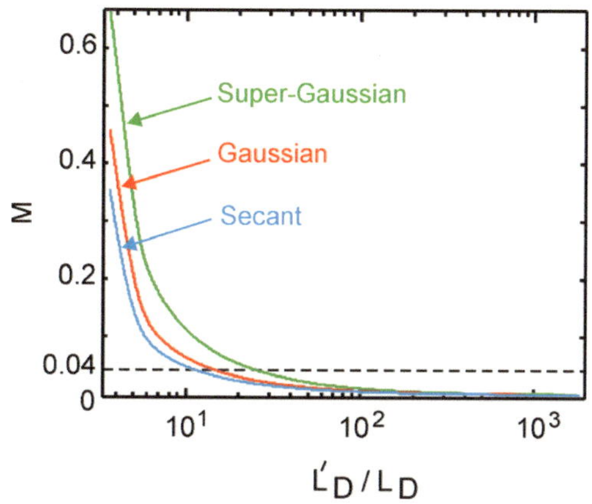

Fig. 7. Evolution of the misfit parameter M versus L'_D / L_D which demonstrates TOD impact on the pulse profile ($N = 2$, $\xi = 8$).

From Fig. 7 we see that super-Gaussian pulse is the most sensitive to the TOD, whereas secant pulse is the least sensitive. Level 0.04 in Fig. 6 indicates the border, where pulse distortions become significant. Less value of misfit parameter can be achieved when $L'_D / L_D > 30$ for all pulse shapes. This condition should be taken into account in order to select suitable fiber for pulse reshaping.

Next we investigate the impact of loss and gain on the pulse reshaping in the steady-state regime. Both effects can be taken into account using the term α in NLSE (1). When $\alpha > 0$ it means the presence of loss in the fiber; $\alpha = 0$ - lossless case; $\alpha < 0$ indicates gain. Fig. 8 shows evolution of misfit parameter M (given by (3)-(4)) versus ξ both for the temporal and spectral pulse shapes. Initial pulse shape here is Gaussian, when initial pulse shape is secant or super-Gaussian pulse evolution is qualitatively similar. Total loss/gain was chosen to be 20 dB, it corresponds to approximately total gain of 2.3 m of Yb3+ doped fiber. Typical attenuation of the photonic crystall fibers (PCF) can exceed tens of dB/km, such that total loss 20 dB correspond to hundreds meters of PCF.

From Fig. 8 we can see that within the short propagation distance both loss and gain don't affect sufficiently pulse evolution, because dotted and dashed lines nearly coincide with appropriate solid lines. However with increasing the propagation distance the loss/gain influence becomes stronger. In the case of loss the dotted spectral and temporal curves are nearly horizontal and they tend asymptotically to some limit without intersection (at least within the calculated propagation distance). Actually it means that the presence of loss damps the strength of the pulse reshaping and the resulted pulse shape differs stronger from the parabolic profile. The influence of the gain is opposite to that one of the loss. Gain enhances pulse reshaping towards the parabolic shape and we obtain actually a similariton propagation regime. Spectral profile (green solid line in Fig. 8) tends to parabolic shape faster as compared to the temporal one. However, one should note that spectral curve doesn't achieve $M = 0$, we can see only some minimum around $\xi = 15$ and then this curve grows up. This behavior is related to the arising the oscillations in the pulse spectral profile in the similariton propagation regime (Kruglov et al., 2002).

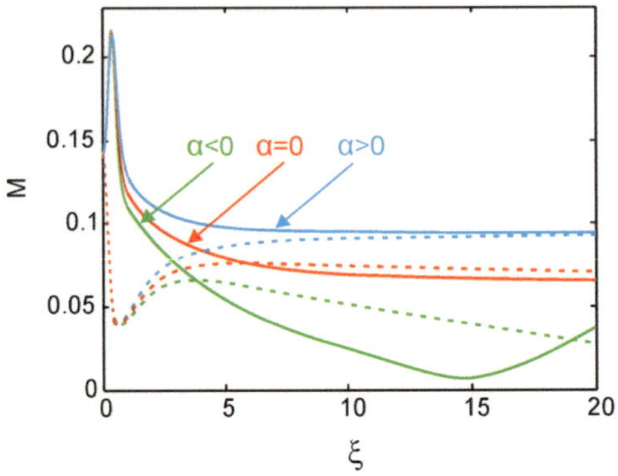

Fig. 8. Evolution of the misfit parameter M versus ξ for the temporal and spectral pulse shapes ($N = 2$). Blue lines denote $\alpha > 0$; red lines - $\alpha = 0$; green lines - $\alpha < 0$. Dotted lines denote temporal curves, solid lines – spectral curves.

Based on the presented results we can estimate some practical conditions for achieving the quasi-parabolic pulses in a passive fiber. One can see that initial super-Gaussian pulse is the best choice to obtain parabolic pulses in a passive optical fiber, but state of the art ultrashort lasers generate usually secant or Gaussian waveform. In this case an initial Gaussian pulse is more preferable than secant, because the resulted pulse shape can be closer to the parabolic one. Next point is that selected fiber should provide suitable dispersion properties at the operation wavelength. Dispersion should be normal and flat enough in order to reduce the impact of high-order dispersion ($L_D' / L_D > 30$), moreover the higher amount of second-order dispersion allows usage of shorter pieces of fiber due to the smaller dispersion length L_D. For example, for realistic dispersion $\beta_2 = 20$ ps^2/km and initial pulse duration 100 fs, the required fiber length ($8L_D$) is about 1.4 m. Application of photonic crystal fibers seems attractive due to the possibility to design the desired dispersion properties by varying geometrical fiber parameters such as the lattice pitch and diameter of air-holes. Because the required fiber length is quite short the influence of attenuation on the pulse reshaping process is expected to be negligible. Next, by varying the initial pulse peak power and/or selecting fiber with appropriate nonlinear coefficient one can achieve operation within the desired range $1.5 < N < 2.5$, where pulse shape in the steady-state regime is closest to the parabolic one (for Gaussian pulse $M \sim 0.07$). Application of pulses with initial positive chirp is desirable to further decrease misfit parameter and required fiber length.

4. Pulse reshaping in the strong nonlinear case

4.1 Optical wave breaking cancellation

Now, we investigate pulse reshaping in the strong nonlinear case ($N > 10$) in the normal dispersive optical fibers. The main difference here is appearing of the optical wave breaking at the initial stage of the pulse propagation. The appearing of optical wave breaking has been investigated theoretically and experimentally (Tomlinson et al., 1985; Rothenberg, 1989; Anderson et al., 1992).

We start our consideration from OWB stage appearing during the pulse propagation in the normal dispersive fiber with $N = 30$ showing in Fig. 9 ($\xi = 0.065$). The optical wave breaking comprises following features. Pulse becomes nearly rectangular with relatively sharp leading and trailing edges and oscillations in the pulse edges arise. The strength of oscillations rises with increasing of N. The chirp function shows that there is a nearly linear chirp over most of the pulse width; however the steep transition regions are developed at the leading and trailing edges. In the spectrum we can see the development of sidelobes. The appearing of the OWB has been explained by the combined action of SPM and normal GVD (Tomlinson et al., 1985; Anderson et al., 1992). Normal GVD tends to change a nonlinear double-peak chirp pattern induced by SPM. A steepening of the chirp function is occurred, because the red-shifted light near the leading edge travels faster and overtakes the unshifted light in the forward tail of the pulse. The opposite scenario occurs for the blue-shifted light near the trailing edge. In both cases, the leading and trailing regions of the pulse contain light at two different frequencies which interfere. Owing to such interference the oscillations near the pulse edges arise. Moreover nonlinear mixing of the overlapped pulse components with different frequencies creates new frequencies. This leads to the development of sidelobes in the spectrum profile.

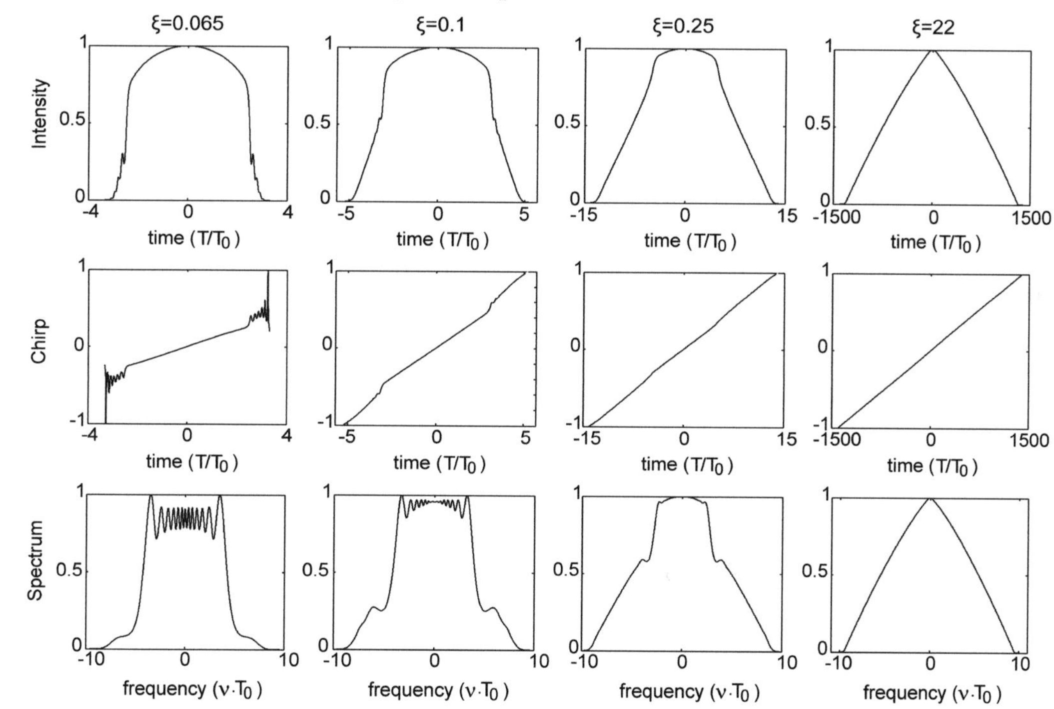

Fig. 9. Evolution of initial Gaussian pulse in a fiber showing optical wave breaking cancellation ($N = 30$). First row comprises pulse temporal intensity over the increasing length ξ; second row - instantaneous frequency (chirp); third row - pulse spectrum. All quantities are normalized to their maxima; time and frequency are normalized to the initial pulse width.

Now let us look on the pulse reshaping at the longer propagation distance. It was shown that oscillations induced by OWB are vanishing at longer distances and pulse transforms to the chirped trapezium-shaped pulse (Karlsson, 1994). Here we investigate this in a more detailed way. One can see from Fig. 9 ($\xi = 0.1$ and $\xi = 0.25$) that oscillations of the chirp function decay at the pulse edges, such that only two bends remain around the pulse center. The magnitudes of these bends decrease with distance, thus actually chirp tends to flattening. At the same time oscillations of the pulse intensity envelope decay as well and we can see also two symmetrical bends on the pulse intensity profile (trapezium-shaped pulse). These bends becomes weaker and moves to the center of the pulse with increasing of propagation distance. A quite long propagation distance is required to complete this process. Only at $\xi = 22$ there are no any bends in the pulse intensity profile. The same behavior one can observe in the pulse spectrum, i.e. the bends indicating spectral sidelobes moves to the center of the spectrum and their magnitude decreases. In addition the SPM induced spectral oscillations decay in the top of the spectrum and SPM-induced broadening of the spectrum is stopping. At the last stage ($\xi = 22$) pulse spectrum is smooth also and it actually repeats temporal intensity profile.
Thus, we can see that OWB fully vanishes at the longer propagation distance and pulse temporal shape and spectrum is fully smooth with linear chirp. One should note that the resulted pulse shape is note trapezium, rather triangular. We explain OWB cancellation by the following way. It is clear that the mentioned above picture of OWB onset doesn't freeze with propagation distance. Owing to the normal dispersion a frequency-shifted light should overtake the unshifted light in the pulse edges more and more with propagation distance. When this frequency-shifted light completely overtakes the unshifted light one can observe actually the formation of low-intensity pulse wings both in the temporal and spectral profile (Fig. 9), the chirp function is linear within these wings. The maximum shifted light from initial nonlinear profile overtakes an unshifted light the fastest, however with increasing the propagation distance the less shifted light overtakes it as well. Owing to that low intensity pulse wings expand up to the pulse peak. The frequency difference between the shifted light and unshifted one continuously reduces during this process, therefore optical shock relaxes and a flattening of chirp function occurs. This is accompanied by reducing the bends at the pulse edges and finally we obtain smooth pulse profile with a linear chirp.
So, we can highlight the main trend of the pulse evolution: normal GVD tends to flatten SPM induced nonlinear chirp pattern with propagation distance. At first this leads to the steepening of the chirp function and appearing the OWB, but then with increasing of propagation distance the chirp flattens and OWB vanishes. However, the question remains: How to find the length where OWB fully vanishes and pulse shape is already smooth? It is clear that in this case the pulse shape and spectrum profile should not include any bends. The straightforward way here is to calculate a second derivative of the temporal intensity function or spectral intensity function. Zeros of those second derivatives will indicate the presence of the bend's point. Figure 10 illustrates application of the second derivative for the analysis of the pulse shape. There are shown Gaussian pulse, parabolic pulse and pulse obtained due to nonlinear reshaping of initial Gaussian pulse after the vanishing of OWB. All shown waveforms in Fig. 10 possess only two symmetrical zeroes of the second derivatives. For Gaussian pulse the bends are located near the center of the pulse edges, whereas bends of parabolic pulse are shifted to the foot of the pulse. Due to the sharp

transitions between zero background and nonzero parabolic profile (resulted from (4)), the appropriate second derivative demonstrates here sharp peaks. In the central part 2-nd derivative of parabolic pulse is horizontal and negative. A pulse obtained due to nonlinear reshaping demonstrates mixed features. Symmetrical bends here are also shifted to the foot of the pulse; however they are not so sharp. In the central part 2-nd derivative is not fully horizontal and moreover a dip in the center remains similar to that one in the Gaussian profile.

Thus, we can see that pulses with smooth edges possess only two symmetrical zeroes indicating the transition between the foot of the pulse and pulse edges. More number of zeroes indicates uneven pulse edges. We can use this in order to characterize the transformation of the pulse shape from the perturbed shape toward the smooth one.

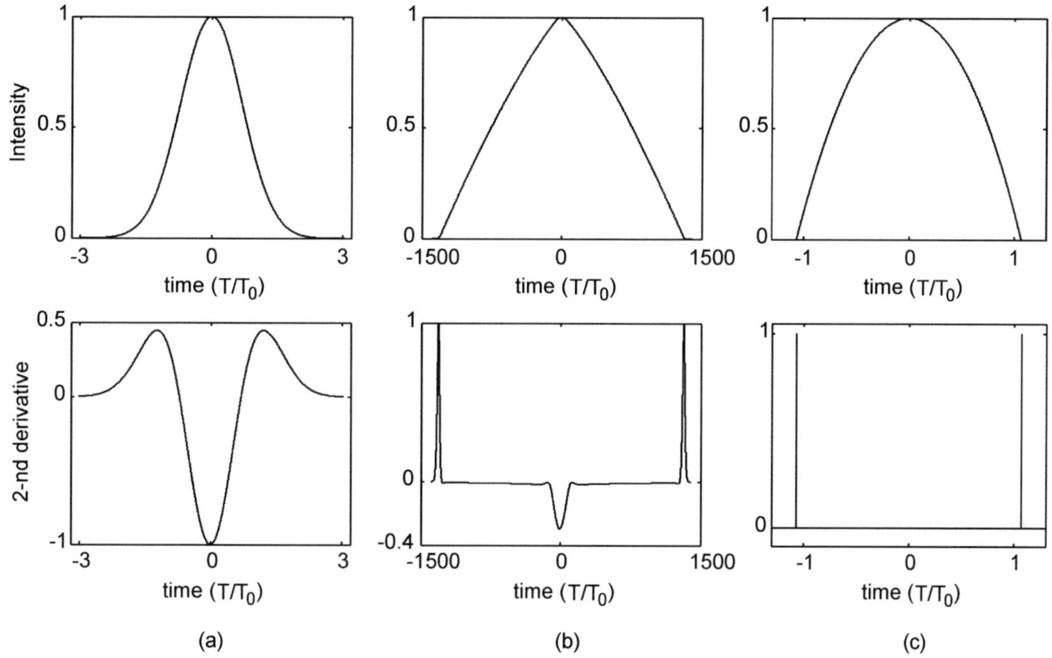

Fig. 10. Pulse waveforms (top raw) and appropriate 2-nd derivative (bottom row). (a) This column shows Gaussian waveform. (b) This column shows pulse waveform obtained after the optical wave breaking cancelled ($N = 30$, $\xi = 22$). (c) This column shows parabolic waveform. Pulse temporal intensities are normalized to its maxima; 2-nd derivatives are normalized to its absolute maxima.

Figure 11 shows how the number of zeroes of 2-nd derivative changes during pulse reshaping in a fiber.

At the start point ($\xi = 0$) pulse possesses only 2 zeroes of the second derivatives associated with initial Gaussian profile. However, the number of zeroes increases very fast with pulse propagation and achieves some maximal values indicating pulse shape perturbations due to OWB and SPM induced spectral oscillations. The greater is soliton number N , the stronger pulse shape perturbations appear. But after that the number of zeroes decreases and we can see the presence of long horizontal plateau, when the number of zeroes is 6. At this stage pulse waveform is already quite smooth without breaking oscillations or spectral oscillations induced by SPM. There are only two weak symmetrical bends at the pulse edges associated with gradual expansion of the low-intensity temporal and spectral wings. When this process is completed, the number of zeroes of the second derivative goes down to 2.

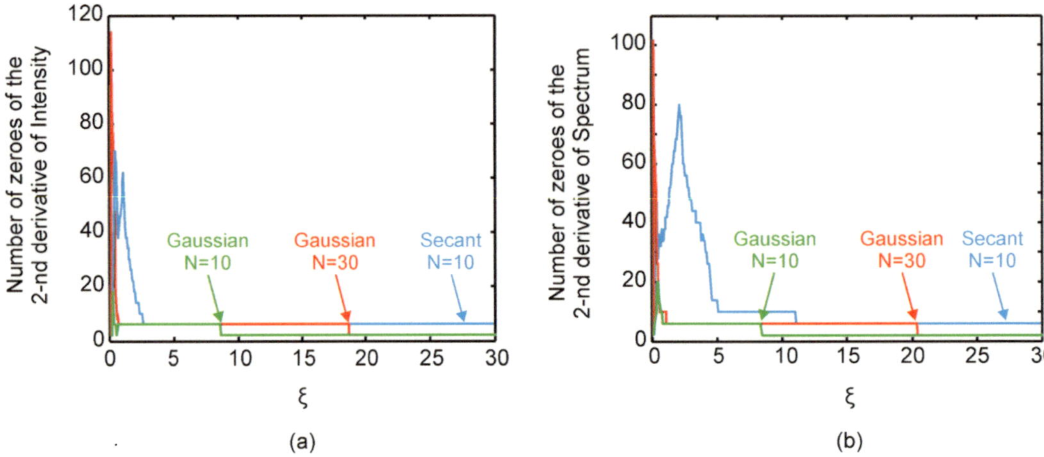

Fig. 11. (a) Number of zeroes of the second derivative of pulse temporal intensity depending on propagation distance. (b) Number of zeroes of the second derivative of pulse spectrum.

From Fig. 11 we can see that initial Gaussian pulse with $N = 10$ becomes smooth the fastest $\xi > 9$, here both the temporal intensity and spectrum have smooth profile. For initial Gaussian pulse with $N = 30$ it is required longer propagation distance $\xi > 21$. In the case of initial secant pulse we can see that bends in the temporal and spectral profile remains until $\xi = 30$. This is related to the stronger distortions due to OWB in the case of initial secant pulse.

4.2 Pulse shape transformations after OWB cancellation

Here we are interested in the pulse shape transformations in the steady-state regime after OWB cancellation. One can see that the resulted pulse shape here strongly differs (see Fig. 9) from the quasi-parabolic shape achieved in the weak nonlinear case, it is rather triangular. Indeed, recently it has been found that one can achieve a pulse with perfect triangular pulse shape from the initial chirped Gaussian pulse with $N = 10$ at $\xi = 0.33$ (Boscolo et al., 2008; Wang et al., 2010). Triangular pulses as well as parabolic pulses are highly desired for a range of photonic applications. Possible applications include: add-drop multiplexing (Parmigiani et al., 2009) or doubling of the optical signals (Latkin et al., 2009).

In order to investigate pulse transformations towards triangular pulse shape we introduce the following triangular fitting function:

$$\begin{cases} A_t(T) = \sqrt{P_t}\sqrt{1 - \left|T / (T_t\sqrt{2.5})\right|}, & |T| \le T_t\sqrt{2.5} \\ A_t(T) = 0, & |T| > T_t\sqrt{2.5} \end{cases}, \qquad (7)$$

where P_t is the peak power of the triangular pulse, T_t is the duration of the triangular pulse (the half-width at 1/e intensity point), $U_t = P_t T_t \sqrt{2.5}$ is the energy of triangular fit pulse.

Applying (3) and (7) we can investigate pulse transformations towards the triangular shape in the same way it was done above for parabolic pulse shape.

Figure 12 shows transformations of the initial chirped Gaussian pulse with ($N = 10$, $C = -4$). We can see that temporal spectral and temporal curves achieves minimum near $\xi = 0.4$. The triangular misfit parameter here is $M \sim 0.03$, whereas for the spectral shape it is slightly higher. However, we can see from Fig. 12 a) that these minima are very sharp and with increasing the propagation distance the pulse shape and spectrum do not remain triangular.

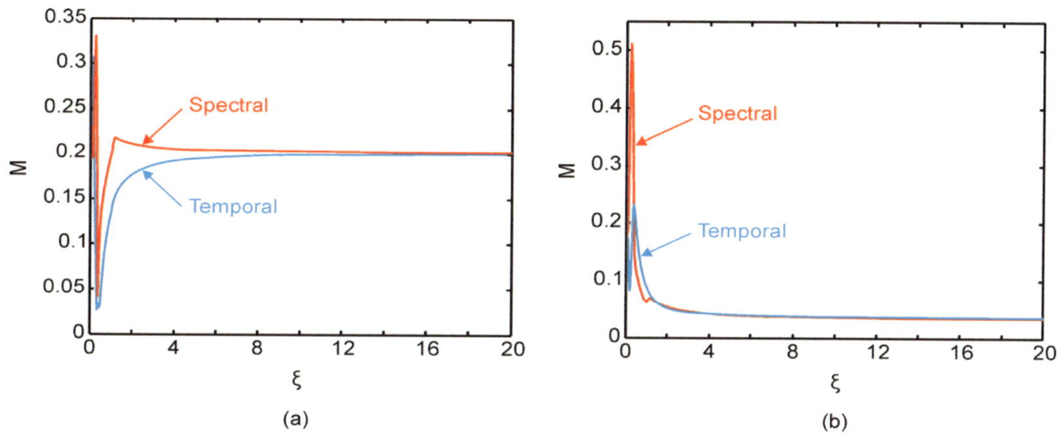

Fig. 12. a) Evolution of the misfit parameter M versus ξ for the temporal and spectral pulse shapes ($N = 10$). a) - triangular fitting; b) - parabolic fitting.

Fig. 12 b) surprisingly shows that instead of triangular shape we obtain here a fully parabolic pulse in the steady-state regime. Parabolic misfit parameter here is $M < 0.04$ from $\xi > 6$ both for temporal and spectral profile and moreover, chirp is linear here (Fig. 13).

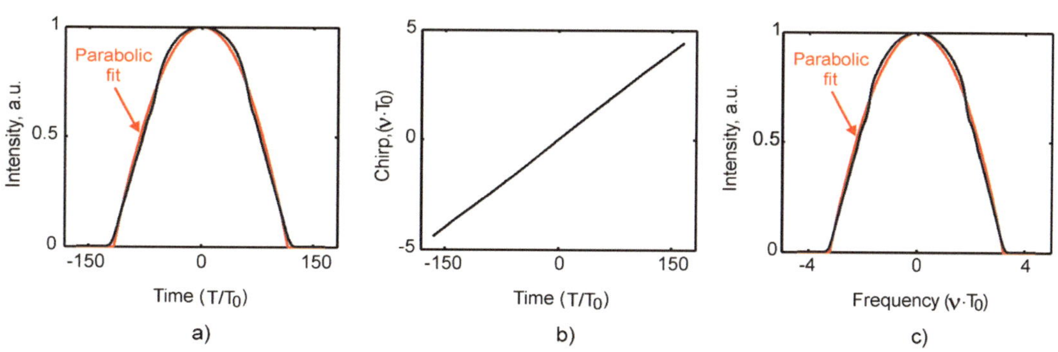

Fig. 13. Normalized pulse temporal intensity ($M = 0.039$) – (a), spectrum ($M = 0.038$) – (c) and chirp (b) in the steady-state regime produced from initial Gaussian pulse with negative chirp ($N = 10$, $\xi = 6$, $C = -4$). Red curves show corresponding parabolic fits.

Thus, in the steady-state regime we can obtain parabolic pulses not only in the weak nonlinear case ($1.5 < N < 2.5$), but also at the higher pulse powers.
In spite of that it is also possible to obtain triangular pulses as well in the steady-state regime. Figure 14 a) clearly shows that applying initial secant pulse without any chirp we can obtain triangular pulses in the steady-state regime. One can see that triangular misfit parameter is $M < 0.04$ from $\xi > 10$ both for temporal and spectral profile. In contrast to that Fig. 14 b) shows that in the case of initial Gaussian pulse the deviation from the triangular fit is sufficiently higher.

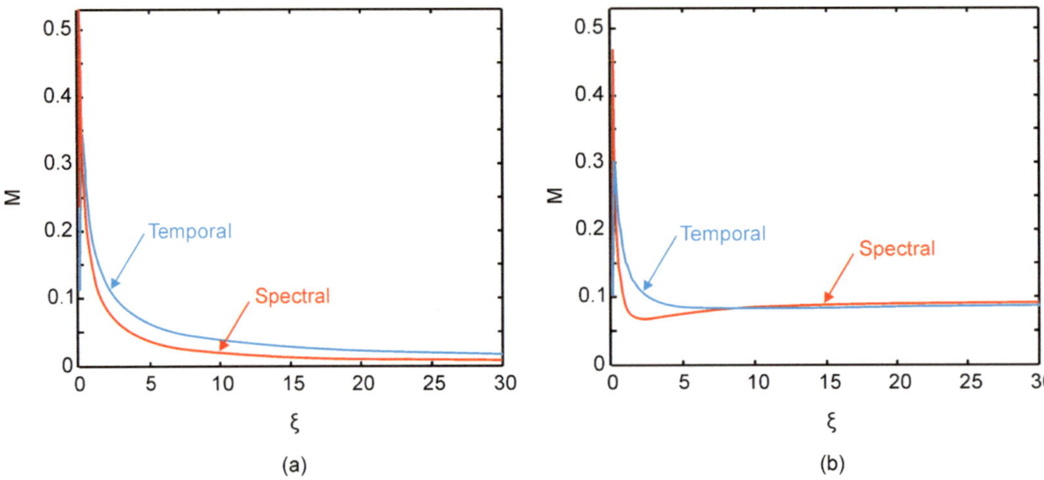

Fig. 14. a) Evolution of the triangular misfit parameter M versus ξ ($N = 10$).a) - initial secant pulse; b) - initial Gaussian pulse.

Figure 15 shows the resulted pulse shape in the steady-state regime obtained from the initial unchirped secant pulse. Pulse shape and spectrum are fully triangular over the whole pulse with the exception of a small flat top. The chirp is also linear here.

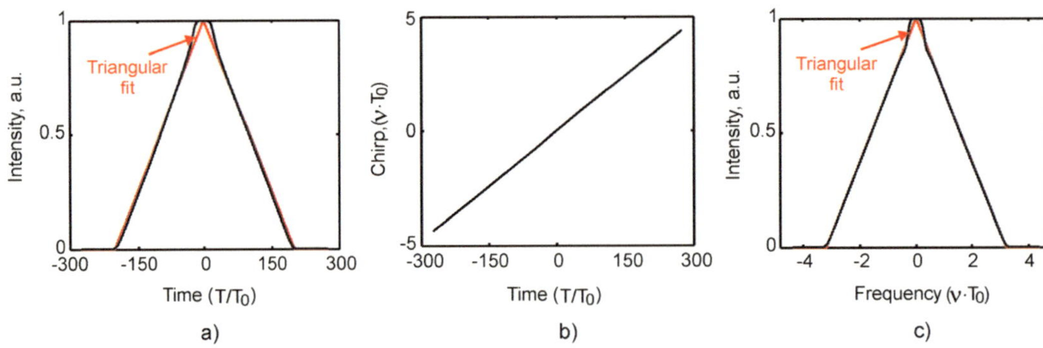

Fig. 15. Normalized pulse temporal intensity ($M = 0.037$) - (a), spectrum ($M = 0.018$) - (c) and chirp (b) in the steady-state regime produced from initial secant pulse ($N = 10$, $\xi = 10$). Red curves show corresponding triangular fits.

5. Conclusion

We have investigated pulse shape transformations in passive optical fibers in the steady-state regime. In the case of super-Gaussian pulse shape (2-nd order) the resulted pulse waveform tends to the fully parabolic very fast. In the case of initial Gaussian pulse quasi-parabolic pulses ($M \sim 0.07$) with linear chirp can be obtained in the weak nonlinear case at fiber length $\xi > 5$. In the case of initial secant pulse the steady-state regime arises at much longer propagation distance and obtained pulse shape differs stronger from the parabolic one. The value of soliton order N strongly affects on the pulse shape in the steady-state. When soliton order lies within the following range $1.5 < N < 2.5$, the shape of quasi-parabolic pulse is closest to the parabolic one. Application of positively chirped pulses is preferable to achieve quasi-parabolic pulses with minimal deviation from the parabolic profile both for temporal and spectral shapes at the shortest propagation distance. Third-order dispersion is able to affect strongly on the pulse shape in the steady-state regime leading to the pulse distortions due to optical shock-type instabilities. It was found that the dispersion length's ratio $L'_D / L_D > 30$ provides undistorted pulse shape. The presence of losses in the fiber damps the strength of pulse reshaping and the resulted pulse shape differs stronger from the parabolic profile, whereas gain enhances pulse reshaping towards the parabolic shape. In the strong nonlinear case ($N \geq 10$) it is also possible to achieve parabolic pulses in the steady-state regime from the initially chirped Gaussian pulses. Moreover, one can obtain also triangular pulses from the unchirped secant pulses.

6. References

Agrawal, G. P. (2007). *Nonlinear Fiber Optics.* Academic Press, Boston.

Akhmediev, N. & Ankiewicz, A. (Ed(s).). (2008). *Dissipative Solitons: From optics to biology and medicine,* Springer, Berlin-Heidelberg.

Anderson, D.; Desaix, M.; Lisak, M. & Quiroga-Teixeiro, M. L. (1992). Wave breaking in nonlinear-optical fibers. *J. Opt. Soc. Am. B,* Vol. 9, No. 8, pp. 1358-1361.

Anderson, D.; Desaix, M.; Karlsson, M.; Lisak, M. & Quiroga-Teixeiro, M. L. (1993). Wave-breaking-free pulses in nonlinear-optical fibers. *J. Opt. Soc. Am. B,* Vol. 10, No. 7, pp. 1185-1190.

Bale, B. G. & Boscolo, S. (2010). Impact of third-order fibre dispersion on the evolution of parabolic optical pulses. *J. Opt.,* Vol. 12, No. 1, pp. 015202.

Barenblatt, G. I. (1996). *Scaling, Self-Similarity, and Intermediate Asymptotics.* Cambridge University Press, Cambridge.

Boscolo, S.; Latkin, A. I. & Turitsyn, S. K. (2008). Passive Nonlinear Pulse Shaping in Normally Dispersive Fiber Systems, *IEEE J. Quantum Electron.,* Vol. 44, No. 12, pp. 1196-1203.

Brabec, T.; Spielmann, Ch. & Krausz, E. (1991). Mode locking in solitary lasers. *Opt. Lett.,* Vol. 16, No. 24, pp. 1961-1963.

Dudley, J. M.; Genty, G.; Coen, S. (2006). Supercontinuum generation in photonic crystal fiber. *Rev. Mod. Phys.,* Vol. 78, No. 4, pp. 1135-1184.

Dudley, J. M.; Finot, C.; Richardson, D. J. & Millot, G. (2007). Self-similarity in ultrafast nonlinear optics. *Nature Physics,* Vol. 3, pp. 597-603.

Duling III, I. N. (1991). Subpicosecond all-fibre erbium laser. Electron. Lett., Vol. 27, No. 6, pp. 544–545.

Fermann, M. E.; Kruglov, V. I.; Thomsen, B. C.; Dudley, J. M. & Harvey, J. D. (2000). Self-similar propagation and amplification of parabolic pulses in optical fibers. *Phys. Rev. Lett.*, Vol. 84, No. 26, pp. 6010-6013.

Finot, C.; Parmigiani, F.; Petropoulos, P. & Richardson, D. J. (2006). Parabolic pulse evolution in normally dispersive fiber amplifiers preceding the similariton formation regime. *Opt. Express,* Vol. 14, No. 8, pp. 3161-3170.

Finot, C.; Provost, L.; Petropoulos, P. & Richardson, D. J. (2007). Parabolic pulse generation through passive nonlinear pulse reshaping in a normally dispersive two segment fiber device," *Opt. Express,* Vol. 15, No. 3, pp. 852-864.

Finot, C.; Dudley, J. M.; Kibler, B.; Richardson, D. J.; Millot, G. (2009). Optical Parabolic Pulse Generation and Applications. *IEEE J. Quantum Electron.,* Vol. 45, No. 11, pp. 1482 - 1489.

French, W. G.; MacChesney, J. B.; O'Connor, P. D.; Tasker, G. W. (1974). Optical waveguides with very low losses, *Bell Syst. Tech. J.,* Vol. 53, No. 5, pp. 951-954.

Gu, C.H. (Ed(s).). (1995). *Soliton Theory and its Applications,* Springer, New York.

Hasegawa, A. & Matsumoto, M. (2003). *Optical Solitons in Fibers* (3rd ed.), Springer-Verlag, Berlin.

Hasegawa, A. & Tappert, F. (1973). Transmission of stationary nonlinear optical pulses in dispersive dielectric fibers. I. Anomalous dispersion. *Appl. Phys. Lett.,* Vol. 23, No.3, pp. 142-144.

Hirooka, T. & Nakazawa, M. (2004). Parabolic pulse generation by use of a dispersion-decreasing fiber with normal group-velocity dispersion," *Opt. Lett.,* Vol. 29, No. 5, pp. 498-500.

Ilday, F. Ö.; Buckley, J. R.; Clark, W. G. & Wise, F. W. (2004). Self-similar evolution of parabolic pulses in a laser. Phys. Rev. Lett., Vol. 92, No. 21, pp. 213902.

Ippen, E. P. & Stolen, R. H. (1972). Stimulated Brillouin scattering in optical fibers. *Appl. Phys. Lett.,* Vol. 21, No. 11, pp. 539-541.

Kao, K. C.; Hockham, G. A. (1966). Dielectric-fibre surface waveguides for optical frequencies. *Proc. IEE,* Vol. 113, No. 7, pp. 1151–1158.

Karlsson, M. (1994). Optical fiber-grating compressors utilizing long fibers. *Opt. Comm.,* Vol. 112, No. 1-2, pp. 48–54.

Kivshar, Y. S. & Agrawal, G. P. (2003). *Optical solitons From Fibers to Photonic Crystals.* Academic Press, San Diego.

Kodama, Y. and Hasegawa, A. (1987). Nonlinear pulse propagation in a monomode dielectric guide. *IEEE J. Quantum Electron.,* Vol. 23, No. 5, pp. 510–524.

Kruglov, V. I.; Peacock, A. C.; Harvey, J. D. & Dudley, J. M. (2002). Self-similar propagation of parabolic pulses in normal dispersion fiber amplifiers. *J. Opt. Soc. Am. B,* Vol. 19, No. 3, pp. 461–469.

Kruglov, V. I., Peacock, A. C. & Harvey, J. D. (2003). Exact self-similar solutions of the generalized nonlinear Schrdinger equation with distributed coefficients, *Phys. Rev. Lett.,* Vol. 90, No. 11, pp. 113902.

Latkin, A. I.; Turitsyn, S. K. & Sysoliatin, A. A. (2007). Theory of parabolic pulse generation in tapered fiber. *Opt. Lett.,* Vol. 32, No. 4, pp. 331–333.

Latkin, A. I.; Boscolo, S.; Bhamber, R. S. & Turitsyn, S. K. (2009) Doubling of optical signals using triangular pulses. *J. Opt. Soc. Am. B*, Vol. 26, No. 8, pp. 1492–1496.

Miya, T.; Terunuma, Y.; Hosaka, T.; Miyashita, T. (1979). Ultimate low-loss single mode fiber at 1.55 μm. *Electron. Lett.*, Vol. 15, No. 4, pp. 106-108.

Mollenauer, L. F.; Stolen, R. H.; & Gordon, J. P. (1980). Experimental Observation of Picosecond Pulse Narrowing and Solitons in Optical Fibers. *Phys. Rev. Lett.*, Vol. 45, No. 13, pp. 1095-1098.

Parmigiani, F.; Finot, C.; Mukasa, K.; Ibsen, M.; Roelens, M. A.; Petropoulos, P. & Richardson, D. J. (2006). Ultra-flat SPM-broadened spectra in a highly nonlinear fiber using parabolic pulses formed in a fiber Bragg grating. *Opt. Express*, Vol. 14, No. 17, pp. 7617-7622.

Parmigiani, F.; Petropoulos, P.; Ibsen, M.; Almeida, P. J.; Ng, T. T. & Richardson, D. J. (2009) Time domain add–drop multiplexing scheme enhanced using a saw-tooth pulse shaper *Opt. Express,* Vol. 17, No. 10, pp. 8362-8369.

Rothenberg, J. E. (1989). Femtosecond optical shocks and wave breaking in fiber propagation. *J. Opt. Soc. Am. B*, Vol. 6, No. 12, pp. 2392-2401.

Schreiber, T.; Nielsen, C. K.; Ortac, B.; Limpert, J. P. & Tünnermann, A. (2006). Microjoule-level all-polarization-maintaining femtosecond fiber source. *Opt. Lett.*, Vol. 31, No. 5, pp. 574-576.

Stolen, R. H. & Askin, A. (1973). Optical Kerr effect in glass waveguide. *Appl. Phys. Lett.*, Vol. 22, No.6, pp. 294-296.

Stolen, R. H.; Bjorkholm, J. E. & Ashkin, A. (1974). Phase-matched three-wave mixing in silica fiber optical waveguides. *Appl. Phys. Lett.*, Vol. 24, No. 7, pp. 308-310.

Stolen, R. H. & Lin, C. (1978). Self-phase-modulation in silica optical fibers. *Phys. Rev. A*, Vol. 17, No. 4, pp. 1448-1453.

Sukhoivanov, I. A.; Iakushev, S. O.; Petrov, S. I.; Shulika, O. V. Optical wave breaking cancellation in the far dispersion field of optical fiber, *Proceedings of Photonics Society Summer Topical Meeting Series*, pp. 86–87, Playa del Carmen, Mexico, July 19-21, 2010.

Tamura, K. & Nakazawa, M. (1996). Pulse compression by nonlinear pulse evolution with reduced optical wave breaking in erbium-doped fiber amplifiers. *Opt. Lett.*, Vol. 21, No. 1, pp. 68-70.

Tomlinson, W. J.; Stolen, R. H. & Johnson, (1985). A. M. Optical wave breaking of pulses in nonlinear optical fibers. *Opt. Lett.*, Vol. 10, No. 9, pp. 457-459.

Wang, H.; Latkin, A. I.; Boscolo, S.; Harper, P. & Turitsyn, S. K. (2010). Generation of triangular-shaped optical pulses in normally dispersive fibre. *J. Opt.*, Vol. 12, No. 3, pp. 035205.

Yakushev, S. O.; Shulika, O. V.; Sukhoivanov, I. A.; Andrade-Lucio, J. A.; Garcia-Perez, A. Quasi-Parabolic Pulses in the Far Field of Dispersion of Nonlinear Fiber, *Proceedings of Frontiers in Optics/Laser Science*, pp. FTuR2, Rochester, NY, USA, October 24-28, 2010.

Zakharov, V. E. & Shabat, A. B. (1972). Exact theory of two-dimensional self-focusing and one-dimensional self-modulation of waves in nonlinear media. *Sov. Phys. JETP*, Vol. 34, pp. 62–69.

Zeytunyan, A.; Yesayan, G.; Mouradian, L.; Kockaert, P.; Emplit, P.; Louradour, F.; Barthélémy, A. (2009). Nonlinear-dispersive similariton of passive fiber. *J. Europ. Opt. Soc. Rap. Public.*, Vol. 4, pp. 09009.

Zeytunyan, A. S.; Khachikyan, T. J.; Palandjan, K. A.; Esayan, G. L. & Muradyan, L. Kh. (2010). Nonlinear dispersive similariton: spectral interferometric study. *Quantum Electron.*, Vol. 40, No. 4, pp. 327.

Part 3

Intense Pulse Propagation Phenomena

11

Linear and Nonlinear Femtosecond Optics in Isotropic Media – Ionization-Free Filamentation

Kamen Kovachev and Lubomir M. Kovachev
Institute of Electronics, Bulgarian Academy of Sciences, Sofia
Bulgaria

1. Introduction

In the process of investigating the filamentation of a power femtosecond (fs) laser pulse many new physical effects have been observed, such as long-range self-channeling (1–3), coherent and incoherent radial and forward THz emission (4–6), asymmetric pulse shaping, super-broad spectra (7–11; 30) and others. The role of the different mechanisms in near zone (up to $1-2$ m from the source) has been investigated experimentally and by numerical simulations, and most processes in this zone are well explained (12–15). When a fs pulse with power of several $P_{cr} = \pi(0.61\lambda_0)^2/(8n_0n_2)$ starts from the laser source a slice-by-slice self-focussing process takes place (16). At a distance of one-two meters the pulse self-compresses, enlarging the k_z spectrum to super-broad asymmetric spectrum $\triangle k_z \approx k_0$. The process increases the core intensity up to tens of $10^{13}W/cm^2$, where different types of plasma ionization, multi-photon processes and higher-order Kerr terms appear (17). Usually, the basic model of propagation in near the zone is a scalar spatio-temporal paraxial equation including all the above mentioned mechanisms (12; 13; 17). The basic model is natural in the near zone because of the fact that the initial fs pulse contains a narrow-band spectrum $\triangle k_z << k_0$. Thus, the paraxial spatio-temporal model gives a good explanation of nonlinear phenomena such as conical emission, X-waves, spectral broadening to the high frequency region and others. In far-away zone (propagation distance more than $2-3$ meters) plasma ionization and higher-order Kerr terms are admitted also as necessary for a balance between the self-focussing and plasma defocussing and for obtaining long range self-channeling in gases.

However, the above explanation of filamentation is difficult to apply in far-away zone. There are basically two main characteristics which remain the same at these distances - the superbroad spectrum and the width of the core, while the intensity in a stable filament drops to a value of $10^{12}W/cm^2$ (12; 17). The plasma and higher-order Kerr terms are too small to prevent self-focussing. The observation of long-range self-channeling (18–20) without ionization also leads to change the role of plasma in the laser filamentation.

In addition, there are difficulties with the physical interpretation of the THz radiation as a result of plasma generation. The plasma strings formed during filamentation should emit incoherent THz radiation in a direction orthogonal to the propagation axis. The nature of the THz emission, measured in (6) is different. Instead of being emitted radially, it is confined to a very narrow cone in the forward direction. The contribution from ionization in far-away zone is negligible (17) and this is the reason to look for other physical mechanism which could

cause THz or GHz radiations. Our analysis on the third order nonlinear polarization of pulses with broadband spectrum indicates that the nonlinear term in the corresponding envelope equation oscillates with frequency proportional to the group and phase velocity difference $\Omega_{nl} = 3(k_0 v_{ph} - v_{gr} \triangle k_z)$. Actually, this is three times the well-known Carrier-to Envelope Phase (CEP) difference (21). This oscillation induces THz generation, where the generated frequency is exactly $\Omega_{THz} = 93GHz$ for a pulse with superbroad spectrum $\triangle k_z \approx k_0$ with carryier wavelength 800 nm.
Physically, one dimensional Schrödinger solitons in fibers appear as a balance between the Kerr nonlinearity and the negative dispersion (22–24). On the other hand, if we try to find 2D+1 and spatio-temporal solitons in Kerr media, the numerical and the real experiments demonstrate that there is no balance between the plane wave paraxial diffraction - dispersion and the Kerr nonlinearity. This leads to instability and self-focusing of a laser beam or initially narrow band optical pulse. Recently Serkin in (25) suggested stable soliton propagation and reducing the 3D soliton problem to one dimensional, with introducing trapping potential in Bose - Einstein condensates.
In this paper we present a new mathematical model, on the basis of the Amplitude Envelope (AE) equation, up to second order of dispersion, without using paraxial approximation. In the non-paraxial zone the diffraction of pulses with superbroad spectrum or pulses with a few cycles under the envelope is closer to wave type (26). For such pulses, a new physical mechanism of balance between nonparaxial (wave-type diffraction) and third order nonlinearity appears. Exact analytical three-dimensional bright solitons in this regime are found.

2. Linear regime of narrow band and broad band optical pulses

The paraxial spatio-temporal envelope equation governs well the transverse diffraction and the dispersion of fs pulses up to $6-7$ cycles under the envelope. This equation relies on one approximation obtained after neglecting the second derivative in the propagation direction and the second derivative in time from the wave equation (27) or from the $3D+1$ AE equation (28). In air, the series of $k^2(\omega)$ are strongly convergent up to one cycle under the envelope and this is the reason why the AE equation is correct up to the single-cycle regime.
The linearized AE, governing the propagation of laser pulses when the dispersion is limited to second order, is:

$$-2ik_0\left(\frac{\partial A}{\partial z}+\frac{1}{v_{gr}}\frac{\partial A}{\partial t}\right)=\Delta A-\frac{1+\beta}{v_{gr}^2}\frac{\partial^2 A}{\partial t^2}, \tag{1}$$

where $\beta = k'' k_0 v_{gr}^2$ is a number representing the influence of the second order dispersion. In vacuum and dispesionless media the following Diffraction Equation (DE) ($v \sim c$) is obtained:

$$-2ik_0\left(\frac{\partial V}{\partial z}+\frac{1}{v}\frac{\partial V}{\partial t}\right)=\Delta V-\frac{1}{v^2}\frac{\partial^2 V}{\partial t^2}. \tag{2}$$

We solve AE (1) and DE (2) by applying spatial Fourier transformation to the amplitude functions A and V. The fundamental solutions of the Fourier images $\hat{A}$ and $\hat{V}$ in $(k_x, k_y, \triangle k_z, t)$ space are:

$$\hat{A} = \hat{A}(k_x, k_y, \triangle k_z, t = 0) \times$$
$$\exp\left\{i\frac{v_{gr}}{\beta+1}\left(k_0 \pm \sqrt{k_0^2 + (\beta+1)\left({k_x}^2 + {k_y}^2 + {\triangle k_z}^2 - 2k_0\triangle k_z\right)}\right)t\right\}, \quad (3)$$

$$\hat{V} = \hat{V}(k_x, k_y, \triangle k_z, t = 0)\exp\left\{iv\left(k_0 \pm \sqrt{{k_x}^2 + {k_y}^2 + (\triangle k_z - k_0)^2}\right)t\right\}, \quad (4)$$

respectively. In air $\beta \simeq 2.1 \times 10^{-5}$, AE (1) is equal to DE (2), and the dispersion is negligible compared to the diffraction. We solve analytically the convolution problem (4) for initial Gaussian light bullet of the kind $V(x, y, z, t = 0) = \exp\left(-(x^2 + y^2 + z^2)/2r_0^2\right)$. The corresponding solution is:

$$V(x, y, z, t) = \frac{i}{2\hat{r}}\exp\left[-\frac{k_0^2 r_0^2}{2} + ik_0(vt - z)\right] \times$$
$$\left\{i(vt + \hat{r})\exp\left[-\frac{1}{2r_0^2}(vt + \hat{r})^2\right] erfc\left[\frac{i}{\sqrt{2}r_0}(vt + \hat{r})\right]\right. \quad (5)$$
$$\left.-i(vt - \hat{r})\exp\left[-\frac{1}{2r_0^2}(vt - \hat{r})^2\right] erfc\left[\frac{i}{\sqrt{2}r_0}(vt - \hat{r})\right]\right\},$$

where $\hat{r} = \sqrt{x^2 + y^2 + (z - ir_0^2 k_0)^2}$. On the other hand, multiplying the solution (5) with the carrier phase, we obtain solution of the wave equation $E(x, y, z, t) = V(x, y, z, t)\exp\left(i(k_0 z - \omega_0 t)\right)$, where ω_0 and k_0 are the carrier frequency and carrier wave number in the wave packet:

$$\Delta E = \frac{1}{v^2}\frac{\partial^2 E}{\partial t^2}, \quad (6)$$

$$E(x, y, z, t) = \frac{i}{2\hat{r}}\exp\left(-\frac{k_0^2 r_0^2}{2}\right) \times$$
$$\left\{i(vt + \hat{r})\exp\left[-\frac{1}{2r_0^2}(vt + \hat{r})^2\right] erfc\left[\frac{i}{\sqrt{2}r_0}(vt + \hat{r})\right]\right. \quad (7)$$
$$\left.-i(vt - \hat{r})\exp\left[-\frac{1}{2r_0^2}(vt - \hat{r})^2\right] erfc\left[\frac{i}{\sqrt{2}r_0}(vt - \hat{r})\right]\right\}.$$

A systematic study on the different kinds of exact solutions and methods for solving wave equation (6) was performed recently in (29). Here, as in (26) we suggest another method: Starting with the ansatz $E(x, y, z, t) = V(x, y, z, t)\exp\left(i(k_0 z - \omega_0 t)\right)$, we separate the main phase and reduce the wave equation to $3D + 1$ parabolic type one (2). Thus, the initial value problem can be solved and exact (5) (or numerical) solutions of the corresponding amplitude equation (2) can be obtained. The solution (5), multiplied by the main phase, gives an exact solution (7) of the wave equation (6). To investigate the evolution of optical pulses at long distances, it is convenient to rewrite AE (1) equation in Galilean coordinate system $t' = t; z' = z - v_{gr}t$:

$$-i\frac{2k_0}{v_{gr}}\frac{\partial A}{\partial t'} = \Delta_\perp A - \beta\frac{\partial^2 A}{\partial z'^2} - \frac{1+\beta}{v_{gr}^2}\left(\frac{\partial^2 A}{\partial t'^2} - 2v_{gr}\frac{\partial^2 A}{\partial t'\partial z'}\right). \tag{8}$$

Pulses governed by DE (2) move with phase velocity and the transformation is $t' = t; z' = z - vt$:

$$-i\frac{2k_0}{v}\frac{\partial V}{\partial t'} = \Delta_\perp V - \frac{1}{v^2}\left(\frac{\partial^2 V}{\partial t'^2} - 2v\frac{\partial^2 V}{\partial t'\partial z'}\right). \tag{9}$$

Here, $\Delta_\perp = \frac{\partial^2}{\partial x^2} + \frac{\partial^2}{\partial y^2}$ denotes the transverse Laplace operator. The corresponding fundamental solution of AE equation (8) in Galilean coordinates is:

$$\hat{A}_G(k_x, k_y, \triangle k_z, t) = \hat{A}_G(k_x, k_y, \triangle k_z, t = 0) \times$$
$$\exp\left\{i\frac{v_{gr}}{\beta+1}\left[k_0 - (\beta+1)\triangle k_z \pm \sqrt{(k_0 - (\beta+1)\triangle k_z)^2 + (\beta+1)(k_x^2 + k_y^2 - \beta\triangle k_z^2)}\right] t\right\}, \tag{10}$$

while the fundamental solution of DE (9) becomes:

$$\hat{V}_G = \hat{V}_G(k_x, k_y, \triangle k_z, t = 0) \times$$
$$\exp\left\{iv\left[k_0 - \triangle k_z \pm \sqrt{(k_0 - \triangle k_z)^2 + k_x^2 + k_y^2}\right] t\right\}. \tag{11}$$

The analytical solution of (11) for initial pulse in the form of Gaussian bullet is the same as (5), but with new radial component $\hat{r} = \sqrt{x^2 + y^2 + (z + vt - ir_0^2 k_0)^2}$ translated in space and time. The numerical and analytical solutions of AE (1) and DE (2) are equal to the solutions of the equations AE (8) and DE (9) in Galilean coordinates with only one difference: in Laboratory frame the solutions translate in z-direction , while in Galilean frame the solutions stay in the centrum of the coordinate system.

The basic theoretical studies governed laser pulse propagation have been performed in so called "local time" coordinates $z = z; \tau = t - z/v_{gr}$. In order to compare our investigation with these results, we need to rewrite AE equation (1) for the amplitude function A in the same coordinate system. Thus Eq. (1) becomes:

$$-2ik_0\frac{\partial A}{\partial z} = \Delta_\perp A + \frac{\partial^2 A}{\partial z^2} - \frac{2}{v_{gr}}\frac{\partial^2 A}{\partial \tau\partial z} - \frac{\beta}{v_{gr}^2}\frac{\partial^2 A}{\partial \tau^2}. \tag{12}$$

Since this is a parabolic type equation with low order derivative on z, we apply Fourier transform to the amplitude function in form: $\hat{A}\left(k_x, k_y, \triangle\omega, z\right) = FFF\left[A\left(x, y, z, t\right)\right]$, where FFF denotes 3D Fourier transform in x, y, τ space and $\triangle\omega = \omega - \omega_0$; $\triangle k_z = \triangle\omega / v_{gr}$ are the spectral widths in frequency and wave vector domains correspondingly. The following ordinary differential equation in $\left(k_x, k_y, \triangle\omega, z\right)$ space is obtained:

$$-2i\left(k_0 - \frac{\triangle\omega}{v_{gr}}\right)\frac{\partial\hat{A}}{\partial z} = -\left(k_x^2 + k_y^2 - \frac{\beta\triangle\omega^2}{v_{gr}^2}\right)\hat{A} + \frac{\partial^2\hat{A}}{\partial z^2}. \tag{13}$$

As can be seen from (13), if the second derivative on z is neglected,then the paraxial spatio-temporal approximation is valid. Equation (13) is more general and we will estimate where we can apply spatio-temporal paraxial optics (PO), and where PO does not works. The fundamental solution of (13) is:

$$\hat{A}\left(k_x, k_y, \triangle\omega, z\right) = \hat{A}\left(k_x, k_y, \triangle\omega, 0\right) \times$$

$$\exp\left\{i\left[\left(k_0 - \frac{\triangle\omega}{v_{gr}}\right) \mp \sqrt{\left(k_0 - \frac{\triangle\omega}{v_{gr}}\right)^2 + k_x^2 + k_y^2 - \frac{\beta\triangle\omega^2}{v_{gr}^2}}\right]z\right\}. \tag{14}$$

The analysis of the fundamental solution (14) of the equation (13) is performed in two basic cases:
a: Narrow band pulses - from nanosecond up to $50 - 100$ femtosecond laser pulses, where the conditions:

$$\frac{\beta\triangle\omega^2}{v_{gr}^2} \le k_x^2 \sim k_y^2 << k_0^2;\ \triangle k_z = \frac{\triangle\omega}{v_{gr}} << k_0 \tag{15}$$

are satisfied, and the wave vector's difference $k_0 - \triangle\omega / v_{gr}$ can be replaced by k_0. Using the low order of the Taylor expansion and the minus sign in front of the square root from the initial conditions, equation (14) is transformed in a spatio - temporal paraxial generalization of the kind:

$$\hat{A}\left(k_x, k_y, \triangle\omega, z\right) = \hat{A}\left(k_x, k_y, \triangle\omega, 0\right)\exp\left[i\left(\frac{k_x^2 + k_y^2 - \frac{\beta\triangle\omega^2}{v_{gr}^2}}{2k_0}\right)z\right]. \tag{16}$$

From (16) the evolution of the narrow band pulses becomes obvious: while the transverse projection of the pulses enlarges by the Fresnel's law, the longitudinal temporal shape will be enlarged in the same away, proportionally to the dispersion parameter β. Such shaping of pulses with initially narrow band spectrum is demonstrated in Fig.1, where the typical Fresnel diffraction of the intensity profile (spot (x, y) projection) is presented. The numerical experiment is performed for 100 femtosecond Gaussian initial pulse at $\lambda = 800$ nm, $\Delta k_z << k_0$, $z_0 = 30\mu m$, $r_0(x, y) = 60\mu m$, with 37.5 cycles under envelope propagating in air ($\beta = 2.1 \times 10^{-5}$). The result is obtained by solving numerically the inverse Fourier transform of the fundamental solution (14) of the AE equation in the local time frame (12). The spot enlarges twice at one diffraction length $z_{diff} = r_0^2 k_0$. Fig. 2 presents the intensity side (x, τ) projection

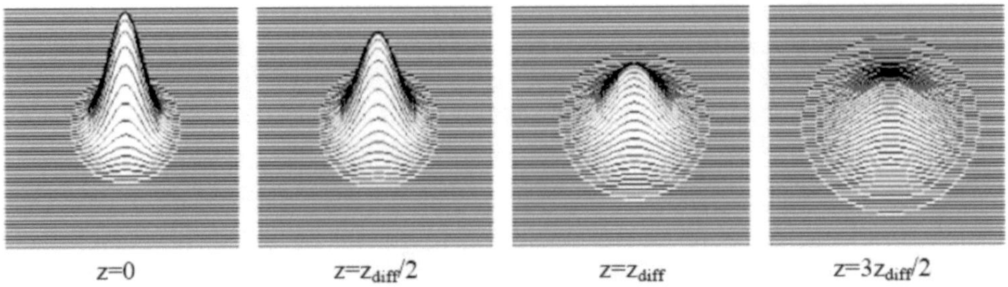

Fig. 1. Plot of the waist (intensity's) projection $|A(x,y)|^2$ of a 100 fs Gaussian pulse at $\lambda = 800\ nm$, with initial spot $r_0 = 60\ \mu m$, and longitudinal spatial pulse duration $z_0 = 30\ \mu m$, as solution of the linear equation in local time (12) on distances expressed by diffraction lengths. The spot deformation satisfies the Fresnel diffraction law and on one diffraction length $z = z_{diff}$ the diameter of the spot increases twice, while the maximum of the pulse decreases with the same factor.

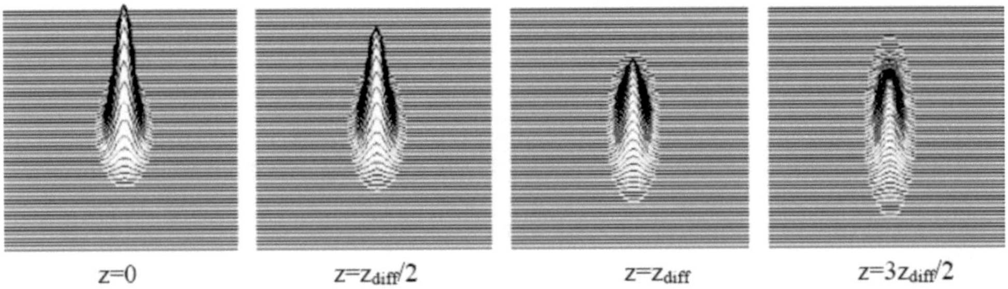

Fig. 2. Side (x, τ) projection of the intensity $|A(x,\tau)|^2$ for the same optical pulse as in Fig. 1. The (x, y) projection of the pulse diffracts considerably following the Fresnel law, while the (τ) projection on several diffraction lengths preserves its initial shape due to the small dispersion. The diffraction - dispersion picture, presented by the side (x, τ) projection, gives idea of what should happen in the nonlinear regime: the plane wave diffraction with a combination of parabolic type nonlinear Kerr focusing always leads to self-focusing for narrow-band $(\triangle k_z << k_0)$ pulses.

of the same pulse. We should note that while the spot ((x, y) projection) enlarges considerably due to the Fresnel law, the longitudinal time shape (the τ projection) remains the same on several diffraction lengths from the small dispersion in air. The diffraction - dispersion picture, presented by the side (x, τ) projection of the pulse, gives idea of what should happen in the nonlinear regime: the plane wave diffraction with a combination of parabolic type nonlinear Kerr focusing always leads to self-focusing for narrow-band $(\Delta k_z << k_0)$ pulses. The same Taylor expansion for narrow band pulses can be performed to fundamental solutions of the equation in Laboratory (3) and Galilean (10) frames.

b: broad band pulses - from attosecond up to $20 - 30$ femtosecond pulses, where the conditions:

$$\frac{\Delta\omega^2}{v_{gr}^2} \sim k_0^2 \propto k_x^2 \sim k_y^2 \tag{17}$$

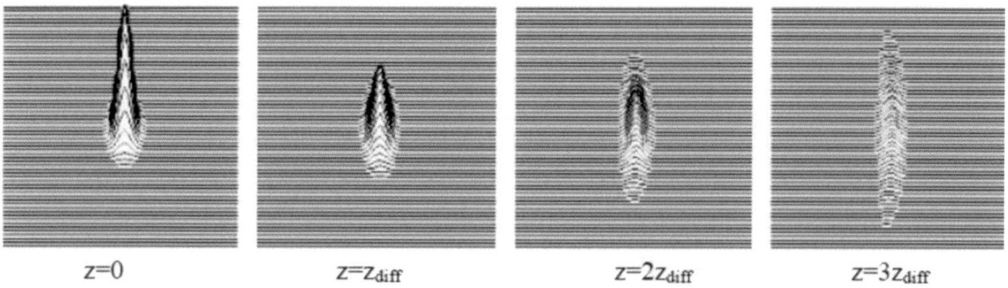

Fig. 3. Side (x,z) projection of the intensity $|A(x,z')|^2$ for a normalized 10 fs Gaussian initial pulse at $\lambda = 800$ nm, $\triangle k_z \simeq k_0/3$, $z_0 = r_0/2$, and only 3 cycles under the envelope (large-band pulse $\triangle k_z \approx k_0$), obtained numerically from the AE equation (8) in Galilean frame. At 3 diffraction lengths a divergent parabolic type diffraction is observed. In nonlinear regime a possibility appear: the divergent parabolic type diffraction for large-band pulses to be compensated by the converged parabolic type nonlinear Kerr focusing.

are satisfied. In this case we can not use Taylor expansion of the spectral kernels in Laboratory (3), Galilean (10) and local time (14) frames. The spectral kernels are in square root and we can expect evolution governed by wave diffraction. That why for broadband pulses we can expect curvature (parabolic deformation) of the intensity profile of the (x,z) or (x,τ) side projection. Fig 3. present the evolution of the intensity (side (x,z) projection) of a normalized 10 fs Gaussian initial pulse at $\lambda = 800$ nm; $\Delta k_z \simeq k_0/3$; $z_0 = r_0/2$; and only 3 cycles under the envelope (broadband pulse), obtained numerically from AE equation (8) in Galilean frame. The solution confirms the experimentally observed parabolic type diffraction for few cycle pulses. And here appears the main physical question for stable pulse propagation in nonlinear regime: Is it possible for the divergent parabolic intensity distribution due to non-paraxial diffraction to be compensated by the converged parabolic type nonlinear Kerr focusing? If this is the case, then a stable soliton pulse propagation exists. As we show below, only for broadband pulses one-directional soliton solution of the corresponding nonlinear equations can be found.

3. Self-focusing of narrow band femtosecond pulses. Conical emission and spectral broadening

The laser pulses in a media acquire additional carrier -to envelope phase (CEP), connected with the group-phase velocity difference. In air the dispersion is a second order phase effect with respect to the CEP. In linear regime the envelope equations contain Galilean invariance, and thus CEP does not influence the pulse evolution. Taking into account the CEP in the expression for the nonlinear polarization of third order, a new frequency conversion in THz and GHz region takes place. In Laboratory frame, the nonlinear polarization of third order for a laser beam or optical pulse, without considering CEP, can be written as follows:

$$n_2 E^3(x,y,z,t)\vec{x} = \vec{x} n_2 \exp\left[i(k_0(z - v_{ph}t)\right] \times \left\{\frac{3}{4}|A|^2 A + \frac{1}{4}\exp\left[2i(k_0(z - v_{ph}t)\right] A^3\right\} + \vec{x} c.c., \tag{18}$$

while in Galilean coordinates ($z' = z - v_{gr}t; t' = t$) the CEP, being an absolute phase (21), is present in the phase of the Third Harmonic (TH) term

$$n_2 E^3 (x, y, z, t) \vec{x} = \vec{x} n_2 \exp\left[i\left(k_0(z' - (v_{ph} - v_{gr})t')\right)\right] \times$$
$$\left\{\frac{3}{4}|A|^2 A + \frac{1}{4}\exp\left[2i\left(k_0(z' - (v_{ph} - v_{gr})t')\right)\right] A^3\right\} + \vec{x}c.c.. \quad (19)$$

Note that we transform the TH term to a frequency shift of $\omega_{nl} = 3k_0(v_{ph} - v_{gr}) \cong 93GHz$ in air of the carrying wave number $\lambda_0 = 800nm$. The nonlinear amplitude equations for power near the critical one for self-focusing in Laboratory and Galilean frame are:

$$-2ik_0\left(\frac{\partial A}{\partial z} + \frac{1}{v_{gr}}\frac{\partial A}{\partial t}\right) = \Delta A - \frac{1+\beta}{v_{gr}^2}\frac{\partial^2 A}{\partial t^2} +$$
$$n_2 k_0^2 \left\{\frac{3}{4}|A|^2 A + \frac{1}{4}\exp\left[2i(k_0(z - v_{ph}t)\right] A^3\right\} + c.c., \quad (20)$$

and

$$-i\frac{2k_0}{v_{gr}}\frac{\partial V}{\partial t'} = \Delta_\perp V - \frac{1+\beta}{v_{gr}^2}\left(\frac{\partial^2 V}{\partial t'^2} - 2v_{gr}\frac{\partial^2 V}{\partial t' \partial z'}\right) +$$
$$n_2 k_0^2 \left\{\frac{3}{4}|V|^2 V + \frac{1}{4}\exp\left[2i\left(k_0(z' - (v_{ph} - v_{gr})t')\right)\right] V^3\right\} + c.c., \quad (21)$$

respectively. We use AE equations (20) and (21) to simulate the propagation of a fs pulse, typical for laboratory-scale experiments: initial power $P = 2P_{kr}$, center wavelength $\lambda = 800$ nm, initial time duration $t_0 = 400\ fs$, corresponding to spatial pulse duration $z_0 = v_{gr}t_0 \cong 120$ μm, and waist $r_0 = 120\ \mu m$.
Fig.4 presents the evolution of the spot $|A(x, y|^2$ of the initial Gaussian laser pulse at distances $z = 0, z = 1/2z_{diff}, z = z_{diff}, z = 3/2z_{diff}$. As a result, we obtain the typical self-focal zone (core) with colored ring around, observed in several experiments (12–14). The $3D + 1$ nonlinear AE equation (21) gives an additional possibility for investigating the evolution of the side projection of the intensity $|A(x, z'|^2$ profile. The side projection $|A(x, z'|^2$ of the same pulse is presented in Fig.5. The initial Gaussian pulse begins to self-compress at about one diffraction length and it is split in a sequence of several maxima with decreasing amplitude. Fig. 6 presents the evolution of the Fourier spectrum of the side projection $|A(k_x, k_{z'}|^2$. At one diffraction length the pulse enlarges asymmetrically towards the short wavelengths (high wave-numbers). It is important to point here, that similar numerical results for narrow band pulses are obtained when only the self-action term in AE equation (21) is taken into account. The TH or THz term (the second nonlinear term in the brackets) practically does not influence the intensity picture during propagation. In conclusion of this paragraph, we should point

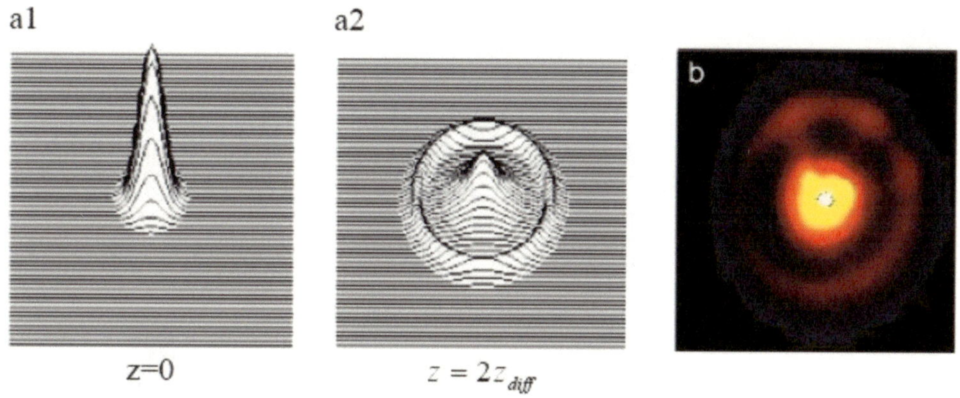

Fig. 4. Nonlinear evolution of the waist (intensity) projection $|A(x,y)|^2$ of a 400 fs initial Gaussian pulse (a1) at $\lambda = 800\ nm$, with spot $r_0 = 120\ \mu m$, and longitudinal spatial pulse duration $z_0 = v_{gr}t_0 \cong 120\ \mu m$ at a distance $z = 2z_{diff}$ (a2), obtained by numerical simulation of the 3D+1 nonlinear AE equation (20). The power is above the critical for self-focusing $P = 2P_{kr}$. Typical self-focal zone (core) surrounded by Newton's ring is obtained. (b) Comparison with the experimental result presented in (12).

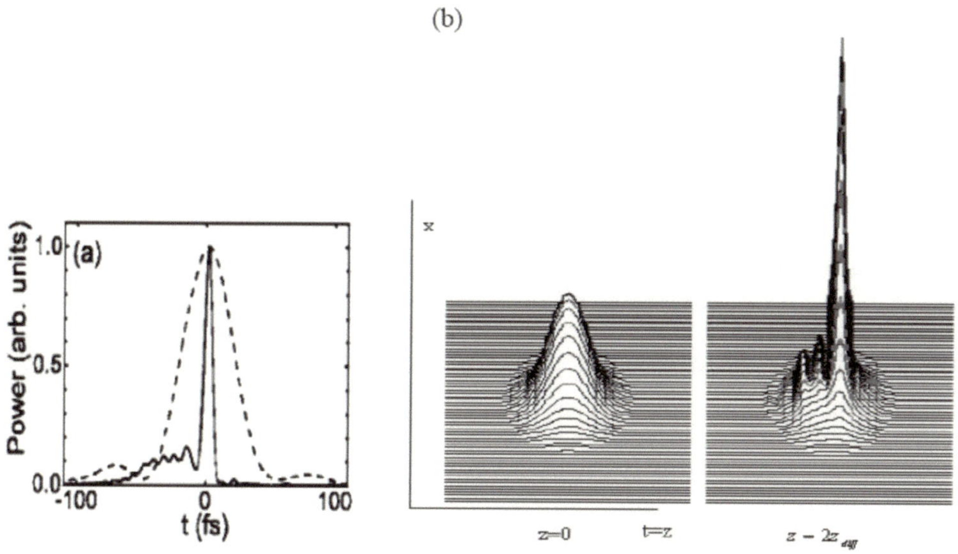

Fig. 5. (a) Experimental result of pulse self compression and spliting of the initial pulse to a sequence of several decreasing maxima (30). (b) Numerical simulation of the evolution of (x,t=z) projection $|A(x,z')|^2$ of the same pulse of Fig. 4 at distances $z = 0, z = z_{diff}$, governed by the (3D+1) nonlinear AE equation (20) and the ionization-free model.

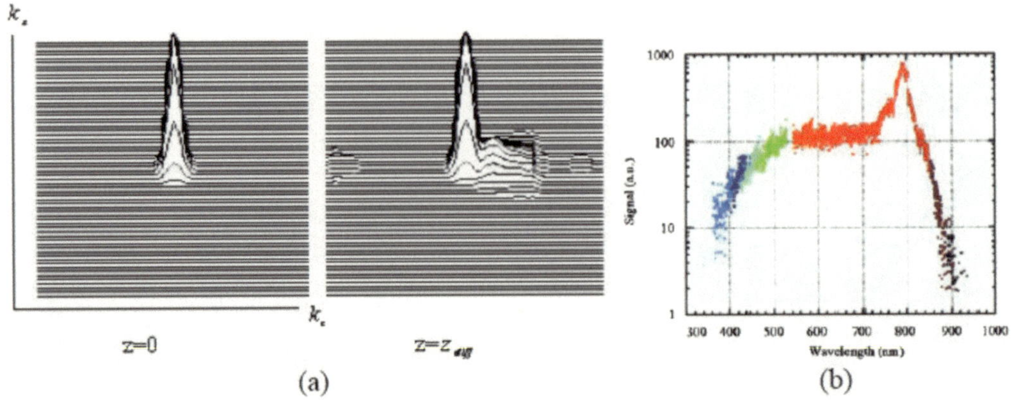

Fig. 6. (a) Fourier spectrum of the same side (x, z) projection of the intensity $|A(k_x, k_z)|^2$ as in Fig.5. At one diffraction length the pulse enlarges asymmetrically forwards the short wavelengths (high k_z wave-numbers),(b) a spectral form observed also in the experiments (12).

out that our non-paraxial ionization-free model (20) and (21) is in good agreement with the experiments on spatial and spectral transformations of a fs pulse in a regime near the critical $P \geq P_{cr}$. Such transformation of the shape and spectrum of the fs pulse is typical in the near zone, up to several diffraction lengths, where the conditions for narrow-band pulse are satisfied $\triangle k_z << k_0$.

4. Carrier-to-envelope phase and nonlinear polarization. Drift from THz to GHz generation

In nonlinear regime the spectrum of the amplitude function becomes large due to different nonlinear mechanisms. The Fourier expression $\hat{A}\left[k_x, k_y, k_0 - k_z, \omega_0 - \omega\right]$ is a function of arbitrary $\Delta k_z = k_0 - k_z$ and $\Delta\omega = \omega_0 - \omega$, which are related to the group velocity $\Delta\omega/\Delta k_z = v_{gr}$ (here, we do not include the nonlinear addition to the group velocity - it is too small for power near the critical one). Let Δk_z denote an arbitrary initial spectral width of the pulse. In the nonlinear regime $\Delta k_z(z)$ enlarges considerably and approaches values $\Delta k_z(z) \simeq k_0$. To see the difference between the evolution of narrow-band $\Delta k_z << k_0$ and broadband $\Delta k_z \simeq k_0$ pulses, it is convenient to rewrite the amplitude function in Laboratory coordinates (the dispersion number $\beta \simeq 2.1 \times 10^{-5}$, being smaller than the diffraction in air, is neglected):

$$A(x, y, z, t) = B_0 B(x, y, z, t) \exp\left(-i(\triangle k_z(z - v_{gr} t)\right), \tag{22}$$

while in Galilean coordinates it is equal to:

$$V\left(x, y, z', t'\right) = B_0 G\left(x, y, z', t'\right) \exp\left(-i\triangle k_z z'\right). \tag{23}$$

The Nonlinear Diffraction Equation (NDE) (20) in Laboratory frame becomes:

$$
\begin{aligned}
& -2i(k_0-\triangle k_z)\left(\frac{\partial B}{\partial z}+\frac{1}{v_{gr}}\frac{\partial B}{\partial t}\right)=\Delta B-\frac{1}{v_{gr}^2}\frac{\partial^2 B}{\partial t^2}+ \\
& n_2k_0^2B_0^2\left\{\frac{3}{4}|B|^2B+\frac{1}{4}\exp\left[2i\left((k_0-\triangle k_z)z-(k_0v_{ph}-v_{gr}\triangle k_z)t\right)\right]B^3\right\}+c.c.,
\end{aligned} \tag{24}
$$

and in Galilean frame, the equation (21) is:

$$
\begin{aligned}
& -i\frac{2(k_0-\triangle k_z)}{v_{gr}}\frac{\partial G}{\partial t'}=\Delta_\perp G-\frac{1}{v_{gr}^2}\left(\frac{\partial^2 G}{\partial t'^2}-2v_{gr}\frac{\partial^2 G}{\partial t'\partial z'}\right)+ \\
& n_2k_0^2B_0^2\left\{\frac{3}{4}|G|^2G+\frac{1}{4}\exp\left[2i\left((k_0-\triangle k_z)z'-k_0(v_{ph}-v_{gr})t'\right)\right]G^3\right\}+c.c.,
\end{aligned} \tag{25}
$$

where $\triangle k_z$ can get arbitrary values. It can be seen that the nonlinear phases in both coordinate systems are equal after the transformation $z'=z-v_{gr}t;\ t'=t$:

$$
(k_0-\triangle k_z)z-(k_0v_{ph}-\triangle k_zv_{gr})t=(k_0-\triangle k_z)z'-k_0(v_{ph}-v_{gr})t'. \tag{26}
$$

On the other hand, the corresponding frequency conversions are different. In Laboratory frame the frequency conversion depends on the spectral width $\triangle k_z$:

$$
\omega_{nl}^{Lab}=k_0v_{ph}-\triangle k_zv_{gr}, \tag{27}
$$

while in Galilean frame the nonlinear frequency conversion is fixed to the offset frequency

$$
\omega_{nl}^{Gal}=k_0(v_{ph}-v_{gr})=31GHz;\ (\lambda=800nm) \tag{28}
$$

in air. The expression of the nonlinear frequency shift in Laboratory frame (27) explains the different frequency arising from pulses with different initial spectral width. When the laser is in ns or ps regime, $\triangle k_z << k_0$ and the nonlinear frequency shift is equal to the third harmonic $3\omega_{nl}^{Lab}=3\omega_0$. In this case (spectral width of the pulse much smaller than the spectral distance to the third harmonic), the phase matching conditions can not be met. Thus, the nonlinear polarization is transformed into a self-action term. The fs pulses on the other hand have initial spectral width of the order $\triangle\omega^{fs}\simeq 10^{13-14}Hz$ and for such pulses at short distances in nonlinear regime the condition $\triangle\omega^{fs}\simeq\omega_{nl}^{Lab}$ can be satisfied. Thus, the nonlinear frequency shift lies within the spectral width of a fs pulse, and from (27) follows the condition for THz and not for TH generation. The self-action enlarges the spectrum up to values $\triangle k_z\simeq k_0$ and thus, following (27), the nonlinear frequency conversion in far field zone drifts from THz to $\sim$ 93 GHz (18). Note that we consider a single pulse propagation, while the laser system generates a sequences of fs pulses. The different pulses have different nonlinear spectral widths when moving from the source to the far field zone. One would detect in an experiment a mix of frequencies from THz up to GHz.

5. Nonlinear sub-cycle regime for $\triangle k_z \approx k_0$

The separation of the nonlinear polarization to self-action and TH, THz or GHz generated terms is appropriate for fs pulses up to several cycles under envelope. For fs narrow-band pulses, as mentioned in the previous section, the pulse shape is changed by the self-action term, while the CEP frequency depending at the spectral width of the pulse $\triangle k_z$ leads to different type of frequency conversion drifts from THz to GHz region. However, when supper-broad spectrum occurs ($\triangle k_z \approx k_0$), the time width of the pulse $\triangle t$ becomes smaller than the period of the nonlinear oscillation ω_{nl}^{Lab}. In this nonlinear sub-cycle regime, the nonlinear term starts to oscillate with ω_{nl}^{Lab} and separation of the self-action and the frequency conversion terms becomes mathematically incorrect, due to the mixing of frequencies (32; 33). For the first time such possibility was discussed in (32), where a correct expression of the nonlinear polarization, including Raman response is presented. In the sub-cycle regime the nonlinear polarization at a fixed frequency and Laboratory frame becomes:

$$n_2 E^3(x,y,z,t) = n_2 \exp\left[i\left((k_0-\triangle k_z)z - (k_0 v_{ph} - \triangle k_z v_{gr})t\right)\right] \times \left\{\exp\left[2i\left((k_0-\triangle k_z)z - (k_0 v_{ph} - \triangle k_z v_{gr})t\right)\right] B^3\right\}, \tag{29}$$

and in Galilean frame it is

$$n_2 E^3(x,y,z',t') = n_2 \exp\left[i\left((k_0-\triangle k_z)z' - k_0(v_{ph} - v_{gr})t'\right)\right] \times \left\{\exp\left[2i\left((k_0-\triangle k_z)z' - k_0(v_{ph} - v_{gr})t'\right)\right] B^3\right\}. \tag{30}$$

In spite of the super-broad spectrum, the dispersion parameter in the transparency region from 400 nm up to 800 nm continues to be small, in the range of $\beta \approx 10^{-4} - 10^{-5}$. The nonlinear amplitude equations for pulses with super-broad spectrum in Laboratory system become:

$$-2i(k_0-\triangle k_z)\left(\frac{\partial A}{\partial z} + \frac{1}{v_{gr}}\frac{\partial A}{\partial t}\right) = \Delta A - \frac{1}{v_{gr}^2}\frac{\partial^2 A}{\partial t^2} + n_2 k_0^2 \exp\left[2i\left((k_0-\triangle k_z)z - (k_0 v_{ph} - \triangle k_z v_{gr})t\right)\right] A^3, \tag{31}$$

and in Galilean frame

$$-i\frac{(k_0-\triangle k_z)}{v_{gr}}\frac{\partial V}{\partial t'} = \Delta_\perp V - \frac{1}{v_{gr}^2}\left(\frac{\partial^2 V}{\partial t'^2} - 2v_{gr}\frac{\partial^2 V}{\partial t' \partial z'}\right) + n_2 k_0^2 \exp\left[2i\left((k_0-\triangle k_z)z' - k_0(v_{ph} - v_{gr})t'\right)\right] V^3. \tag{32}$$

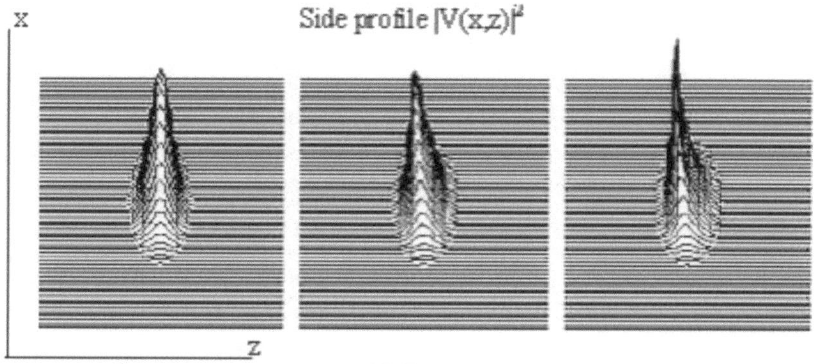

Fig. 7. Numerical simulations for an initial Gaussian pulse with super-broad spectrum $\triangle k_z \approx k_0$ governed by the nonlinear equation (32). The power is slightly above the critical $P = 2P_{kr}$. The side projection $|V(x,z')|^2$ of the intensity is plotted. Instead of splitting into a series of several maxima, the pulse transforms its shape into a Lorentzian of the kind $V(x,y,z') \simeq 1/[1+x^2+y^2+(z'+ia)^2+a^2]$.

Fig. 7 shows a typical numerical solution of the nonparaxial nonlinear equation (31) (or (32)) for an initial Gaussian pulse with super-broad spectrum $\triangle k_z \approx k_0$. It is obtained by using the split step method (4 step Runge-Kutta method for the nonlinear part). These results are the same both in Laboratory and Galilean coordinate frames differing only by a translation. The side projection $|V(x,z'|^2$ of the intensity profile is plotted for different propagation distances. Instead of splitting into to a series of several maxima, the pulse transforms its shape in a Lorentzian type form of the kind $V(x,y,z) \simeq 1/[1+x^2+y^2+(z'+ia)^2+a^2]$. Here, the number a accounts for compression in z' direction and a spatial angular distribution. Fig. 8 presents the evolution of the spectrum $|V(k_x,k_{z'}|^2$ of the side intensity projection for the same pulse. The spectrum enlarges forwards the small k_z wave-numbers (long wavelengths) - typical for Lorentzian type profiles. To compare with Fig. 8, Fig. 9 gives a plot of the side projection $|V(k_x,k_z'|^2$ of the spectrum of a Lorentzian profile $V(x,y,z') = 1/[1+x^2+y^2+(z'+ia)^2+a^2]$, $a = 2$ increases toward the small wave-numbers. The numerical experiments lead to the conclusion that a possible shape of the stable $3D+1$ soliton can be in the form of a Lorentzian profile. Thus, if we take as an initial condition Lorentzian, instead Gaussian one, a relative stability in the shape and spectrum can be expected. Fig. 10 shows the evolution of the $|V(x,z'|^2$ profile of a pulse with initial Lorentzian shape $V(x,y,z',t=0) = 1/[1+x^2+y^2+(z'+ia)^2+a^2]$, $a = 2$. The pulse propagates at distance of one diffraction length, preserving its initial shape.

6. Spectrally asymmetric 3D+1 soliton solution

The numerical simulations in the previous section for broad band spectrum pulses demonstrate a stable soliton propagation with a specific initial Lorentzian shape. To find an exact soliton solution, we require that $\triangle k_z = k_0$ and $\triangle\omega \cong \omega_0$ be satisfied. In air $\beta \cong 0$ and the amplitude equation (31) can be rewritten as:

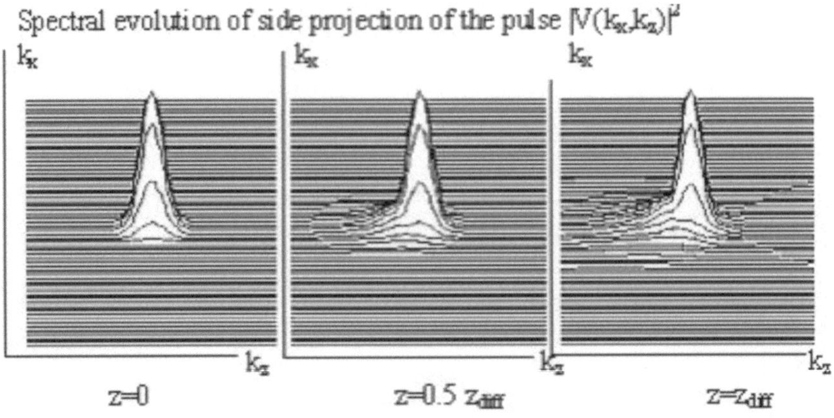

Fig. 8. The evolution of the spectrum $|V(k_x, k_z)|^2$ of the same side intensity projection $|V(x, z')|^2$. The spectrum enlarges towards small k_z wave-numbers (long wavelengths) - typical for Lorentzian profiles.

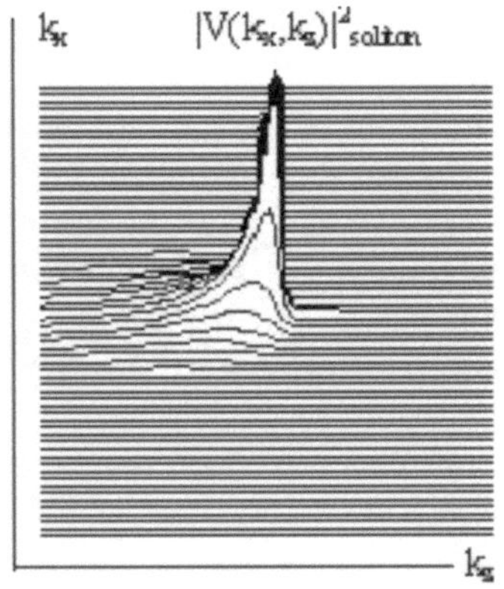

Fig. 9. Plot of the side projection $|V(k_x, k_z|^2$ of the spectrum of a Lorentzian profile $V(x, y, z) = 1/[1 + x^2 + y^2 + (z + ia)^2 + a^2]$, $a = 2$ increasing towards the small k_z wave-numbers (compare with Fig. 8).

$$\Delta B - \frac{1}{v_{gr}^2}\frac{\partial^2 B}{\partial t^2} + k_0^2 n_2 B_0^2 \exp\left[i\left(2\triangle\omega_{nl}t\right)\right]B^3 = 0. \tag{33}$$

To minimize the influence of the GHz oscillation ω_{nl}, we use an amplitude function with a phase opposite to CEP:

$$B(x, y, x, t) = C(x, y, z, t)\exp(-i\triangle\omega_{nl}t). \tag{34}$$

This corresponds to an oscillation of our soliton solution with frequency $\omega_{nl} \simeq 31$ GHz. The equation (33) becomes:

$$\Delta C - \frac{1}{v_{gr}^2}\frac{\partial^2 C}{\partial t^2} + k_0^2 n_2 B_0^2 C^3 = 2i\frac{\triangle\omega_{nl}}{v_{gr}^2}\frac{\partial C}{\partial t} - \frac{\triangle\omega_{nl}^2}{v_{gr}^2}C \tag{35}$$

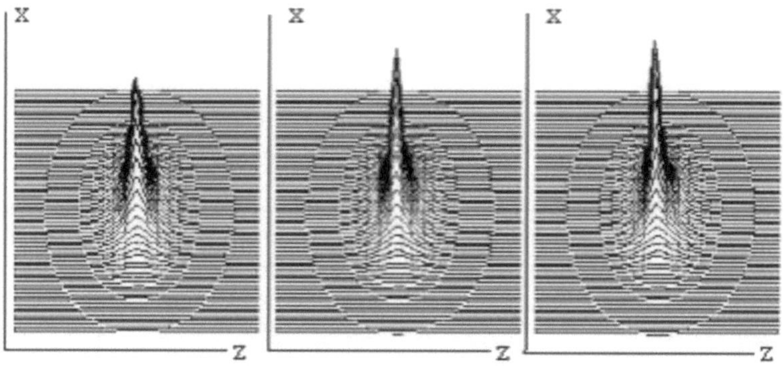

Fig. 10. Evolution of the $|V(x,z')|^2$ profile of a pulse with super-broad spectrum $\triangle k_z \approx k_0$ and initial Lorentzian shape $V(x,y,z',t=0) = 1/([1+x^2+y^2+(z'+ia)^2+a^2]$, $a=2$, governed by the nonlinear equation (32). The pulse propagates over one diffraction length with relatively stable form.

To estimate the influence of the different terms on the propagation dynamics we rewrite equation (35) in dimensionless form. Substituting:

$$t = t_0 t;\ z = z_0 z;\ x = r_0 x;\ y = r_0 y; \tag{36}$$

$$r_0/z_0 = \delta \sim 1;\ z_0 = v_{gr} t_0;\ t_0 \cong 2 \times 10^{-13} - 10^{-14} sec, \tag{37}$$

we obtain the following normalized equation:

$$\Delta C - \frac{\partial^2 C}{\partial t^2} + \gamma C^3 = i\alpha \frac{\partial C}{\partial t} - \beta C, \tag{38}$$

where $\gamma = r_0^2 k_0^2 n_2 B_0^2$ is the nonlinear constant, $\alpha = 2\triangle\omega_{nl} r_0^2 / v_{gr}^2 t_0$ and $\beta = \triangle\omega_{nl}^2 r_0^2 / v_{gr}^2$. For typical fs laser pulse at carrier wavenumber 800 nm with spot $r_0 = 100\ \mu m$, the constants of both terms in the r.h.s of equation (38) are very small ($\alpha \sim 10^{-2}$ and $\beta \sim 10^{-4}$) and can be neglected. Thus, equation (38) becomes:

$$\Delta C - \frac{\partial^2 C}{\partial t^2} + \gamma C^3 = 0. \tag{39}$$

Furthermore, we shall assume that the new envelope wave equation (39) has solutions in the form:

$$C\,(x,y,z,t) = C(\tilde{r}), \tag{40}$$

where $\tilde{r} = \sqrt{x^2+y^2+(z+ia)^2-(t+ia)^2}$. From the nonlinear wave equation (39), using (40), the following ordinary nonlinear equation is obtained:

$$\frac{3}{\tilde{r}}\frac{\partial C}{\partial \tilde{r}} + \frac{\partial^2 C}{\partial \tilde{r}^2} + \gamma C^3 = 0. \tag{41}$$

The number a counts for the longitudinal compression and the phase modulation of the pulse. When the nonlinear coefficient is slightly above the critical and reaches the value $\gamma = 2$, equation (41) has exact particle-like solution of the form:

$$C = \frac{sech(ln(\tilde{r}))}{\tilde{r}}. \tag{42}$$

Using the fact that $exp(ln(\tilde{r})) = \tilde{r}$ and $exp(-(ln\tilde{r})) = \frac{1}{\tilde{r}}$, the solution (42) is simplified to the following algebraic soliton:

$$C(\tilde{r}) = \frac{2}{1 + x^2 + y^2 + (z + ia)^2 - (t + ia)^2} \tag{43}$$

The solution (43) gives the time evolution of our Lorentz initial form, investigated in the previous section. As be seen from equation (39), the solution appears as a balance between the parabolic (not paraxial) wave type diffraction of broad band pulse $\triangle k_z = k_0$ and the nonlinearity of third order. The maxima of this solution are at the points where $\tilde{r}^2 = 0$. If we turn back to standard, not normalized coordinates, and solve the second order equation $z^2 + 2iaz - 2iav_{gr}t - v_{gr}^2 t^2 = 0$, only one real solution $z = v_{gr}t$ can be obtained. It corresponds to one-directional propagation with position of the maximum on the z - coordinate $z = v_{gr}t$. As it was pointed above, Fig. 8 presents the initial k_x, k_z spectrum of the soliton (43). While the k_x, k_y spectrum is symmetric, the k_z projection is fully asymmetric, enlarging forwards to low k_z wave-numbers (long wavelengths), and has typical Lorentz shape. Recently, in experiments with $2 - 3$ cycle pulses long range filaments with similar spectral profile (34) are observed. We suppose that in this experiment a $3D + 1$ Lorentz type soliton was found experimentally for the first time.

7. Conclusions

In this paper we investigate femtosecond pulse propagation in air, governed by the AE equation, in linear and nonlinear regime. The equation allows to solve the problem of propagation of pulses with super-broad spectrum. Note that this problem can not be studied in paraxial optics. In linear regime the fundamental solutions of AE (1) and DE (2) are obtained and different regimes of diffraction are analyzed. The typical fs pulses up to 50 fs diffract by the Fresnel law, in a plane orthogonal to the direction of propagation, while their longitudinal shape is preserved in air or is enlarged a little, due to the dispersion. Broad-band pulses (only a few cycles under envelope) at several diffraction lengths diffract in a parabolic form. We solve the convolution problem of the diffraction equation DE (2) for an initial pulse in the form of a Gaussian bullet, and obtain an exact analytical solution (5). A new method for solving evolution problems of the wave equation is also suggested. We investigate precisely the nonlinear third order polarization, including the CEP into account. This additional phase transforms TH term to THz or GHz terms, depending on the spectral width of the pulse. Thus, we suggest a new mechanism of THz and GHz generation from fs pulses in nonlinear regime. For pulses with power a little above the critical for self-focusing, we investigate two basic cases: pulses with narrow-band spectrum and with broad-band spectrum. The numerical simulation of the evolution of narrow-band pulses (standard 100 fs pulses), gives a typical conical emission and a spectral enlargement to the short wavelengths. Our study of broad-band pulses leads to the conclusion that their propagation is governed by the nonlinear wave equation with third order nonlinear term (39), when the THz oscillation is neglected as small term. An exact soliton solution of equation (39), with $3D + 1$ Lorentz shape is also obtained. The soliton appears as a balance between parabolic divergent type diffraction and parabolic convergent type of nonlinear self-focusing. Numerically, we demonstrate a relative stability of the soliton pulse with respect to the THz oscillations.

8. Acknowledgements

This work is partially supported by the Bulgarian Science Foundation under grant DO-02-0114/2008.

9. References

[1] A. Braun, G. Korn, X. Liu, D. Du, J. Squier, and G. Mourou, "Self-channeling of high-peak-power femtosecond laser pulses in air", Opt. Lett. , 20(1), 73-75 (1995).
[2] L. Wöste, C. Wedekind, H. Wille, P. Rairoux, B. Stein, S. Nikolov, C. Werner, S.Nierdermeier, F. Ronneberger, H. Schillinger, and R. Sauerbrey, "Femtosecond atmospheric lamp", AT-Fachverlag, Stuttgard, Laser and Optoelectronik 29, 51-53 (1997).
[3] G. Méchain, C. D'Amico, Y.-B. André, S. Tzortzakis, M. Franco, B. Prade, A. Mysyrowicz, A. Couairon, E. Salmon, R. Sauerbrey, "Length of plasma filaments created in air by a multiterawatt femtosecond laser", Opt. Commun. , 247, 171-108 (2005).
[4] S. Tzortzakis, G. Méchain, G. Patalano, Y.-B. André, B. Prade, M. Franco, A. Mysyrowicz, J. M. Munier, M. Gheudin, G. Beaudin, and P. Encrenaz, "Coherent subterahertz radiation from femtosecond infrared filaments in air", Opt. Lett. , 1944-1946, (2002).
[5] C. D'Amico, A. Houard, M. Franco, B. Prade, A. Mysyrowicz, "Coherent and incoherent THz radiation emission from femtosecond filaments in air", Optics Express, 15, 15274-15279 (2007).
[6] C. D'Amico, A. Houard, S. Akturk, Y. Liu, J. Le Bloas, M. Franco, B. Prade, A. Couairon, V. T. Tikhonchuk, and A. Mysyrowicz, "Forward THz radiation emission by femtosecond filamentation in gases: theory and experiment", New J. of Phys., 10, 013015 (2008).
[7] C. P. Hauri, W. Kornelis, F. W. Helbing, A. Couairon, A. Mysyrowicz, J. Biegert, U. Keller, "Generation of intense, carrier-envelope phase locked few-cycle laser pulses through filamentation", Appl. Phys. B, 79, 673-677 (2004).
[8] C. P. Hauri, A. Guandalini, P. Eckle, W. Kornelis, J. Biegert, U. Keller. "Generation of intense few cycle laser pulses through filamentation - parameter dependence", Optics Express, 13, 7541 (2005).
[9] A. Couairon, J. Biegert, C. P. Hauri, W. Kornelis, F. W. Helbing, U. Keller, A. Mysyrowicz, "Self-compression of ultrashort laser pulses down to one optical cycle by filamentation", J. Mod. Opt., 53, 75-85 (2006).
[10] S.L. Chin, A. Brodeur, S. Petit, O. G. Kosareva, V. P. Kandidov, " Filamentation and supercontituum generation during the propagation of powerful ultrashort laser pulses in optical media (white light laser)", J. Nonlinear Opt. Phys. Mater., 8, 121-146 (1998).
[11] J. Kasparian, R. Sauerbrey, D. Mondelain, S. Niedermeier, J. Yu, Y. P. Wolf, Y.-B. André, M. Franco, B. S. Prade, S. Tzortzakis, A. Mysyrowicz, H. Wille, M. Rodriguez, L. Wöste, "Infrared extenstion of the supercontinuum generated by femtosecond terrawattlaser pulses propagating in the atmosphere", Opt. Lett. , 25, 1397-1399 (2000).
[12] A. Couairon, and A. Mysyrowicz, "Femtosecond filamentation in transparent media", Physics Reports, 441, 47-189 (2007).
[13] S. L. Chin, S. A. Hosseini, W. Liu, Q. Luo, F. Théberge, N. Aközbek, A. Becker, V. P. Kandidov, O. G. Kosareva, and H. Schoeder, "The propagation of powerful femtosecond laser pulses in optical media: physics, applications, and new challenges", Can. J. Phys. 83, 863-905 (2005).

[14] Daniele Faccio, Alessandro Averhi, Antonio Lotti, Paolo Di Trapani, Arnaud Couairon, Dimitris Papazoglou, Stelios Tzortzakis, "Ultrashort laser pulse filamentation from spontaneous X Wave formation in air", Optics Express, 16 1565-1569 (2008)
[15] M. Kolesik and J. V. Moloney, "Perturbative and non-perturbative aspects of optical filamentation in bulk dielectric media.", Optics Express, 16, 2971-2986 (2008).
[16] Y. R. Shen, *The Principles of Nonlinear Optics*, Wiley-Interscience, New York, 1984.
[17] P. Béjot, J. Kasparian, S. Henin, V. Loriot, T. Viellard, E. Hertz, O. Faucher, B. Lavorel, and J.-P. Wolf, "Higher-Order Kerr Terms Allow Ionization-Free Filamentation in Gases", Phys. Rev. Lett., 104, 103903 (2010).
[18] G. Méchain, A. Couairon, Y.-B. André, C. D'Amico, M. Franco, B. Prade, S. Tzortzakis, A. Mysyrowicz, R. Sauerbrey, "Long-range self-channeling of infrared laser pulses in air: a new regime without ionization", Appl. Phys B, 79, 379-382 (2004).
[19] A. Dubietis, E. Gaižauskas, G. Tamošauskas, P. Di Trapani, "Light filaments without self-channeling", Phys. Rev. Lett 92, 253903 (2004).
[20] Todd A. Pitts, Ting S. Luk, James K. Gruetzner, Thomas R. Nelson, Armon McPherson, Stewart M. Cameron and Aaron C. Bernstein, "Propagation of self-focused laser pulse in atmosphere: experiment versus numerical simulation", J. Opt. Soc. Am. B , 21, 2006-2016 (2004).
[21] Martin Wegener, *Extreme Nonlinear Optics*, (Springer-Verlag, Berlin Heidelberg, 2005).
[22] A. Hasegawa, *Optical Solitons in Fibers*, (Springer, Berlin,1989).
[23] G. P. Agrawal, *Nonlinear Fiber Optics*, (Academic, San Diego, 2001).
[24] E. M. Dianov, P.V. Mamyshev, A. M. Prokhorov, and V. N. Serkin, *Nonlinear Effects in Fibers*, (Harwood Academic, NewYork, 1989).
[25] V. N. Serkin, A. Hasegawa, and T. L. Belyaeva, "Nonautonomous Solitons in External Potentials", Phys. Rev. Lett. 98, 074102 (2007).
[26] Lubomir M. Kovachev, Kamen Kovachev, "Diffraction of femtosecond pulses: nonparaxial regime",J. Opt. Soc. Am. A , 25, 2232-2243 (2008); "Erratum " , 25, 3097-3098 (2008).
[27] I. P. Christov, "Propagation of femtosecond light pulses", Opt. Comm., 53, 364-366 (1985).
[28] T. Brabec, F. Krausz, "Nonlinear Optical Pulse Propagation in the Single-Cycle Regime", Phys. Rev. Lett. 78, 3282-3285 (1997).
[29] A. P. Kiselev, "Localized Light Waves: Paraxial and Exact Solutions of the Wave Equation", Optics and Spectroscopy, 102, 603-622 (2007).
[30] Stefan Skupin, Gero Stibenz, Luc Berge, Falk Lederer, Thomas Sokollik, Matthias Schnurer, Nickolai Zhavoronkov, and Gunter Steinmeyer ,"Self-compression by femtosecond pulse filamentation: Experiments versus numerical simulations" Phys. Rev. E 74, 056604 (2006).
[31] J. Kasparian, R. Sauerbrey, and S. L. Chin, "The critical laser intensity of self-guided light filaments in air", Appl. Phys. B 71, 877-879 (2000).
[32] M. Kolesik, E. M. Wright, A. Becker, and J. V. Moloney, "Simulation of third-harmonic and supercontinuum generation for femtosecond pulses in air", Appl. Phys. B, 85, 531-538 (2006).
[33] L. M. Kovachev, "New mechanism for THz oscillation of the nonlinear refractive index in air: particle-like solutions", J. Mod. Opt., 56, 1797 - 1803 (2009).
[34] E. Schulz and M. Kovacev, private communication.

12

Dispersion of a Laser Pulse at Propagation Through an Image Acquisition System

Toadere Florin
INCDTIM Cluj Napoca
Romania

1. Introduction

The purpose of this chapter is to analyze different lenses and fibers, in order to find the best solution for the compensation of the laser pulse dispersion. We generate a laser pulse which is captured by an image acquisition system. The system consists of a laser, an optical fiber and a CMOS senor. In figure 1, we use a confocal resonator to generate the laser pulse; then the generated light is focused into an optical fiber using a lens; the light is propagated through the fiber and at the output of the fiber the light is projected on a CMOS sensor. For the same system, we propose three different combinations in which different lenses and fibers are used in order to compensate the dispersion of a laser pulse at propagation through the image acquisition system. Laser generates Hermite Gaussian modes. We use the fundamental mode which is the Gaussian pulse. This pulse spreads at propagation through the free space. In order to avoid the spreading, we focus the pulse into an optical fiber using different lenses. Also the lenses suffer of chromatic dispersion. In order to decrease the effect of the chromatic dispersion, we design and analyze the functionality of a singlet, an achromatic doublet and an apochromat. At the output of the lens the pulse is focalized into an optical fiber. We take in consideration the step index fiber, the graded index fiber and self phase modulation fiber. The step index fiber suffers of intermodal dispersion, an alternative solution is to use the grade index fiber and the best solution is provided by the self phase modulation fiber. Finally, at the output of the fiber the light spreads on the CMOS sensor. During the functionality of the senor it introduces different temporal and spatial noises which degrade the quality of the pulse. Consequently, we have to reconstruct the image of the pulse using the Laplace, the amplitude and the bilateral filters.

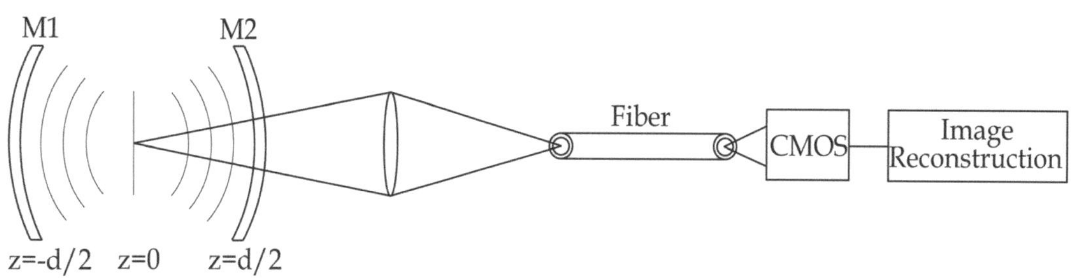

Fig. 1. A schematic of the image capture system

2. The laser modes

In order to find the laser modes we consider a confocal resonator system like that in figure 1. The optical axis is noted with z, and the light propagates from left to right in report with the optical axis. The resonator is made by two concave mirrors of equal radii of curvature $R=\frac{d}{2}$ separated by a distance d, and one mirror is a partially refractive mirror M_2. We consider the middle of the resonator in the point $z=-\frac{d}{2}+\frac{d}{2}=0$. After certain calculus (Poon & Kim, 2006), the modes in the middle of the resonator can be express as:

$$\psi(x,y,z=0)=E_0 \exp\left[-\frac{\left(x^2+y^2\right)}{w_0^2}\right]\cdot H_m\left(\frac{\sqrt{2}x}{w_0}\right)H_n\left(\frac{\sqrt{2}y}{w_0}\right) \tag{1}$$

where:
w_0 is the waist of the beam,
H_m is the Hermite Gaussian polynomial,

$$H_m(x)=(-1)^m e^{x^2}\frac{d^m}{dx^m}e^{-x^2}. \tag{2}$$

We have the 2D solution represented in figure 2 (Toadere & Mastorakys 2009, 2010).

Fig. 2. The fundamental Hermite Gaussian mode TEM_{00}

Each set (m,n) corresponds to a particular transverse electromagnetic mode of the resonator. The electric (and magnetic) field of the electromagnetic wave is orthogonal in the middle of the resonator in point $z=0$. The lowest-order Hermite polynomial H_0 is equal to unity; hence the mode corresponding to the set (0,0) is called the TEM_{00} mode and has a Gaussian radial profile. The laser output comprises a small fraction of the energy in the resonator that is coupled out through a partially refractive mirror. The width of the Gaussian beam monotonically increases in function of propagation on direction z, and reaches $\sqrt{2}$ times its original width at Rayleigh range. For a circular beam, this means that the mode area is doubled at this point (Poon & Banarje, 2001), (Poon & Kim, 2006).
In this paper we consider that the laser generates a pulse with a Gaussian radial profile (TEM_{00}). To avoid the spreading of the pulse, in the Rayleigh range at 20mm, we focus the

pulse in to a fiber using a lens. In order to attenuate the chromatic dispersion we use the singlet, the doublet, the apochromat, and the step fiber, the graded index fiber and the non linear index fiber.

3. The optical system analysis

When we work with optical components, the most important problem is that it is impossible to image a point object as a perfect point image. An optical system is made by a set of components (surfaces) through which the light passes. The optical sensor is analyzed in space by the point spread function (PSF) and in the spatial frequency by the modulation transfer function (MTF). These are the most important integrative criterions of imaging evaluation for the optical system. The PSF gives the 2D intensity distribution of the image of a point source. PSF gives the physically correct light distribution in the image plane including the effects of aberrations and diffraction. Errors are introduced by design (geometrical aberrations), optical and mechanical fabrication or alignment. MTF characterize the functionality of the optical system in spatial frequencies. Most optical systems are expected to perform a predetermined level of image integrity. A method to measure this quality level is the ability of the optical system to transfer various levels of details from the object to the image. This performance is measured in terms of contrast or modulation, and is related to the degradation of the image of a perfect source produced by a lens. MTF describe the image structure as a function of spatial frequency and is specified in lines per millimeter. It is obtained by Fourier transform of the image spatial distribution (Goodmann, 1996), (Yzuka, 2008).

When an optical system process an image using incoherent light, then the function which describe the intensity in the image plane produced by a point in the object plane is called the impulse response function:

$$g(x,y) = H\left[f(x,y)\right] \tag{3}$$

H is an operator representing a linear, position (or space) invariant system. The input object intensity pattern and the output image intensity pattern are related by a simple convolution equation:

$$g(x,y) = \int\int_{-\infty}^{+\infty} f(\alpha,\beta) H\left[\delta\left((x-\alpha, y-\beta)\right)\right] d\alpha d\beta,\ g(x,y) = \int\int_{-\infty}^{+\infty} f(\alpha,\beta) h(x-\alpha, y-\beta) d\alpha d\beta, \tag{4}$$

α and β are spatial frequencies (line/mm) which are defined as the rate of repetition of a particular pattern in unit distance.

$$h(x-\alpha, y-\beta) = H\left[\delta(x-\alpha, y-\beta)\right] \tag{5}$$

is the impulse response of H; in optics, it is called the point spread function (PSF). The net PSF of the optical part of the image acquisition system is a convolution between the individual responses of the optical components: the lens, the fiber and the optical part of the CMOS:

$$PSF = PSF_{lens} * PSF_{fiber} * PSF_{CMOS}\ . \tag{6}$$

We work with multiple convolutions, and we focus our attention on space analysis using the point spread function, which is specific to each component of the optical sensor. The optical fiber is analyzed from the spatial resolution point of view (Toadere & Mastorakis, 2009, 2010).

The PSF characterize the image analyses in space but also we can characterize the image in frequency using the optical transfer function (OTF) (Yzuka, 2008). The optical transfer function is the normalized autocorrelation of the transfer function and has the formula:

$$H(\alpha,\beta)=\frac{\iint P\left(x+\frac{\alpha}{2},y+\frac{\beta}{2},\right)P\left(x-\frac{\alpha}{2},y-\frac{\beta}{2},\right)}{\iint P(x,y)^2\,dxdy}. \tag{7}$$

The numerator represents the area of overlap of two pupil functions, one of which is displaced by $\frac{\alpha}{2},\frac{\beta}{2}$ and $-\frac{\alpha}{2},-\frac{\beta}{2}$ in directions x an y and the other in opposite directions $-x$ and $-y$. OTF is defined as the rapport between the area of the overlap of displace pupil function and complete area of the pupil function.

The changes in contrast that happens when an image passes trough an optical system is expected to have a lot to do with the optical transfer function (Goodmann, 1996) (Yzuka, 2008), (Toadere & Mastorakis, 2010). The definition of the modulation transfer function (MTF) is:

$$MTF=\frac{contrast\ \ of\ \ output\ \ image}{contrast\ \ of\ \ input\ \ image} \tag{8}$$

which represent the ratio of the contrast of the output image to that of the input image.

The relation between OTF and MTF is:

$$MTF=|OTF|. \tag{9}$$

The modulation transfer function is identical to the absolute value of the optical transfer function. The net sensor MTF is a multiplication between the transfer functions of the individual components:

$$MTF=MTF_{lens}\times MTF_{fiber}\times MTF_{CMOS}. \tag{10}$$

In general, the contrast of any image which has propagated through an image acquisition system is worse then the contrast of the original input image.

3.1 The PSF and MTF with aberrations

When we work with real optical systems, which have aberrations, the point spread function the optical transfer function and the modulation transfer function suffers modifications due to a phase distortion term $W(x,y)$ (Goodmann, 1996):

$$PSF=\frac{1}{\lambda^2 d^2 A_p}\left\|FT\left\{p(x,y)\cdot e^{-\frac{2x}{\lambda}W(x,y)}\right\}\Bigg|_{f_x=\frac{x}{\lambda d},f_y=\frac{y}{\lambda d}}\right\| \tag{11}$$

where:
λ is the wavelength,
FT is the Fourier transform,
d is the distance from the aperture to the image plane,
A_p is the area of the aperture,
$W(x,y)$ is the aberration of the pupil,
$p(x,y)$ is the pupil function,

$$P(x,y) = p(x,y) \cdot e^{-\frac{2x}{\lambda}W(x,y)}. \tag{12}$$

The optical transfer function is:

$$OTF(f_x, f_y) = \frac{FT(PSF)}{FT(PSF)\big|_{f_x=0, f_y=0}} \tag{13}$$

and the modulation transfer function is:

$$MTF(f_x, f_y) = \left|OTF(f_x, f_y)\right|. \tag{14}$$

3.2 The monochromatic aberrations

Aberrations are the failure of light rays emerging from a point object to form a perfect point image after passing through an optical system. Aberrations lead to blurring of the image, which is produced by the image-forming optical system. The wave front emerging from a real lens is complex because has error in the design, fabrication and lens assembly. Nevertheless, well made and carefully assembled lenses can possess certain inherent aberrations. To describe the primary monochromatic aberrations, of rotationally symmetrical optical systems, we specify the shape of the wave front emerging from the exit pupil. For each object point, there will be a quasi-spherical wave front converging toward the paraxial image point (Goodmann, 1996), (Kidger, 2001).

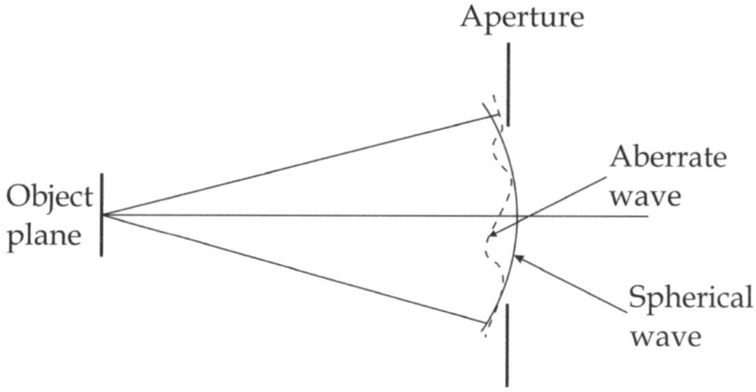

Fig. 3. The wavefront aberrations

In figure 3 the wave aberration function, W(x,y), is the distance, in optical path length, from the reference sphere to the wavefront in the exit pupil measured along the ray as a function

of the transverse coordinates (x,y) of the ray intersection with a reference sphere centered on the ideal image point.
To specify the aberrations we use the Siedel field aberration formula:

$$W(r,\theta) = W_{020}r^2 + W_{040}r^4 + W_{131}hr^3\cos(\theta) + \\ + W_{220}h^2r^2 + W_{311}h^3r\cos(\theta) + (higher \quad order \quad terms) \tag{15}$$

where:
W_{klm} are the wave aberration coefficients of the modes,
h is the height of the object,
r^2 is the defocus,
r^4 is the spherical aberration,
$hr^3\cos(\theta)$ is the coma,
$h^2r^2\cos^2(\theta)$ is the astigmatism,
h^2r^2 is the field curvature,
$h^3r\cos(\theta)$ is the distortion.

This Seidel aberration formula represents orthogonal polynomials which have the next properties: field aberrations describe the wavefront for a single object point as a function of pupil coordinates (*x*,*y*) and field height *h*. The aberrations are described functionally as a linear combination of polynomials. Point aberrations depend only on pupil coordinates and each polynomial term represents a single aberration. The aberration polynomial may be extended to higher order; these aberrations presented in equation (15) are up to fourth order. (Kidger 2001).
The Siedel aberrations for thin lenses can be express in function of bending and magnification (Geary, 2002), (Kidger, 2001). The bending can be express in function of the thin lens curvature:

$$B = \frac{c_1 + c_2}{c_1 - c_2}. \tag{16}$$

From the formula of the Lagrange invariant, the transverse magnification is given by:

$$m = \frac{y'}{y} = \frac{nu}{n'u'} \tag{17}$$

and the magnification is:

$$M = \frac{m+1}{m-1}. \tag{18}$$

Consequently, the Siedel aberrations are: W_{040} is the spherical aberration, W_{131} is the coma, W_{222} the astigmatism, W_{220} the field curvature, W_{311} is the distortion, W_{020} is the axial color and W_{111} is the lateral color:

$$W_{040} = \frac{1}{16} y_a^2 \phi^3 (a_1 + a_2(B - a_3M)^2 - a_4M^2), \tag{19}$$

$$W_{131} = \frac{1}{4\lambda} y_a^2 \phi^2 L(a_5 B - a_6 M), \tag{20}$$

$$W_{222} = \frac{1}{2\lambda} L^2 \phi, \tag{21}$$

$$W_{220} = \frac{1}{4\lambda} L^2 \phi \frac{n_g + 1}{n_g}, \tag{22}$$

$$W_{311} = 0, \tag{23}$$

$$W_{020} = \frac{1}{2\lambda} y_a^2 \frac{\phi}{\nu}, \tag{24}$$

$$W_{111} = 0 \tag{25}$$

where:
y_a is the aperture,
ϕ is the lens power,
ν is the Abbe number,
n_g are the glass refraction indices,
$L = -nu_a y_c$ is the Lagrange invariant,
$B = \frac{c_1 + c_2}{c_1 - c_2}$ is the bending, $M = \frac{1+m}{1-m}$ is the magnification,

$$a_1 = \left(\frac{n_g}{n_g - 1}\right)^2, \; a_2 = \frac{n_g + 2}{n_g(n_g - 1)^2}, \; a_3 = \frac{2(n_g^2 - 1)}{n_g + 2}, \; a_4 = \frac{n_g}{n_g + 2}, \; a_5 = \frac{n_g + 1}{n_g(n_g - 1)}, \; a_6 = \frac{2n_g + 1}{n_g}.$$

3.3 The correction of the aberrations

In the paragraph 3.2, we presented the mathematical relations that are used in the optical design which implies Seidel aberrations (Kidger, 2004), (Toadere & Mastorakis, 2010). In order to optimize the defects produced by the aberrations we use the defect vector *f* which is a set of *m* functions f_i that depend on a set on *n* variables. The function is of the type:

$$\sigma^2 = f^t \cdot f. \tag{26}$$

A is a $(n \times m)$ matrix of first derivatives:

$$A_{ij} = \frac{\partial f_i}{\partial x_j} \tag{27}$$

and *f* are changes in the variables from the current design. The gradient *g* is a $(n \times 1)$ vector given by:

$$g = \frac{1}{2}\nabla\sigma^2 \quad (28)$$

its components are:

$$g_i = \frac{\partial\sigma^2}{\partial x_i} = 2\left(f_1\frac{\partial f_1}{\partial x_i} + f_2\frac{\partial f_2}{\partial x_i} + \ldots f_m\frac{\partial f_m}{\partial x_i}\right), \quad (29)$$
$$g = A^t f.$$

Method of Least Squares:

$$g = A^t(f_0 + As),$$

$$g = g_0 + A^t As,$$

$$C = A^t A,$$

$$g_0 + Cs = 0.$$

is a set of simultaneous linear equations known as the normal equations of least-squares. Providing that the matrix C is not singular, these equations can always be solved, and the formal solution s may be written:

$$s = -C^{-1}g_0. \quad (30)$$

The basic idea of the damped least-squares is to start with the basic equation for the least squares condition. g_0 is the gradient at the starting point and augment the diagonal of the matrix C by the addition or factoring of a damping coefficient. Modifications of the form $c_{ii} + p$ for example, are called additive damping. In the case of additive damping, the equation for the damped least-squares solution reduces to:

$$g_0 + ps + Cs = 0. \quad (31)$$

As the damping factor p increases, the third term in the equation above becomes small and the solution vector becomes parallel to the gradient vector:

$$s = -\frac{1}{p}g_0 \quad (32)$$

3.4 The lens design

Lens design refers to the calculation of lens construction parameters that will meet a set of performance requirements and constraints. Construction parameters include surface profile types and the parameters such as radius of curvature, thickness, semi diameter, glass type and optionally tilt and decenter. Before we proceed, we notice that the human eye can only distinguish aberrations up to the fourth or fifth order. When we design the lens we have to take in consideration the aberrations, the aberration correction and the design considerations.
We design a singlet, a doublet and an apchromat. We are interested about resolution of these lenses configurations. A singlet has chromatic aberration; a doublet can focus two wavelengths and an apochromat can focus three wavelengths (Geary, 2002). Therefore, the

type of the lenses that are used in our analysis has significant impact on the shape and resolution of the pulse at the output of these lenses.

3.4.1 The design of the singlet

The singlet has the lens focal length 20*mm* and $f/2$ aperture. We use the glass BK 7, and we assume the object is at infinity (M = 1). The merit functions are the axial color and the coma (Kidger, 2001, 2004), (Toadere & Mastorakis, 2010). To solve this problem we must solve the equation system (figure 4):

$$\begin{cases} f_1 = \phi - \phi_1 \\ f_2 = \frac{1}{2}\phi^2 L(a_5 B - a_6 M) \end{cases} \tag{33}$$

where:

ϕ is the power of the lens.

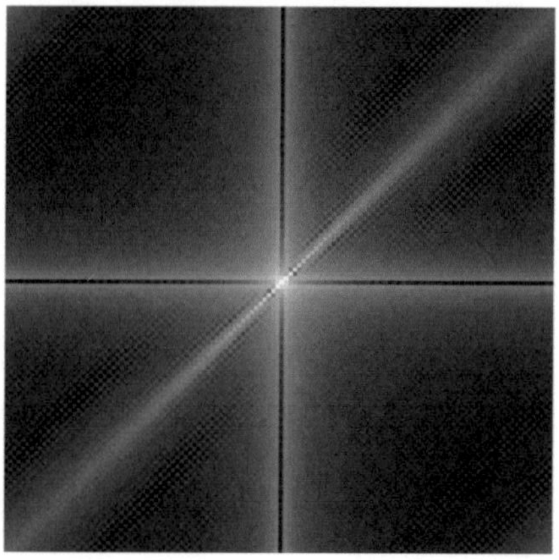

Fig. 4. The log of the PSF for the singlet

3.4.2 The design of an achromatic doublet

The achromatic doublet has the focal length 23*mm* with an *f/2* aperture. Assume the object is at infinity (M =1). We use the glasses BK 7 and SF 2. The merit functions are coma and spherical aberrations (Kidger 2001, 2003), (Geary, 2002). To solve this problem we must solve the equations system (figure 5):

$$\begin{cases} f_1 = \phi - (\phi_1 + \phi_2) \\ f_2 = \frac{\phi_1}{\nu_1} + \frac{\phi_2}{\nu_2} \end{cases} \tag{34}$$

where:

ϕ_1 is the power of the first lens,

ϕ_2 is the power of the second lens,

v_1 , v_2 are the corresponding Abbe numbers.

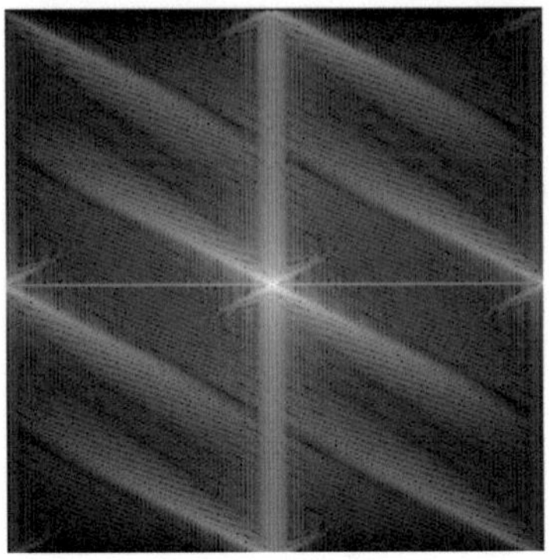

Fig. 5. The log of the PSF for the achromatic doublet

3.4.3 The design of an apochromat

The apochromat has the lens focal length 20*mm* with an $f/2$ aperture. We use the glass F2, KZFSN5, FK51, and we assume the object is at infinity (M = 1). The merit functions are spherical aberration and the axial color (Kidger 2001, 2004), (Geary, 2002). To solve this problem we must solve the equation system (figure 6):

$$\begin{cases} f_1 = \phi - \phi_1 + \phi_2 + \phi_3 \\ f_2 = \left(\dfrac{1}{\nu_1}\right)\phi_1 + \left(\dfrac{1}{\nu_2}\right)\phi_2 + \left(\dfrac{1}{\nu_3}\right)\phi_3 \\ f_3 = \left(\dfrac{P_1}{\nu_1}\right)\phi_1 + \left(\dfrac{P_2}{\nu_2}\right)\phi_2 + \left(\dfrac{P_3}{\nu_3}\right)\phi_3 \end{cases} \tag{35}$$

where:

ϕ_1, ϕ_2, ϕ_3 are the powers of the elements,
v_1, v_2, v_3 are the Abbe numbers,
P_1, P_2, P_3 are the partial dispersions.
The first equation determines the power, the second equation the axial color and the third equation the longitudinal color.

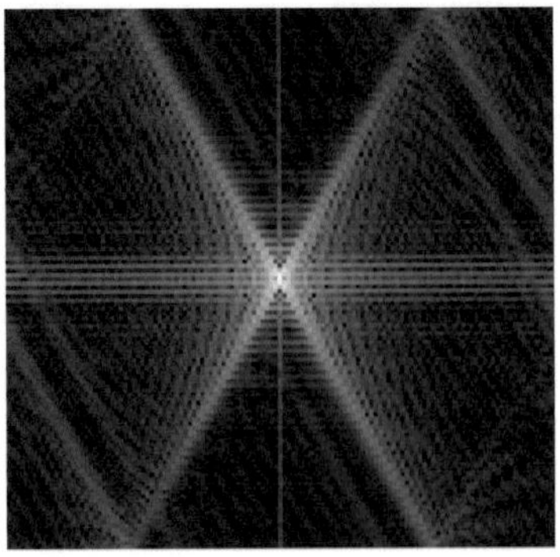

Fig. 6. The log of PSF for the apochromat

4. The optical fiber

An optical fiber is a thin, flexible and transparent fiber that acts as a waveguide in order to transmit light between the two ends of the fiber. During the radiation propagation trough the optical fiber it suffers of material dispersion, modal dispersion and polarization dispersion. Happily there are different types of fibers which allow us to reduce the modal dispersion and the polarization dispersion. Material dispersion is a problem that can be solved only by the designer and the producer of the fiber. When we make the physical model of the refraction index of the fibers we take in consideration the modal dispersion and the polarization dispersion. Modal dispersion happens in multimode fibers. Usually, the waveguide effect is achieved using in the core of the fiber a refractive index that is slightly higher than the refraction index of the surrounding cladding.
In order to reduce the effect of the modal dispersion, we analyze the functionality of the graded index fiber, the step index fiber and the fiber based on self caring effect. The step and graded index fibers use a linear refractive index and the fiber with self caring effect use a non linear refractive index. Polarization gives us information about linear and nonlinear comportment of the refractive index of the fibers. The polarization is deduced from the Maxwell equations.

4.1 The Maxwell equations

The Maxwell equations are (Mitsche, 2009), (Poon & Banarje, 2001), (Poon & Kim, 2006):

$$\nabla \cdot \vec{D} = 0 \tag{36}$$

$$\nabla \cdot \vec{B} = 0 \tag{37}$$

$$\nabla \times \vec{B} = \mu_0 \frac{\partial \vec{D}}{\partial t}, \tag{38}$$

$$\nabla \times \vec{E} = -\frac{\partial \vec{B}}{\partial t} \tag{39}$$

and:

$$\vec{D} = \varepsilon_0 \vec{E} + \vec{P}, \tag{40}$$

$$\vec{B} = \mu_0 \left(\vec{H} + \vec{M} \right), \tag{41}$$

$$\vec{j} = \sigma \vec{E} \tag{42}$$

where:

$\vec{E}$ is the electric field strength (V / m),

$\vec{H}$ is the magnetic field strength (A / m),

$\vec{D}$ is dielectric displacement (As / m^2),

$\vec{B}$ is the magnetic induction (Vs / m^2),

$\vec{J}$ is the current density (A/m^2),

$\vec{P}$ is the polarization,

$\vec{M}$ is the magnetization,

σ is the conductivity.

We rearrange the equation (39) using the equation (40):

$$\nabla \times \nabla \times \vec{E} = \nabla \times \left(-\frac{\partial \vec{B}}{\partial t} \right), \tag{43}$$

$$\nabla\left(\nabla \cdot \vec{E}\right) - \nabla^2 \vec{E} = -\frac{\partial}{\partial t}\left(\nabla \times \vec{B}\right),$$

$$\nabla\left(\nabla \cdot \vec{E}\right) - \nabla^2 \vec{E} = -\frac{\partial}{\partial t}\left(\mu_0 \frac{\partial \vec{D}}{\partial t}\right),$$

$$\nabla\left(\nabla \cdot \vec{E}\right) - \nabla^2 \vec{E} = -\mu_0 \frac{\partial^2 \vec{D}}{\partial t^2},$$

$$\nabla\left(\nabla \cdot \vec{E}\right) - \nabla^2 \vec{E} = -\mu_0 \frac{\partial^2}{\partial t^2}\left(\varepsilon_0 \vec{E} + \vec{P}\right),$$

$$\nabla\left(\nabla \cdot \vec{E}\right) - \nabla^2 \vec{E} = -\mu_0 \varepsilon_0 \frac{\partial^2 \vec{E}}{\partial t^2} - \mu_0 \frac{\partial^2 \vec{P}}{\partial t^2}. \tag{44}$$

If $\vec{E} \| \vec{P}$ and $D \| E$, it fallows that $\nabla \vec{D} = \nabla \vec{E} = 0$ and the equation (44) becomes:

$$\nabla^2 \vec{E} = \mu_0 \varepsilon_0 \frac{\partial^2 \vec{E}}{\partial t^2} + \mu_0 \frac{\partial^2 \vec{P}}{\partial t^2}. \tag{45}$$

The polarization is express as (Mitsche, 2009), (Poon & Banarje, 2001), (Poon & Kim, 2006):

$$P = \varepsilon_0 \left(\chi^{(1)} \vec{E} + \chi^{(2)} \vec{E}^2 + \chi^{(3)} \vec{E}^3 + \ldots \right). \tag{46}$$

4.2 The linear refractive index

For the linear case we take from equation (46) only the linear term:

$$P = \varepsilon_0 \chi^{(1)} \vec{E}. \tag{47}$$

Using equation (47) we rewrite the equation (40):

$$\vec{D} = \varepsilon_0 \vec{E}\left(1 + \chi^{(1)}\right). \tag{48}$$

In equation (48) the term inside the brackets represents the dielectric constant:

$$1+\chi^{(1)}=\varepsilon=\left(n+i\frac{C}{2\omega}\alpha\right)^2 \tag{49}$$

where:
n is the index of refraction,
α is the coefficient of absorption.
In equation (49) if $\alpha \approx 0$ then:

$$\varepsilon = n^2 . \tag{50}$$

Having these conditions we insert the equation (47) in to equation (45) and we obtain the linear wave equation (Mitsche, 2009), (Poon & Banarje, 2001):

$$\nabla^2\vec{E}=\frac{n^2}{c^2}\frac{\partial^2}{\partial t^2}\vec{E}, \tag{51}$$

and equivalently for the magnetic field:

$$\nabla^2\vec{H}=\frac{n^2}{c^2}\frac{\partial^2}{\partial t^2}\vec{H}. \tag{52}$$

4.2.1 Optical propagation through the step index fiber

Step-index fibers are optical fibers with the simplest possible refractive index profile: a constant refractive index n_1 in the core with some radius *r*, and another constant value n_2 in the cladding (Mitsche, 2009):

$$n_1 > n_2 , \tag{53}$$

$$n_2 \cong n_1(1-\Delta), \tag{54}$$

where:
$\Delta=\frac{n_1-n_2}{n_1}$ is the fractional change in the index of refraction,
n_1 is the refractive index in the core,
n_2 is the refractive index in the cladding.
By construction, this type of optical fiber has constant index of refraction in the core. This fact leads to the apparition of the modal dispersion during the propagation of the Gaussian pulse trough the step index fiber. At the output of the fiber the shape of the pulse is spread which produce intensity attenuation. Consequently, this type of optical fiber has modest performances.

4.2.2 Optical propagation through the graded index fiber

A graded-index fiber is an optical fiber whose core has a refractive index that decreases with increasing radial distance from the fiber axis. The index profile is very nearly parabolic. The advantage of the graded-index is the considerable decrease in modal dispersion ensuring a constant propagation velocity for all light rays (Mitsche, 2009):

$$n(x,y) = n_0\left[1 + \Delta(x,y)\right] \tag{55}$$

where:
n_0 is the intrinsic refractive index of the medium,
$n(x,y)$ is the medium index of refraction in the location (x,y),
$\Delta(x,y)$ is the variation of $n(x,y)$.
In reference (Poon & Kim, 2006) is presented a beautiful demonstration in which a plane wave propagates trough a graded index fiber. After the plane wave is substitute in the wave equation, the equation is solved and the results are the Hermite Gaussian polynomials. Since we have total mathematical compatibility with the equation (1), the only concern should be related to the propagation trough the refractive index. Due to the periodic focusing by the graded index, the distribution of the Gaussian pulse does not deform during its propagation through the fiber. This means that the Gaussian spatial confining of the light wave is preserved as the light propagates through the fiber. Therefore, the fiber preserves the spatial resolution of the original Gaussian pulse.

4.3 The nonlinear refractive index

For the nonlinear case (Mitsche, 2009), (Poon & Banarje, 2001), (Poon & Kim, 2006) using the equation (46), the polarization is express taking in consideration the first nonlinear and non zero term:

$$P = \varepsilon_0\left(\chi^{(1)}\vec{E} + \chi^{(3)}\vec{E}^3\right) \tag{56}$$

the second term in the expression (46) vanishes due to the statistical glass structure.
Using equation (50) we express the nonlinear term as:

$$\varepsilon = n^2 = 1 + \chi^{(1)} + \chi^{(3)}E^2 = \varepsilon_{linear} + \chi^{(3)}E^2, \tag{57}$$

$$\varepsilon = \varepsilon_{linear}\left(1 + \frac{\chi^{(3)}E^2}{\varepsilon_{linear}}\right). \tag{58}$$

In this condition the refractive index is:

$$n = n_0\sqrt{1 + \frac{\chi^{(3)}E^2}{\varepsilon_{linear}}} \approx n_0\left(1 + \frac{\chi^{(3)}E^2}{2n_0^2}\right) \tag{59}$$

where:
n_0 is the refractive index at zero intensity.
We note the term:

$$n_2 = \frac{\chi^{(3)}}{2n_0}, \tag{60}$$

$$n = n_0 + n_2 E^2 , \quad (61)$$

or:

$$n = n_0 + n_2 I , \quad (62)$$

where:
n_0 is the refractive index at zero intensity,
n_2 is the Kerr coefficient.

4.3.1 Wave propagation in a nonlinear inhomogeneous medium

The wave propagation in a nonlinear inhomogeneous medium (Poon & Kim, 2006) is governed by the combination of self phase modulation due to the Kerr effect and the group velocity dispersion which balance out each others and can lead to solitons (self sustaining pulses). The optical pulse propagates into a fiber whose index of refraction depends on the pulse intensity. The index of refraction is given by the equation (62). This type of fiber ensures the best propagation conditions. At the output of the fiber the pulse preserves its shape and also it is amplified in intensity.

5. The CMOS sensor

The image at the output of the optical fiber is projected on the image sensor. In this analysis we use a passive pixel complementary metal oxide semiconductor (PPS CMOS). We analyze the modulation transfer function (MTF) of the CMOS and the electrical part of the CMOS considering the photon shot noise and the fixed pattern noise (FPN). Finally, we use a Lapacian filter, an amplitude filter and a bilateral filter in order to reconstruct the noisy blurred image.

5.1 The optical part of a PPS CMOS sensor

The PPS CMOS image capture sensors it is a complex device which converts the focalized light in to numerical signal. CMOS image sensors consists of a $m \times n$ array of pixels; each pixel contains: the photodetector that converts the incident light in to photocurrent, the circuits for reading out photocurrent; part of the readout circuits are in each pixel, the rest are placed at the periphery of the array. CMOS sensors integrate on the same chip the capture and processing of the signal (Holst & Lomheim, 2007).

In our analyses we use a pixel made in $0.5\mu m$ technologies. To model the sensor response as a linear space invariant system, we assume the *n+/p-sub* photodiode with very shallow junction depth, and therefore we can neglect generation in the isolated *n+* regions and only consider generation in the depletion and *p*-type quasi-neutral regions. We assume a uniform depletion region. The parameters values of the pixel are: $z = 5.4\mu m$, $L_d = 4\mu m$, $L = 10\mu m$, $w = 4\mu m$, $\lambda = 550nm$. 1/2 inch CMOS with C optical interface is selected, i.e. its back working distance is 23 ± 0.18 mm. The visual band optical system has 60° field of view (FOV), f/number 2.5 (Toadere, 2010).

In figure 7 we have the cross section of a pixel and we can see that it is part of a periodic structure of pixels. The picture presents a structure of a complex device compound from the lenses , the colors filters and the analog part responsible with the conversion from

photons to charges and then in to voltage. Supplementary, not represented in the figure, we have conversion from analog signal to digital signal and numeric colors processing on the same chip.
The photodiodes are semiconductor devices responsive with capture of photons. They absorb photons and convert them in to electrons. The collected photons increase the voltage across the photodiode, proportional with the incident photon flux. The photodiodes work by direct integration of the photocurrent and dark current. They should have appropriate FOV, fill factor, quantum efficiencies and pixel dimension for the sensitive array. A good light capture allows sensor to obtain a high dynamic range scene.

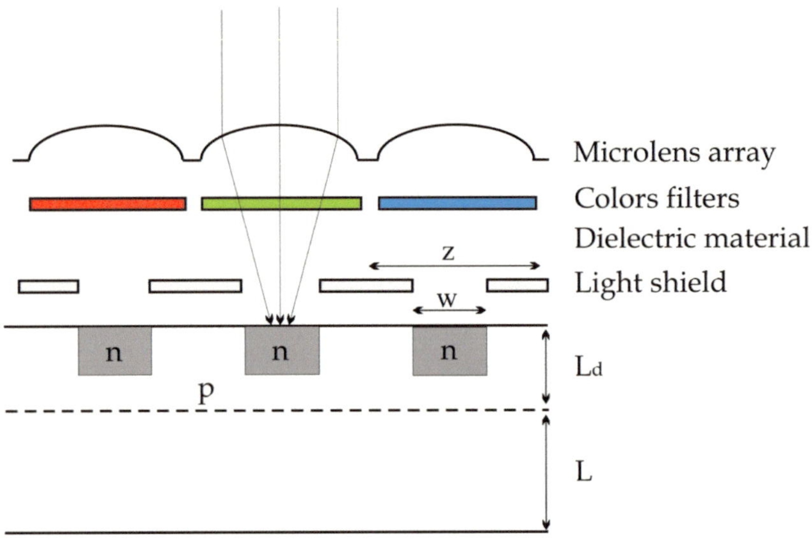

Fig. 7. The view of the simplified pixel cross section

z is the distance between pixel,
w is the pixels width,
L is the quasi neutral region,
L_d is the depletion length.

5.1.1 The modulation transfer function of the CMOS image sensors

The sharpness of a photographic imaging system or of a component of the system (the lens and the optical part of CMOS) is characterized by the MTF, also known as spatial frequency response. The optical part of the CMOS is characterized by its afferent MTF (Holst & Lomheim 2007). The contrast in an image can be characterized by the modulation:

$$M = \frac{s_{max} - s_{min}}{s_{max} + s_{min}} \tag{63}$$

where:
s_{max} and s_{min} are the maximum and minimum pixel values over the image.
Note that $0 \leq M \leq 1$. Let the input signal to an image sensor be a 1D sinusoidal monochromatic photon flux:

$$F(x,f) = F_0\left[1+\cos(2\pi fx)\right] \tag{64}$$

for $0 \le f \le f_{Nyquist}$.

The sensor modulation transfer function is defined as:

$$MTF(f) = \frac{M_{out}(f)}{M_{in}(f)} \tag{65}$$

from the definition of the input signal $M_{in} = 1$. MTF is difficult to find analytically and is typically determined experimentally. For the beginning we made a 1D analysis for simplicity and at the end we generalize the results to 2D model, which we will use in our analyses.

By making several simplifying assumptions, the sensor can be modeled as a 1D linear space-invariant system with impulse response $h(x)$ that is real, nonnegative, and even. In this case the transfer function (Toadere & Mastorakis, 2010):

$$H(f) = F\left[h(x)\right] \tag{66}$$

is real and even, and the signal at x is:

$$S(x) = F(x,f) * h(x), \tag{67}$$

$$S(x) = F_0\left[1+\cos(2\pi fx)\right] * h(x),$$

$$S(x) = F_0\left[H(0) + H(f)\cos(2\pi fx)\right],$$

therefore:

$$S_{\max} = F_0\left[H(0) + \left|H(f)\right|\right], \tag{68}$$

$$S_{\min} = F_0\left[H(0) - \left|H(f)\right|\right], \tag{69}$$

and the sensor MTF is given by:

$$MTF(f) = \frac{\left|H(f)\right|}{H(0)}. \tag{70}$$

In figure 7 we have a 1-D doubly infinite image sensor. To model the sensor's response as a linear space-invariant system, we assume *n+/p-sub* photodiode with very shallow junction depth, and therefore we can neglect generation in the isolated *n+* regions and only consider generation in the depletion and *p*-type quasi-neutral regions. We assume a uniform depletion region (from $-\infty$ to ∞). In figure 8, the monochromatic input photon flux $F(x)$ to the pixel current $iph(x)$ can be represented by the linear space invariant system. $iph(x)$ is sampled at regular intervals z to get the pixels photocurrents.

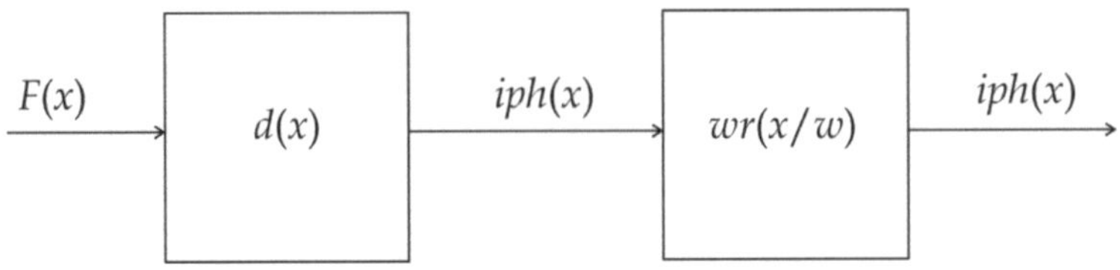

Fig. 8. The process of photogeneration and integration

$$r\left(\frac{x}{w}\right)=\begin{cases}1 & |x|\leq\frac{w}{2}\\ 0 & otherwise\end{cases} \tag{71}$$

$d(x)$ is the spatial impulse response corresponding to the conversion from photon flux to photocurrent density. We assume a square photodetector and the impulse response of the system is thus given by:

$$h(x)=d(x)*\varpi r\left(\frac{x}{\varpi}\right) \tag{72}$$

and its Fourier transform (transfer function) is given by:

$$H(f)=D(f)\varpi^2\,\text{sinc}(\varpi f) \tag{73}$$

where:

$D(0)=n(\lambda)$,

$n(\lambda)$ is the spectral response.

The spectral response is a fraction of the photon flux that contributes to photocurrents as a function of wave length. $D(f)$ can be viewed as a generalized spectral response (function of spatial frequency as well as wavelength).
After some calculus we get $D(f)$ as:

$$D(f)=\frac{q\left(1+\alpha L_f-e^{\alpha L_d}\right)}{1+\alpha L_f}-\frac{qL_f\alpha e^{\alpha L_d}\left(e^{\alpha L}-e^{\frac{L}{L_f}}\right)}{\left(1-\left(\alpha L_f\right)^2\right)\sinh\left(\frac{L}{L_f}\right)}, \tag{74}$$

$$H(f)=D(f)w^2\,\text{sinc}(wf), \tag{75}$$

the modulation transfer functions for $|f|\leq\frac{1}{2p}$ is:

$$MTF(f)=\frac{|H(f)|}{H(0)}=\frac{D_f}{D_0}w^2\,\text{sinc}(wf) \tag{76}$$

$\frac{D_f}{D_0}$ is called the diffusion MTF and $\sin c(wf)$ is called the geometric MTF.

Consequently, we have:

$$MTF_{CMOS} = MTF_{diffu\sin} \cdot MTF_{geometric} \,. \tag{77}$$

But in our analyses we use 2D signals so we must generalize 1D case to 2D case. We know that we have square aperture with length w for each photodiode:

$$MTF\left(f_x, f_y\right) = \frac{\left|H\left(f_x, f_y\right)\right|}{H(0)}, \tag{78}$$

$$MTF\left(f_x, f_y\right) = \frac{D_{\left(f_x, f_y\right)}}{D_0} w^2 \sin c\left(wf_x\right) \sin c\left(wf_y\right), \tag{79}$$

where:

f_x is the spatial frequency on x direction,

f_y is the spatial frequency on y direction.

Spatial frequency (lines/mm) is defined as the rate of repetition of a particular pattern in unit distance. It is indispensable in quantitatively describing the resolution power of a lens. The first level in a CMOS image sensor is a lens which focuses the light on each pixel photodiode.

In figure 9 we have the graphical representation of the *MTF(f)* calculated in equation (79).

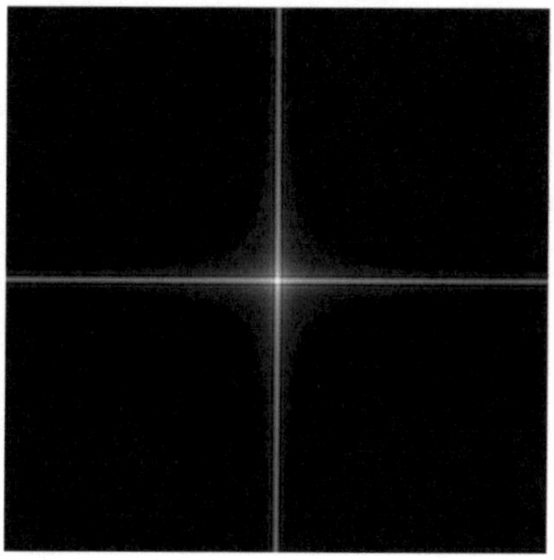

Fig. 9. The log of the PSF for the CMOS sensor

Diffusion MTF decreases with the wavelength. The reason is that the quasi-neutral region is the first region of absorption, and therefore photogenerated carriers due to lower wavelength photons (which are absorbed closer to the surface) experience more diffusion than those generated by higher wavelengths.

5.2 The electrical part of the PPS CMOS sensor

The PPS CMOS image sensor consists of a $n \times m$ PPS array. They are based on photodiodes without internal amplification. In these devices each pixel consists of a photodiode and a transistor in order to connect it to a readout structure (figure 10.). Then, after addressing the pixel by opening the row-select transistor, the pixel is reset along the bit line. The readout is performed one row at a time. At the end of integration, charge is read out via the column charge to voltage amplifiers. The amplifiers and the photodiodes in the row are then reset before the next row readout commences. The main advantage of PPS is its small pixel size. In spite of the small pixel size capability and a large fill factor, they suffer from low sensitivity and high noise due to the large column's capacitance with respect to the pixel's one. Also during the signal propagation trough the bit configuration it suffers of temporal noises perturbations (Holst & Lomheim 2007), (Toadere & Mastorakis, 2010), (Toadere, 2010).

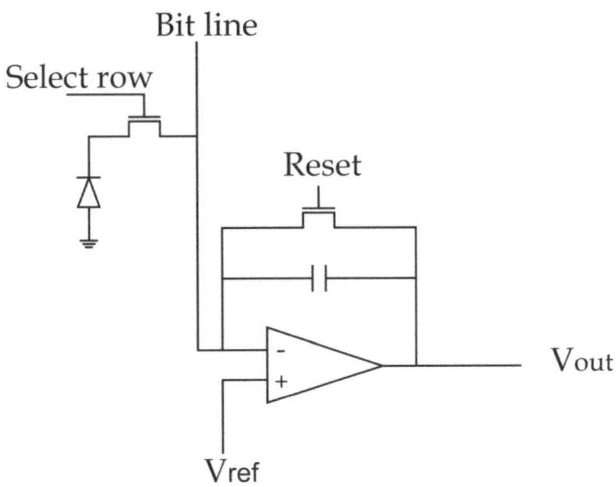

Fig. 10. A schematic of a passive pixel sensor

The pixel photodiode works by direct integration of the photocurrent and dark current on the photodiode condenser during the integration time. At the end of the integration time the condenser charge is read out by the next electronic block.

$$Q_{t\text{int}} = \left(i_{ph} + i_{dc}\right)t_{\text{int}}, \tag{80}$$

$$i_{ph} = q\int f(\lambda)\eta(\lambda)d\lambda, \tag{81}$$

where:

$q = 1.6 \times 10^{-19}C$ is the electron charge,

i_{ph} is the photodiode current,

i_{dc} is the dark current.

Dark current i_{dc} is the leakage current and it corresponds to the photocurrent under no illumination. It can not be accurately determined analytically or using simulation tools. Fluctuate with temperature and introduces unavoidable photon shot noise. The photon shot noise, dark current noise and thermal noise are signal dependent noises; reset and offset noises are signal independent noises.

5.2.1 The electrical noises

Image noise is a random, usually unwanted, variation in brightness or color information in an image. In a CMOS sensor image noise can originate in electronic noise that can be divided in temporal and FPN or in the unavoidable shot noise of an ideal photon detector. Image noise is most apparent in image regions with low signal level, such as shadow regions or underexposed images (Holst & Lomheim 2007).
The photon shot noise is generated by fluctuations in static *dc* current flow through depletion regions of a *pn* junction, resulted after the photons to electrons conversion process. The diode also suffers of dark current noise. Thermal noise is generated by thermally induced motion of electrons in the resistive regions of a MOS transistor channel in strong inversion polarization. Some time the photon shot noise and thermal noise can be considered as white Gaussian noise. In addition we have the reset, read and FPN noises due to other components electronics. Noises represent an additive process (Toadere, 2010).
Shot Noise is associated with the random arrival of photons at any detector. The lower the light levels the smaller the number of photons which reach our detector per unit of time. Consequently, there will not be continuous illumination but a bombardment by single photons and the image will appear granulose. The signal intensity, i.e. the number of arriving photons per unit of time, is stochastic and can be described by an average value and the appropriate fluctuations. The photon shot noise has the Poisson distribution:

$$P(k,\lambda)=\frac{e^{-\lambda}\lambda^{k}}{k!} \tag{82}$$

where:
$k=1\div n$, n is a non-negative integer,
λ is a positive real number.
The readout noise of a PPS CMOS is generated by the electronics and the analog-to-digital conversion. Readout noise is usually assumed to consist of independent and identically distributed random values; this is called white noise. The noise is assumed to have the normal white Gaussian distribution with mean zero and a fixed standard deviation proportional to the amplitude of the noise. The analog to digital convertor produces quantization errors. Whose effect can be approximated by uniformly distributed white noise whose standard deviation is inversely proportional to the number of bits used.

5.2.2 The fixed pattern noise

In a perfect image sensor, each pixel should have the same output signal when the same input signal is applied, but in image sensors the output of each sensor is different. The FPN is defined as the pixel-to-pixel output variation under uniform illumination due to device and interconnect mismatches across the image sensor array. These variations cause two types of FPN: the offset FPN, which is independent of pixel signal, and the gain FPN or photo response non uniformity, which increases with signal level. Offset FPN is fixed from frame to frame but varies from one sensor array to another. The most serious additional source of FPN is the column FPN introduced by the column amplifiers. In general PPS has

FPN, because PPS has very large operational amplifier offset at each column. Such FPN can cause visually objectionable streaks in the image. Offset FPN caused by the readout devices can be reduced by correlated double sampling (CDS). Each pixel output is readout twice, once right after reset and a second time at the end of the integration. The sample after reset is then subtracted from the one after integration (figure 11).
For a more detailed explanation, check out the paper by Abbas El Gammal (El Gamal et al. 1998). In this paper we focus our attention in FPN effects on image quality and we do not compute the FPN, we accept the noises as they are presented in references.

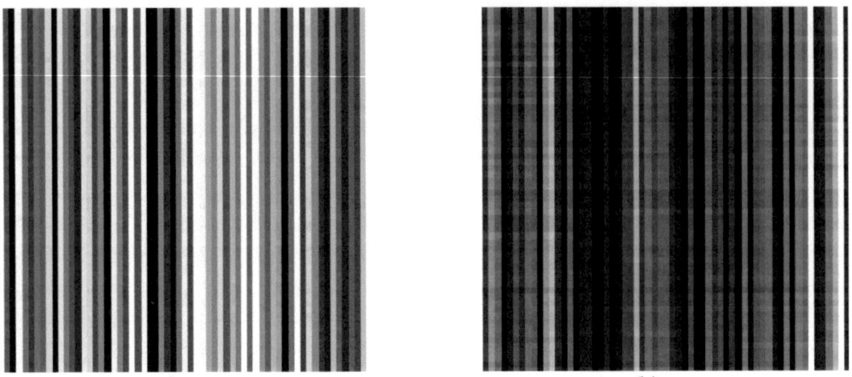

a) b)

Fig. 11. a) the FPN of the PPS without CDS, b) the FPN of the PPS with CDS

5.2.3 The dynamic range

Dynamic range is the ratio of the maximum to minimum values of a physical quantity. For a scene, the ratio is between the brightest and darkest part of the scene. The dynamic range of a real-world scene can be 100000:1. Digital cameras are incapable of capturing the entire dynamic ranges of scenes, and monitors are unable to accurately display what the human eye can see. The sensor dynamic range (*DR*) quantifies its ability to image scenes with wide spatial variations in illumination. It is defined as the ratio of a pixel's largest nonsaturating photocurrent $i_{\max}$ to its smallest detectable photocurrent $i_{\min}$ or the ratio between full-well capacity and the noise floor. The maximum amount of charge that can be accumulated on a photodiode capacitance is called full-well capacity. The initial and maximum voltages are V_{reset} and $V_{\max}$, they depend on the photodiode structures and operating conditions (Holst & Lomheim, 2007), (Toadere, 2010). The largest saturating photocurrent is determined by the well capacity and integration time t_{int} as:

$$i_{\max} = \frac{qQ_{\max}}{t_{\text{int}}} - i_{dc} \tag{83}$$

the smallest detectable signal is set by the root mean square of the noise under dark conditions. *DR* can be expressed as:

$$DR = 20\log_{10}\frac{i_{\max}}{i_{\min}} = 20\log_{10}\frac{qQ_{\max} - i_{dc}t_{\text{int}}}{\sqrt{qt_{\text{int}}i_{dc} + q(\sigma_{read}^2 + \sigma_{DNSU}^2)}} \tag{84}$$

where:

$Q_{\max}$ is the effective well capacity,

σ_{read}^2 is the readout circuit noise.

σ_{DNSU}^2 is the offset *FPN* due to dark current variation, commonly referred to as *DSNU* (dark signal non-uniformity).

5.2.4 The analog to digital conversion

The analog to digital conversion is the last block of the analog signal processing circuits in the CMOS image sensor. In order to convert the analog signal in to digital signal we compute the: analog to digital curve, the voltage swing and the number of bits. The quality of the converted image is good and the image seams to be unaffected by the conversion (Holst & Lomheim, 2007), (Toadere, 2010).

6. The image reconstruction

In the process of radiation capture with our proposed image acquisition system, the input photon flux is deteriorated by the combined effects of the optical aberrations and electrical noises. The optics is responsible for colors fidelity and spatial resolution; the electronics introduce temporal and spatial electrical noises. At the output of the electrical part the image is corrupted by the optical blur and the combined effect of the FPN and the photon shot noise. In order to reduce the blur we use a Laplacian filter, to reduce the FPN we use a frequencies amplitude filter which block the spikes spectrum of the FPN. Finally we reject the remains noise using a bilateral filter.

6.1 The Laplacian filter

In order to correct the blur and to preserve the impression of depth, clarity and fine details we have to sharp the image using a Laplacian filter. A Laplacian filter is a 3x3 pixel mask:

$$L = \begin{bmatrix} 0 & -1 & 0 \\ -1 & 4 & -1 \\ 0 & -1 & 0 \end{bmatrix}. \tag{85}$$

To restore the blurred image we subtract the Laplacian image from the original image (Toadere, 2010), (Toadere & Mastorakis, 2010).

6.2 The amplitude filter

The FPN is introduced by the sensor's column amplifiers and consists of vertical stripes with different amplitudes and periods. Such type of noise in the Fourier plane produces a set of spikes periodic orientate. A procedure to remove this kind of noise is to make a transmittance mask in Fourier 2D logarithm plane. The first step is to block the principal components of the noise pattern. This block can be done by placing a band stop filter $H(u,v)$ in the location of each spike. If $H(u,v)$ is constructed to block only components associated with the noise pattern, it fallows that the Fourier transform of the pattern is given by the relation (Yzuka, 2008), (Toadere, 2010), (Toadere & Mastorakis, 2010):

$$P(u,v) = H(u,v)\log\left[G(u,v)\right] \tag{86}$$

$G(u,v)$ is Fourier transform of the corrupted image $g(x,y)$.
After a particular filter has been set, the corresponding pattern in the spatial domain is obtained making the inverse Fourier transform:

$$p(x,y) = F\left\{\exp\left[P(u,v)\right]\right\}. \tag{87}$$

6.3 The bilateral filter

In order to reduce the remains noise, after the amplitude filter, we use a bilateral filter. It extends the concept of Gaussian smoothing by weighting the filter coefficients with their corresponding relative pixel intensities. Pixels that are very different in intensity from the central pixel are weighted less even though they may be in close proximity to the central pixel. This is effectively a convolution with a non-linear Gaussian filter, with weights based on pixel intensities. This is applied as two Gaussian filters at a localized pixel, one in the spatial domain, named the domain filter, and one in the intensity domain, named the range filter (Toadere, 2010), (Toadere & Mastorakis, 2010).

7. The result of simulations

All the blocks presented in this chapter are taken in consideration in our simulations. Although the CMOS sensor has the Bayer color sampling and interpolation, we did not take in consideration these blocks because we work with black and white images. The figure 12 presents the propagation of the laser pulse through the singlet, the step index fiber and the CMOS sensor. The figure 13 presents the propagation of a laser pulse through the achromatic doublet, the graded index fiber and the CMOS sensor. The figure 14 presents the propagation of a laser pulse through the apochromat, the self phase modulation fiber and the CMOS sensor

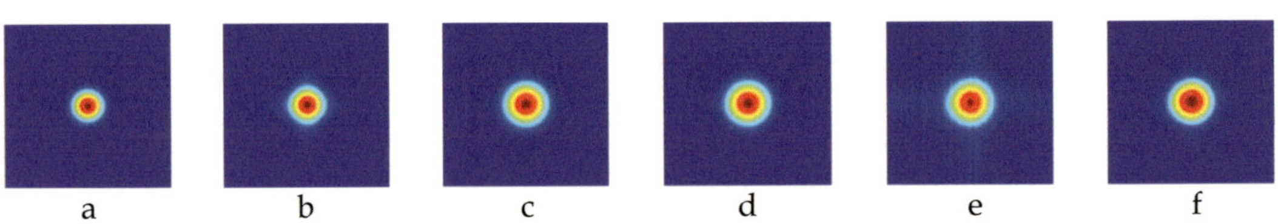

Fig. 12. The image at the output of the a) laser resonator, b) singlet, c) step fiber, d) optical part of the CMOS, e) electrical part of the CMOS, f) filtered image

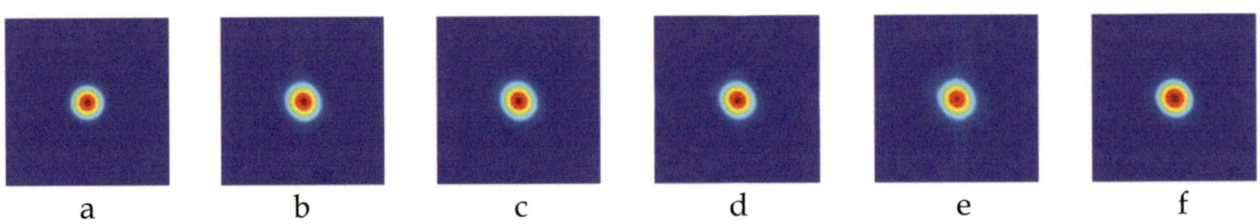

Fig. 13. The image at the output of the a) laser resonator, b) achromatic doublet, c) graded index fiber, d) optical part of the CMOS, e) electrical part of the CMOS, f) filtered image

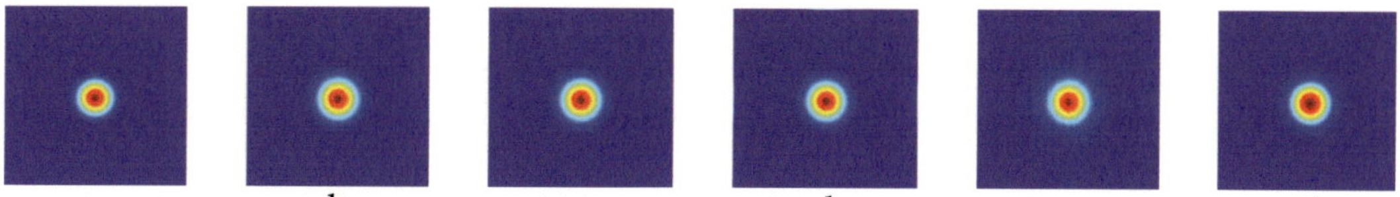

Fig. 14. The image at the output of the a) laser resonator, b) apochromat, c) self phase modulation fiber, d) optical part of the CMOS, e) electrical part of the CMOS, f) filtered image

8. Conclusions

In this paper we simulate the propagation of a Gaussian laser pulse through different image capture systems in order to find the best configuration that preserve the shape of the pulse during its propagations. We simulate the image characteristics at the output of each block from our different systems configurations. We simulate the functionality of the singlet, the achromatic doublet and the apochromat in order to reduce the chromatic dispersion. We simulate the functionality of the step index fiber, the graded index fiber and the self phase modulation fiber in order to reduce the modal dispersion. We simulate some properties of the CMOS sensor. The sensor suffers of different noises. The purpose of this paper was to put to work together, in the same system, optical and electrical components and to recover the degraded signal. In these types of complex systems, a controlled simulation environment can provide the engineer with useful guidance that improves the understanding of design considerations for individual parts and algorithms.

9. References

Geary J. M. (2002). *Introduction to lens design with practical zemax example*, Willmann Bell, ISBN 0-943396-75-1, Richmond, USA

Goodmann J. (1996). *Introduction to Fourier optics*, McGraw-Hill, ISBN 0-07-024254-2, New York, USA

El Gamal A.; Fowler B.; Min H. & Liu X. (1998). Modeling and Estimation of FPN Components in CMOS Image Sensor, *Proceedings of SPIE, vol. 3301*, pp.168-177, San Jose, California, USA, Aprill, 1998

Holst G. C. & Lomheim T. S. (2007). *CMOS/CCD sensors and camera systems*, Spie Press, ISBN 9780970774934, Bellingham, USA

Kidger M. (2001). *Fundamental optical design*, Spie Press, ISBN 0-8194-3915-0, Bellingham, USA

Kidger M. (2004). *Intermediate optical design*, Spie Press, ISBN 0-8194-5217-3, Bellingham, USA

Mitschke F. (2009). *Fiber optics physics and technology*, Springer, ISBN 978-3-642-03702-3, Berlin, Germany

Poon T. C. & Banarje P.P. (2001). *Contemporary optical image processing with matlab*, Elseiver, ISBN: 0-08-043020-1, Oxford, UK

Poon T. C. & Kim T. (2006). *Engineering optics with matlab*, World Scientific, ISBN 981-256-872-7, Singapore

Toadere F. & Mastorakis N. (2009). Imaging a laser pulse propagation trough an image acquisition system, *Recent Advances in Circuits, Systems, Electronics, Control and Signal Processing*, pp. 93-98, ISBN: 978-960-474-139-7, Tenerife, Spain, December 14-16, 2009

Toadere F. & Mastorakis N. (2010). Simulation the functionality of a laser pulse image acquisition system, *WSEAS transaction on circuits and systems*, Issue 1, Volume 9, (January 2010), pp. 22-31, ISSN 1109-2734

Toadere F. (2010). Conversion from light in to numerical signal in a digital camera pipeline, *Proceedings of SPIE on CD ROM, Volume 7821*, Constanta, Romania, 26-28 August, 2010

Yzuka K. (2008). *Engineering Optics, Springer*, ISBN 978-0-387-75723-0, New York, USA

13

Third-Order Optical Nonlinearities of Novel Phthalocyanines and Related Compounds

Zhongyu Li[1,2], Zihui Chen[2], Song Xu[1], Xinyu Zhou[2] and Fushi Zhang[2]
[1]*Changzhou University, Changzhou*
[2]*Tsinghua University, Beijing*
PR China

1. Introduction

The field of nonlinear optics has been developing for a few decades as a promising field with important applications in the domain of optoelectronics and photonics. Materials that exhibit nonlinear optical (NLO) behavior are useful because they allow manipulation of the fundamental properties of laser light beams, and are hence of great technological importance in areas such as photonic switching, optical computing and other optical data processing systems (Perry et al., 1994, 1996; Shirk et al., 2000). NLO activity was first found in inorganic crystals (Zyss, 1994), such as $LiNbO_3$, but the choice of these materials is rather limited. Also, most of them have either low NLO responses or important drawbacks for processing into thin films and being incorporated into micro-optoelectronic devices. By the mid-1980s, organic materials emerged as important targets of choice for nonlinear optical applications because they exhibit large and fast nonlinearities and are, in general, easy to process and integrate into optical devices. Moreover, organic compounds offer the advantage of tailorability: a fine-tuning of the NLO properties can be achieved by rational modification of the chemical structure. Finally, they are ideal to achieve the ultimate goal of device miniaturization by going into the molecular level.

Large optical nonlinearities in organic molecules usually arise from highly delocalized π-electron systems. During the last two decades, phthalocyanines (Pcs) have been intensively investigated for their third-order NLO properties both in solutions and as thin films because of their extensively delocalized two dimensional 18π-electron system (Ho, 1987; Shirk, 1989; Kambara, 1996; Ma, 2003; Zhou, 2004; He, 2007). They also exhibit other additional advantages, namely, exceptional stability, versatility, and processability features. The architectural flexibility of phthalocyanines is well exemplified by the large number of metallic complexes described in the literature, as well as by the huge variety of substituents that can be attached to the phthalocyanine core. Furthermore, some of the four isoindole units can be formally replaced by other heterocyclic moieties, giving rise to different phthalocyanine analogues. All these chemical variations can alter the electronic structure of the macrocyclic core, and therefore, they allow the fine-tuning of the nonlinear response.

Aside from their practical interest, Pc-related molecules present very attractive features for fundamental NLO studies. Since the unsubstituted and many substituted compounds are planar (2D, two dimensional), they offer the possibility of investigating of the role of dimensionality on the NLO response. Moreover, by introducing peripheral substituents or by adding axial substituents, one may obtain three-dimensional (3D) structures with pyramidal shape and examine the effect of a third-dimension on the NLO response. Owing to the extended π system, it is well-known that these Pc-related compounds exhibit a high aggregation tendency, and the aggregates usually display outstanding nonlinear optical properties. In order to improve the nonlinear optical properties of materials, the third-order optical nonlinearities of novel phthalocyanines and related compounds continue attracting attention.

This chapter arises from the need to compile the important advances obtained in the field of the NLO properties of phthalocyanines and analogues. Far from giving an exhaustive description of all the work that have been done in this area. We will focus on the third-order optical nonlinearities of some novel phthalocyanines and related compounds which were studied by our group using femtosecond degenerate four-wave mixing (DFWM) technique or picosecond Z-scan method.

2. Theory of nonlinear optics

Nonlinear optics is a material phenomenon in which intense light induces a nonlinear response in a medium, and in return the medium modifies the optical fields in a nonlinear way. In fact, all media are nonlinear to a certain degree of an applied optical field. The effect of a light wave on a material is usually described through the induced electrical polarization ***P***. At lower irradiation intensities, one can assume that this polarization is a linear function of the applied electric field ***E***,

$$P = \chi^{(1)} \cdot E \tag{1}$$

where $\chi^{(1)}$ is the linear susceptibility. However, when the material is subjected to an intense applied electric field (a laser light), one must take into account the deviation of ***P*** from a linear dependence on ***E***. Then the nonlinear polarization can be expressed by

$$\boldsymbol{P} = \chi^{(1)} \cdot \boldsymbol{E} + \chi^{(2)} \cdot \boldsymbol{EE} + \chi^{(3)} \cdot \boldsymbol{EEE} + \ldots \tag{2}$$

where $\chi^{(2)}$ and $\chi^{(3)}$ the quadratic (first-order) and cubic (second-order) susceptibilities, respectively. These two parameters determine the magnitude of the second- and third-order nonlinear optical responses. At the molecular level, a similar equation can be written for the microscopic polarization ***P*** induced in an atom or a molecule,

$$\boldsymbol{P} = \alpha \cdot \boldsymbol{E} + \beta \cdot \boldsymbol{EE} + \gamma \cdot \boldsymbol{EEE} + \ldots \tag{3}$$

with the coefficients α, β, and γ being the linear polarizability, the first (quadratic) hyperpolarizability, and the second (cubic) hyperpolarizability, respectively. The corresponding susceptibilities and hyperpolarizabilities are second rank ($\chi_{ij}^{(1)}$ or α_{ij}), third rank ($\chi_{ijk}^{(2)}$ or β_{ijk}), and fourth rank ($\chi_{ijkl}^{(3)}$ or γ_{ijkl}) tensors, respectively. In general, geometrical symmetries may reduce the number of independent nonzero components. The odd rank tensor $\chi^{(2)}$ is zero in centrosymmetric media whereas the even rank tensor $\chi^{(3)}$ does not have any symmetry restrictions and can take place in any materials. $\chi^{(n)}$ is frequency dependent and, as a result, resonant and non-resonant parameters differ significantly depending upon the measurement frequencies. The macroscopic third-order susceptibility $\chi^{(3)}$ is directly related to the microscopic averaged hyperpolarizability $\gamma(-\omega_4;\omega_1,\omega_2,\omega_3)$ through local field correction factors $f(\omega_i)$ at the frequencies of the applied electric fields through the relation,

$$\chi^{(3)}(-\omega_4;\omega_1,\omega_2,\omega_3) = Nf(\omega_4)f(\omega_1)f(\omega_2)f(\omega_3)\gamma(-\omega_4;\omega_1,\omega_2,\omega_3) \quad (4)$$

where ω_4 is the output frequency, ω_1, ω_2 and ω_3 are the input ones, relating with each other by the expression $\omega_4 = |\omega_1 \pm \omega_2 \pm \omega_3|$; N is the number of molecules per volume unit which can be obtained from the product of mass density D and Avogadro's constant A, divided by the molar mass M, i.e., $N = A \times D/M$; $f(\omega_i)$ is the local field factor at irradiation frequency ω_i. The local field factor at optical frequencies for a pure liquid can be estimated by the expression derived by Lorentz

$$f(\omega_i) = [n(\omega_i)^2 + 2)]/3 \quad (5)$$

where $n(\omega_i)$ is the linear refractive index for a liquid at frequency ω_i. The local field is the actual electric field acting on the microscopic species in the material. It should be emphasized that Lorentz's local field factor is an approximation since it considers the species to occupy a spherical cavity in the material and the local environment of the species is treated as a continuum. Also note that the indices of refraction used in the local field approximations for optical frequencies are generally assumed to be independent of the applied electric field.

3. Measurement of third-order NLO properties

The initial stages in the development of novel third-order NLO materials are the synthesis of molecules with large γ value and their incorporation into materials with substantial $\chi^{(3)}$ coefficient. Although many additional factors must be considered for practical device applications (e.g., thermal and photochemical stability, processability, etc.) the NLO properties of molecules and molecular materials can be adequately described here by reference to values of γ and $\chi^{(3)}$. Each of two parameters can be measured by using various experimental techniques. In most such measurements, the effects of resonance must always be considered; γ and $\chi^{(3)}$ are all wavelength dependent, being strongly resonance enhanced whenever the fundamental or harmonic frequencies are close to an electronic excitation. Although large NLO responses are required, any significant absorption of light is obviously highly undesirable for most application of NLO materials. This aspect has led to the use of expression "transparency/efficiency trade-off" when discussing the merit of such materials.

For studies of cubic NLO properties, values of γ and $\chi^{(3)}$ are usually obtained from third-harmonic generation (THG) (Hermann et al., 1973) or degenerate four-wave mixing (DFWM) (Linder et al., 1982). The electric-field induced second-harmonic generation (EFISHG) technique may also be used to determine γ. THG studies simply involve measurement of the light produced at the third harmonic (TH), but are often complicated by the fact that many NLO materials absorb strongly in the TH region, even when using a 1907 nm fundamental which gives a TH at 636 nm. THG occurs almost instantaneously through purely electronic interactions that do not depend on the population of the excited state. The disadvantage of the THG technique is that dynamic nonlinearities are not probed and no information on the time response of the optical nonlinearity can be given.

DFWM is a convenient method for measuring both electronic and dynamic nonlinearities. It involves three laser beams of the same frequency interacting in a materials to produce a forth degenerate beam, the intensity of which allows determination of $\chi^{(3)}$. DFWM may hence circumvent the effects of resonance, but has the additional aspect that vibrational and orientational mechanisms can contribute substantially to the observed optical nonlinearity. In contrast, only the electronic hyperpolarizability is fast enough to be probed by THG measurements. Hence, THG-derived γ values are generally considerably smaller than those determined via DFWM, and are more useful for deriving structure activity correlations (SACs) for the purely electronic part of γ, i.e., the only part of interest for potential practical applications. Polarization and time-dependent DFWM measurements can be used to distinguish the electronic (parametric) part of γ from other (nonparametric) components, but several experimental subtleties must be considered if this is to be achieved reliably. It is important to remember that $\chi^{(3)}$ values which are most often determined for solution will always be concentration dependent.

Another experimental method that has become popular for characterization of cubic NLO materials is the relatively simple, but highly sensitive Z-scan technique which measures the phase change induced in a single laser beam on propagation through a material (Sheik-Bahae et al., 1989, 1990). The sample is moved along the propagation path (z) of a focused Gaussian beam whilst its transmittance is measured through a finite aperture in the far filed, and the sign and magnitude of the nonlinear refractive index n_2 are deduced from the resulting transmittance curve (Z-scan). If the value of n_2 is positive, then the material has a tendency to shrink a laser pulse and is termed "self-focusing" (SF), whilst a negative n_2 characterized "self-defocusing" (SDF) behaviour. Although the ultrafast, electronic n_2 arises from the real part of $\chi^{(3)}$, published n_2 values often include, and indeed may be dominated by, nonparametric (thermal) contributions. In cases where nonlinear refraction is accompanied by nonlinear absorption, the nonlinear absorption coefficient α_2 can be determined by performing a second Z-scan with the aperture removed. Effective $\chi^{(3)}$ and γ values can also be determined via this method, but must be interpreted with great care in term of their actual origins which may be largely thermal effects.

Ultrafast, time-resolved optical Kerr effect (OKE) measurements (which involve another type of four-wave mixing process) may also be occasionally be used to determine $\chi^{(3)}$ and γ values. The OKE is the optically-induced birefringence caused by a nonlinear phase shift. There are various mechanisms (electronic deformation, molecular reorientation, molecular libration, molecular redistribution, electrostriction, thermal change) that are responsible for the change of nonlinear refractive index in the OKE and each mechanism has different strengths and response times. In OKE, a strong beam is used as a pump beam and a weak beam is used as a probe beam. The intensity of the probe beam transmitted through a Kerr cell is measured as a function of the delay time between pump and probe.

In each of the THG, DFWM, Z-scan and OKE techniques, γ is derived from the measured $\chi^{(3)}$ values by using the solute number density of the solution. In a few cases, $\chi^{(3)}$ values have also been determined by using Stark spectroscopy. It should be remembered that comparisons of $\chi^{(3)}$ or γ values obtained using different techniques with different experimental conditions are generally of little utility.

4. Novel phthalocyanines and analogues with third-order NLO properties

The third-order NLO properties of phthalocyanines and its analogues have been extensively investigated since 1990, and this area has been reviewed (Torre et al., 2004). The discussion here will be limited to selected highlights of our studies and more recent developments.

4.1 Novel phthalocyanines

The electronic structure of phthalocyanine molecules can be tailored by either metal substitution at the central binding site or by altering the peripheral and axial functionalities, thus affording great versatility in controlling their electro-physical properties. Incorporation of a metal atom in the center of the macrocycle results in two types of charge-transfer transitions: metal-to-ligand and ligand-to-metal. The molecular hyperpolarizability could be enhanced by the metal-ligand bonding through the transfer of the electron density between the metal atom and the conjugated ligand systems. Furthermore, peripheral substituents could also influence the nonlinearity through intermolecular interactions. Therefore, the third-order NLO susceptibility $\chi^{(3)}$ could vary by several orders of magnitude through chemical modification of the macrocycle structure (Nalwa et al., 1993, 1999; He et al., 2007).

4.1.1 Diarylethene-phthalocyanine dyads

In general, the absorption spectra of monomeric phthalocyanines are dominated by two intense bands, a Soret band in the near ultraviolet region (at around 350 nm) and a Q band in the visible region (at around 670 nm), with a molar extinction coefficient in the range of 10^5 M^{-1} cm^{-1}. When electron-donating alkoxy/aryloxy groups are introduced at the periphery closest to the phthalocyanine ring, a large bathochromic shift occurs in the Q band. These large red-shifted phthalocyanines are very soluble in common organic solvents and can be easily be prepared thickness-controlled thin films for practical applications to photonic devices. However, little attention has been paid to the third-order nonlinear properties of these molecules. Herein, we report the third-order nonlinear optical properties of aryloxy substituted novel diarylethene-phthalocyanine dyads **1-4** (the structures are shown in Figure 1) measured by femtosecond DFWM technique at 800 nm under off-resonant condition (Li et al., 2007).

Figure 2 shows the temporal response of the DFWM signal as a function of the probe delay in the diarylethene-phthalocyanine dyads **1-4**. The DFWM signal of each sample is well-fitted by a Gaussian function (solid curve), and the half-width of fitting curve is similar to the autocorrelated pulse duration. All the signal profiles almost exhibit symmetry about the maximum signal (the zero time delay), which indicate that the responses time of the third-order optical nonlinearities are shorter than the experimental time resolution (50 fs). Such instantaneous response means that the Kerr effect (electronic component) from the distortion of the large π-conjugated electron charge distribution of diarylethene-phthalocyanine dyad molecules is main reason for generating the DFWM signal.

The evaluated values of $\chi^{(3)}$, γ and nonlinear refractive index for the diarylethene-phthalocyanine dyads **1-4** in DMF solution are summarized in Table 1. Among the samples, diarylethene-phthalocyanine dyad **4** possesses the highest γ_e value of 1.12×10^{-30} esu. It was found that the second-order hyperpolarizability values of the diarylethene-phthalocyanine

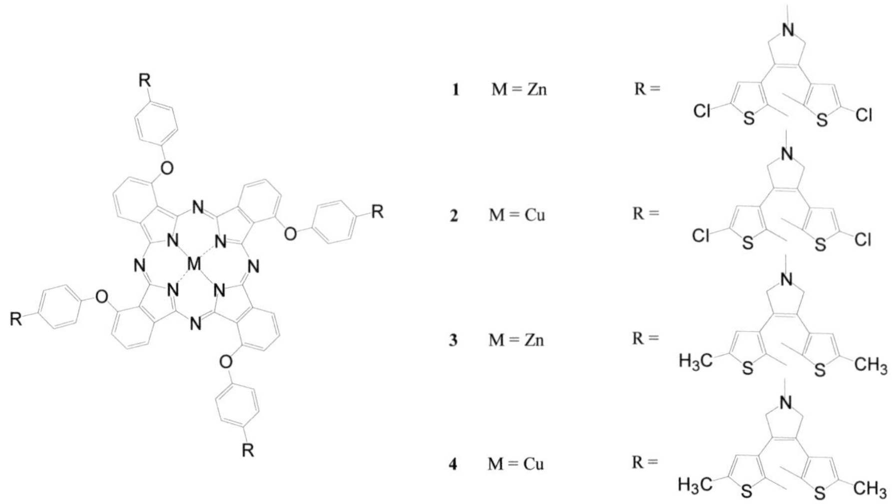

Fig. 1. Structures of diarylethene-phthalocyanine dyads **1-4**

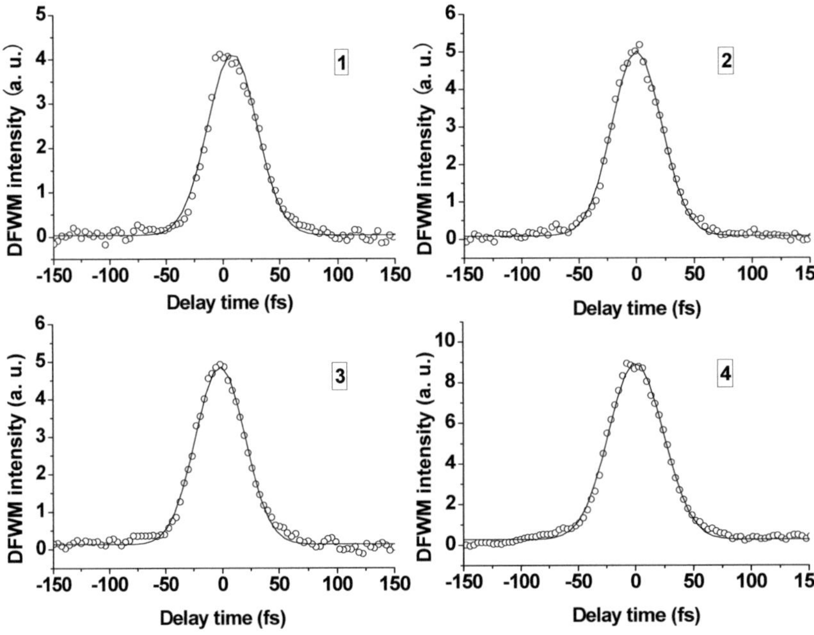

Fig. 2. Temporal profiles of DFWM signal of the diarylethene-phthalocyanine dyads **1-4**

dyads **1-4** are 2-3 times larger than that of tetra-aryloxy substituted metal-free phthalocyanine. This enhancement of γ mainly comes from the d valence orbital contribution of central metal atoms (Zn and Cu) of the diarylethene-phthalocyanine dyads. Furthermore, it is observed that the γ values of the copper substituted diarylethene-phthalocyanine dyads (Cu-DE-Pcs) are larger than those of the zinc substituted diarylethene-phthalocyanine dyads (Zn-DE-Pcs). This behavior can explain from the electronic structures of the upper occupied and lower vacant molecular orbitals for the ground states of metal substituted phthalocyanines (MPcs). In the case of Zn-DE-Pcs with completely filled d-shell, we can suppose that the probability for the charge transfer mechanism is very less. The 3d subshell of Zn-DE-Pcs is filled and deep enough to form rather pure molecular orbitals. The Zn-DE-Pcs exhibit a large gap between the HOMO and LUMO. In the case of Cu-DE-Pcs, the unfilled d valence orbital can be split into serials level due to the interaction between the d electrons and π-conjugation electrons of Pc ring, this result will lower the transition energy in low-lying d orbital-ligand or d-d transition. The existence of excited state with low transition energy will enhance the nonlinear optical susceptibilities of the material. The unfilled d orbit of Cu atoms will couple with the conjugated electrons of Pc ring leading to the extension of conjugated systems. As a result, the Cu-DE-Pcs with larger conjugated systems will show larger optical nonlinearities than the Zn-DE-Pcs. Moreover, the γ value of the methyl substituted diarylethene-phthalocyanine dyad **3/4** is larger than that of the relative chlorine substituted diarylethene-phthalocyanine dyad **1/2**. This is probably attributed to electron-pushing effect of the methyl group of the diarylethene-phthalocyanine dyad **3** and **4**, which leads to large polarization of molecules.

Sample	$\chi^{(3)}$ ($\times 10^{-14}$ esu)	γ ($\times 10^{-31}$ esu)	n_2 ($\times 10^{-13}$ esu)
1	4.50	7.60	8.33
2	5.05	8.53	9.35
3	4.92	8.32	9.12
4	6.62	11.2	12.3

Table 1. Evaluated values of $\chi^{(3)}$, γ and nonlinear refractive index for the diarylethene-phthalocyanine dyads **1-4** in DMF solution

4.1.2 Azobenzene-phthalocyanine dyads

In 2008, we reported the photo-responsive J-aggregation behavior of a novel α-aryloxy-substituted zinc phthalocyanine (azobenzene substituted zinc phthalocyanine, hereafter, abbreviated as azo-ZnPc dyad) and its third-order optical nonlinearity (Chen et al., 2008). The azo-ZnPc dyad was synthesized through a rather facile route as shown in Scheme 1. The third-order optical nonlinearities of the photo responsive J-aggregates of the azo-ZnPc dyad (before and after irradiation conditions) were measured using a Z-scan technique at 532 nm with pulse duration of 25 ps.

In our previous studies, it was found that α-aryloxy-substituted Zinc phthalocyanine could form J-type self-aggregate in noncoordinating solvents through the complementary coordination of the peripheral oxygen atom of one phthalocyanine to the central Zn^{2+} in another phthalocyanine (Zhang et al., 2007a, 2007b). Moreover, the α-aryloxy-substituted Zinc phthalocyanine formed J-aggregate showed another characteristic, that is the addition of methanol could break the J-aggregate, as a consequence, typical Q-band restored.

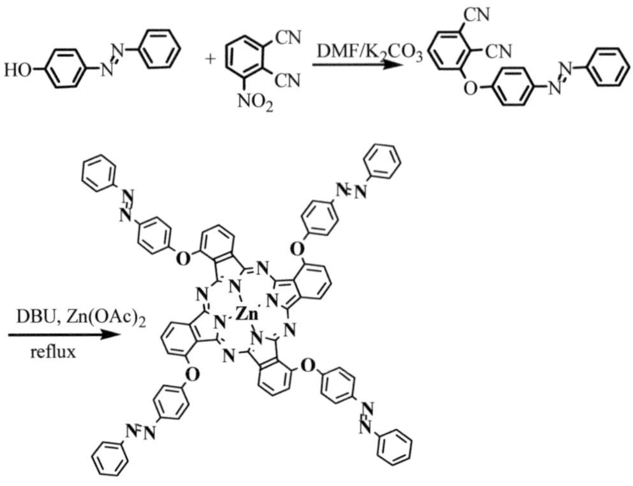

Scheme 1. Synthesis route of a novel α-aryloxy-substituted zinc phthalocyanine

The studied azo-ZnPc dyad shows same absorption behaviors as that of the reported α-aryloxy-substituted Zinc phthalocyanine. Figure 4 shows the spectra change of initial and UV-illuminated solutions of azo-ZnPc in chloroform with the addition of methanol. As methanol was titrated, the absorption at 740 nm of both of the solutions decreased gradually and finally disappeared completely with the increase in the absorption at 698 nm. The unusual red-shifted peak at 740 nm can be attributed to the formation of J-aggregates. For azo-ZnPc, there is a stronger tendency to form J-aggregate when the azobenzene units is in the cis-conformation, and the irradiation of UV light will cause enhancement of J-aggregation.

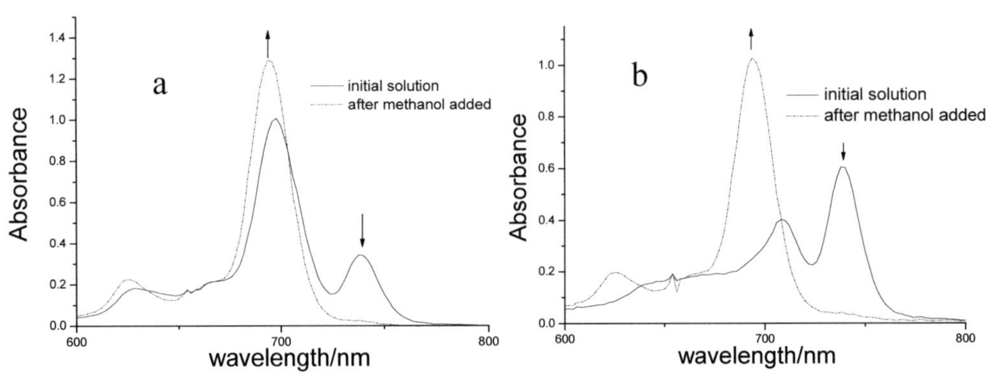

Fig. 3. Absorption spectral changes of azo-ZnPc (a: before UV light irradiation; b: after UV light irradiation for 3 min, c = 1.19 × 10^{-5} M) in chloroform .

The open aperture curves (Figure 4) of Z-scan measurement exhibit the normalized valleys, indicating the presence of reverse saturable absorption with a positive coefficient β. And the normalized transmission for the closed aperture of Z-scan measurement is shown in Figure 5. The large valley-to-peak configurations of closed aperture curves suggest that the refractive index changes are negative, exhibiting a strong defocusing effect.

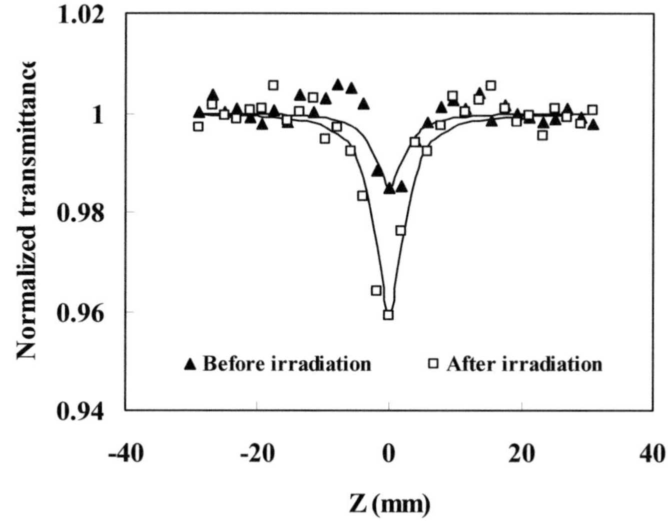

Fig. 4. Normalized transmission without aperture at 532 nm (open aperture) as a function of distance along the lens axis. The filled triangles and open squares are measured data for before irradiation and after irradiation of azo-phthalocyanine, respectively. Each point corresponds to the average of 5 pulses. The solid line is the theoretical fit.

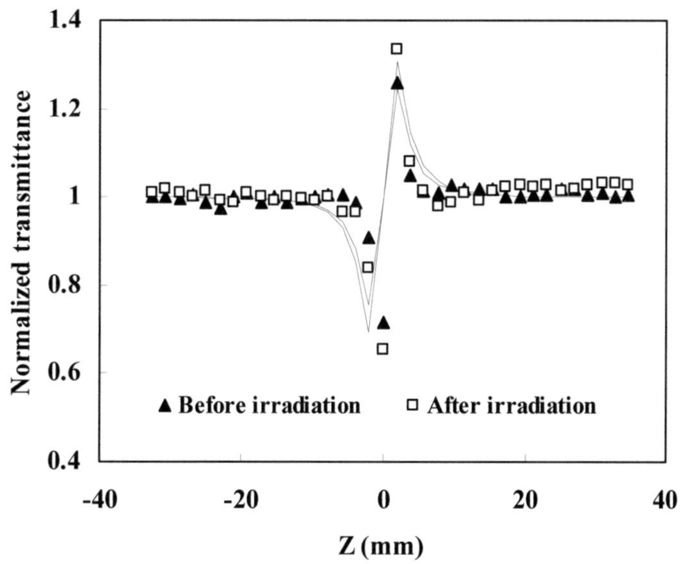

Fig. 5. Normalized transmission for the closed aperture of Z-scan measurement

The nonlinear absorption coefficient (β, m/W), the nonlinear refraction coefficient (n_2, m^2/W), the third-order nonlinear susceptibility ($\chi^{(3)}$, esu) and the molecular second hyperpolarizability (γ, esu) are calculated and listed in Table 2. It is found that the studied azo-phthalocyanine dyad shows large second-order molecular hyperpolarizabilities which are of the order of 10^{-30} esu both before and after irradiation conditions. The value of γ of before irradiation is 1.25 times larger than that of after irradiation condition. This enhancement may be attributed to the increase of J-aggregation degree of azo-ZnPc dyad after UV light irradiation.

Sample	Before irradiation	After irradiation
n_2 / × 10^{-19} m^2 W^{-1}	4.26	5.30
β / × 10^{-9} m W^{-1}	1.48	4.03
$\chi^{(3)}$ / × 10^{-13} esu	2.26	2.82
$\gamma\gamma$ / × 10^{-30} esu	3.87	4.82

Table 2. Values of The nonlinear absorption coefficient, the nonlinear refraction coefficient, the third-order nonlinear susceptibility and the molecular second hyperpolarizability of Azo-ZnPc dyad.

4.1.3 Azobenzene-containing water soluble unsymmetrical phthalocyanines

In 2009, we reported the photoswitching of the third-order nonlinear optical properties of unsymmetrical azobenzene-containing metal phthalocyanines (structures are shown in Figure 6) based on reversible host-guest interacions (Chen, 2009).

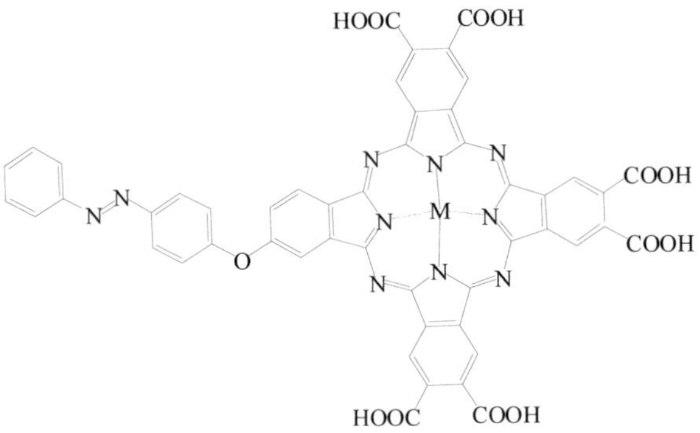

MPc7

Fig. 6. Structures of unsymmetrical azobenzene-containing metal phthalocyanines

It has been well established that azobenzene could reversibly assemble with α-cyclodextrin through host-guest interaction under suitable external photo-stimuli, and this phenomenon has been exploited as the basis of some molecular shuttles and motors (Breslow & Dong, 1998; Dugave & Demange, 2003). However, their applications in phthalocyanine chemistry have rarely been studied yet. We believe that this reversible host-guest interaction can be used to modulate the NLO properties of phthalocyanines if the phthalocyanines were judicious designed. Therefore, we prepared for the first time two azobenzene containing water soluble unsymmetrical metal phthalocyanines. Their reversible host-guest interaction with α-cyclodextrin in aqueous media and the resulting effects on the NLO properties of such molecules were also investigated. Scheme 2 shows the structures and the synthesis of target azobenzene containing water soluble unsymmetrical zinc (II) and copper (II) phthalocyanines (abbreviated as Zn-Pc7 and Cu-Pc7, respectively) and their inclusion complexes with α-cyclodextrin (Zn-Pc8, Cu-Pc8)

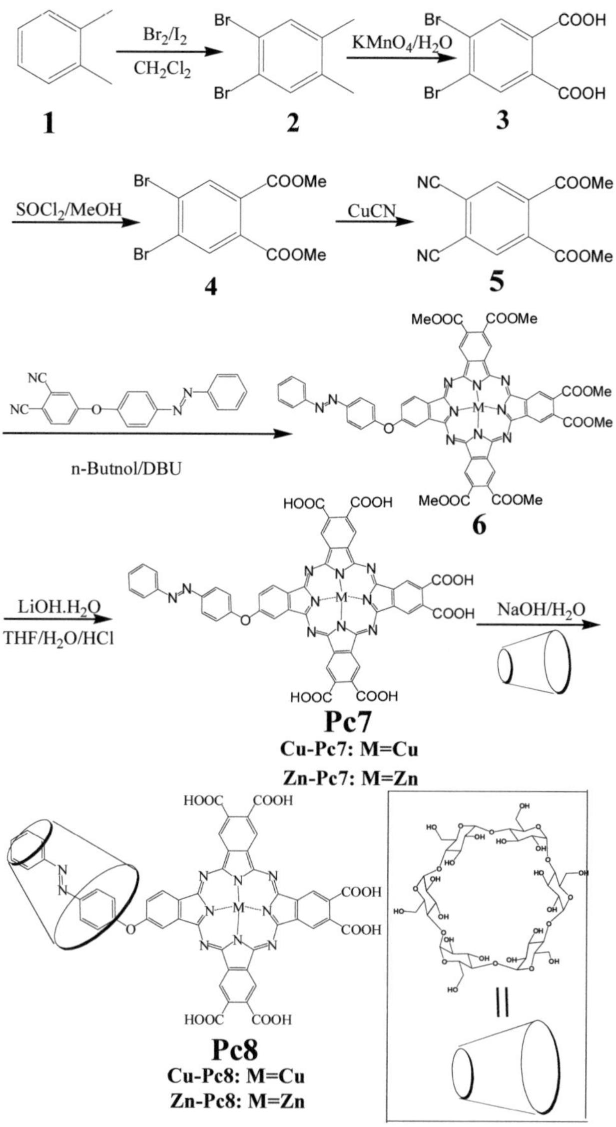

Scheme 2. Synthesis of target azobenzene containing water soluble unsymmetrical zinc (II) and copper (II) phthalocyanines and their inclusion complexes with α-cyclodextrin (Zn-Pc8, Cu-Pc8)

Z-scan studies show that each azobenzene containing water soluble unsymmetrical phthalocyanine consists of an electro donating phenylazophenoxy group (D) and six electron withdrawing carboxyl (A) forming a D-π-A alignment along the x axis. As a result of such unique chemical structure, all the samples showed very large molecular cubic hyperpolarizabilities which are of the order 10^{-30} esu. The Azobenzene moieties of these compounds could reversibly associate with α-CD to form inclusion complexes through host-guest interaction in aqueous media upon alternating illumination of UV and visible light, resulting apparent influences to the third-order NLO properties of these phthalocyanines.

This influence is especially striking for the phthalocyanine whose central metal atom is Cu^{2+}. The molecular cubic hyperpolarizability γ of its inclusion complex with α-CD is 2.10 $\times 10^{-30}$ esu. When the inclusion complex disassociated under the illumination of 365 nm light, the γ value was 4.2×10^{-30} esu, which is an 100% increase. Taking account of the large molecular cubic hyperpolarizabilities of these compounds, our endeavors toward ideal third-order NLO photoswitching systems is very promising, with sufficient room for improvement. This work suggested that reversibly control either the chemical structure or the molecular packing arrangement of excellent third-order NLO materials is an attractive strategy for constructing ideal third-order NLO photoswitching systems. Moreover, the present study emphasized the reversible host-guest interaction between azobenzene and α-CD on the packing style of phthalocyanines, which may provide new insights to the host-guest chemistry.

4.1.4 Novel copper phthalocyanine-ferrocene dyad

González-Cabello (González-Cabello et al., 2003) reported that the interplanar distance between the two ciclopentadiynyl rings combined with the rigid, stereochemically well defined, π-conjugated linkers between the ferrocene and the Pc-subunits provide an excellent situation for the cofacial stacking of the Pc macrocycles, thus allowing potential NLO-favorable through space interactions between the individual Pc subunits. Moreover, the combination of an electron acceptor moiety such as phthalocyanine and electron donor unit such as ferrocene may give rise to intramolecular charge transfer that may enhance the nonlinear optical response. Herein, third-order nonlinear optical property of a novel copper phthalocyanine-ferrocene dyad (the structure is shown in Figure 7) measured by femtosecond degenerate four–wave mixing technique under off-resonant condition (Bin et al., 2008).

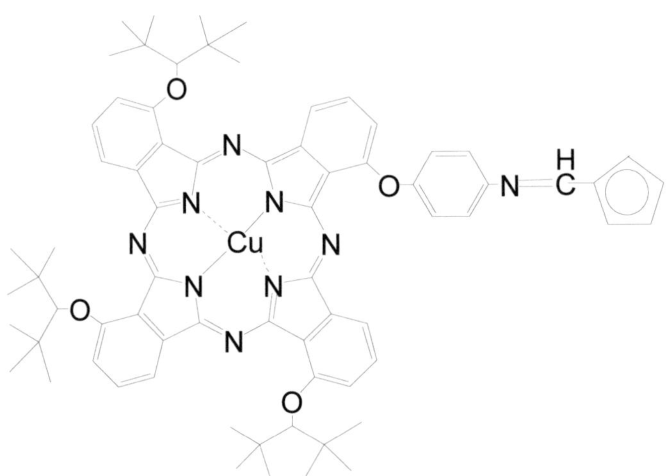

Fig. 7. Structure of novel copper phthalocyanine-ferrocene dyad

The DFWM measurement shows that the second-order molecular hyperpolarizability of this compound was measured to be 1.74×10^{-30} esu, and its response time was also obtained and no more than 50 fs. This large and ultrafast third-order optical nonlinear response is mainly enhanced by the formation of intramolecular charge-transfer which can enhance the delocalized movements of the large π-electrons in the molecules.

4.1.5 Novel thiophene-bearing phthalocyanines

Recently, we have measured the third-order nonlinear optical properties of two thiophene-bearing phthalocyanines (Ni-TPc and Cu-TPc, their structures are shown in Figure 8) using Z-scan technique at 532 nm (Chen et al., 2011). Both Ni-TPc and Cu-TPc were found to show large molecular cubic hyperpolarizabilities whose values are of the order of 10^{-30} esu. The γ value of Cu-TPc is 1.5 times larger than that of Ni-TPc, mainly as a result of their notably different nonlinear absorptions. Most notably, the nonlinear absorption and the nonlinear refraction contribute almost equally to their molecular cubic hyperpolarizabilities, while for our previous studied Pcs, the nonlinear refraction always plays an absolutely predominant role. It is assumed that the incorporation of thiophene rings into phthalocyanines could notably increase the multi-photo absorption cross-section of Pc.

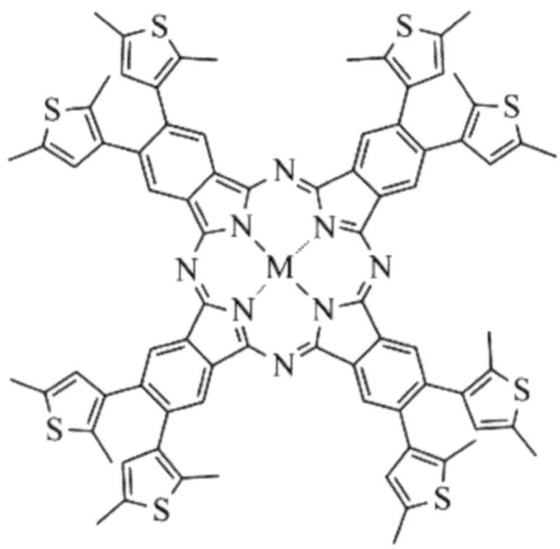

M= Cu^{2+}, Ni^{2+}

Fig. 8. Chemical structures of two thiophene-bearing phthalocyanines, Ni-TPc and Cu-TPc

4.2 Novel phthalocyanine related compounds

Phthalocyanines are special organic systems in a way that they offer tremendous opportunities in tailoring their photophysical and optical properties over a wide range either by substituting different metal atoms into the central binding site or by altering the peripheral and axial functionalities. It is possible to incorporate a variety of peripheral substituents around the phthalocyanines core as well as replace some of the isoindole units by other heterocyclic moieties, giving rise to different phthalocyanine analogues. For example, in the previous studie (Liu et al., 1999), it was shown that the phthalocyanine may lose a bridging nitrogen atom when complexed with phosphorus which was proved to be a phthalocyanine analogue–dihydroxy phosphorus (V) tetrabenzotriazacorrole (TBC).

4.2.1 Dihydroxy phosphorus (V) tetrabenzotriazacorroles

Due to its special three-dimensional π-electron structure and features which are different from phthalocyanines, the TBC macrocycle should have a potential application in photonic devices. However, to our knowledge, there is no report on third-order optical nonlinearity of dihydroxy phosphorus (V) tetrabenzotriazacorrole. Herein, we report the third-order nonlinear optical properties of a series phthalocyanine analogues, non-sulfonated {$P(OH)_2TBC$}, sulfonated {$P(OH)_2TBCSn$} and isopropoxyl substituted {$P(OH)_2TBC(O^iPr)_4$} dihydroxy phosphorus (V) tetrabenzotriazacorroles (the structures are shown in Figure 9) measured by femtosecond (50 fs) degenerate four-wave mixing technique (Huang et al., 2008).

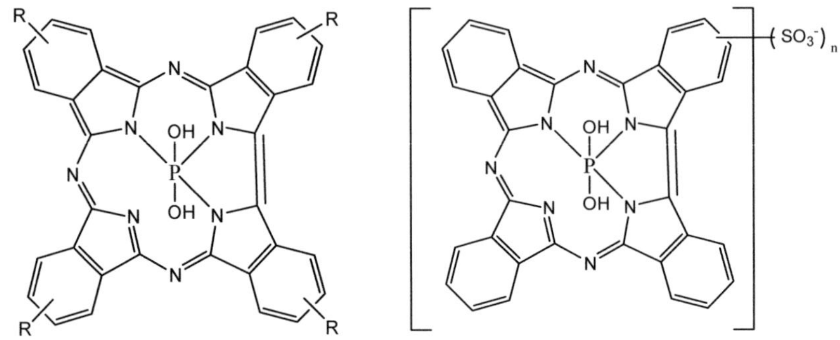

$P(OH)_2TBC$ R=H
$P(OH)_2TBC(O^iPr)_4$ R=O^iPr

$P(OH)_2TBCS_n$

(a) (b)

Fig. 9. Structural formulae of **(a)** $P(OH)_2TBC$ and $P(OH)_2TBC(O^iPr)_4$; **(b)** $P(OH)_2TBCSn$.

The evaluated values of $\chi(3)$ and γ for these phthalocyanine analogues are summarized in Table 3 together with the visible absorption maxima (λ_{max}) and nonlinear refractive index in DMF solutions. The γ of samples show high values as large as 10^{-31} esu, and the highest value of 2.26×10^{-31} esu for the $P(OH)_2TBC(O^iPr)_4$ was observed, which is almost two times larger than that of phthalocyanine analogue $P(OH)^2TBC$. This is probably due to the isopropoxyl group is attached to the benzene rings of the phthalocyanine analogue $P(OH)_2TBC(O^iPr)_4$, and isopropoxyl group is a electron-donating group which leads to large polarization of molecules. Considering molecular structure of invested phthalocyanine analogues, two hydroxyl moieties complexed with central phosphorus of molecule in axial direction, and form a three-dimensional configuration, which can enhance third-order optical nonlinearity of molecule. Moreover, their response times are also obtained and no more than 50 fs, which are commonly accepted to the contribution from the transient motion of the conjugate electron distribution.

Sample	λ_{max} (nm)	$\chi^{(3)}$ (×10-14esu)	γ (×10-14esu)	n_2(×10-14esu)
$P(OH)_2TBC$	655	4.40	1.21	8.15
$P(OH)_2TBCSn$	657	4.27	1.11	7.98
$P(OH)^2TBC(OiPr)_4$	677	4.91	2.26	9.01

Table 3. The evaluated values of $\chi^{(3)}$, γ, and n_2 for the dihydroxy phosphorus (V) tetrabenzotriazacorroles together with the visible absorption maxima (λ_{max}) in DMF solution.

4.2.2 Rare earth polymeric phthalocyanines

In 2008, we (Zhao et al., 2008) have synthesized three novel tri-dimensional phthalocyanine polymers with lanthanum (LaPPc), gadolinium (GdPPc) and ytterbium (YbPPc) as centric atoms from a tetranuclear phthalonitrile (the structures and synthesis route of rare earth polymeric phthalocyanines are shown in Scheme 3). And third-order optical nonlinearities of these compounds in DMF solution were measured by a picosecond Z-sacn technique at 532 nm.

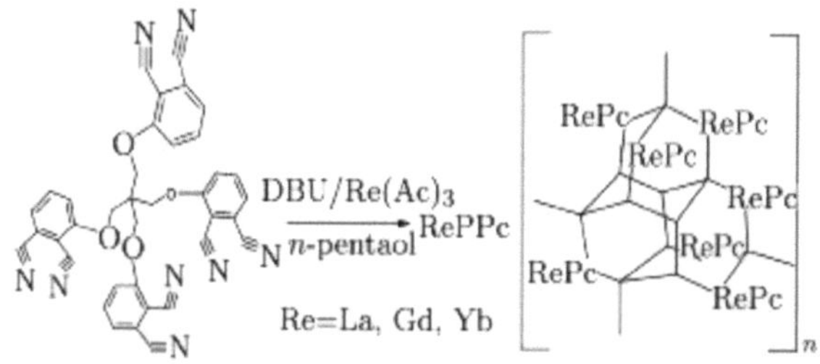

Scheme 3. Structures and synthesis route of rare earth polymeric phthalocyanines

Based on the Z-scan measurements, it is found that these phthalocyanine polymers show large third-order nonlinear susceptibilities which are of the order of 10^{-12} esu. Both the nonlinear absorption β and nonlinear refraction n_2 decreases with the order of LaPPc>GdPPc>YbPPc. For all the compounds, the values of $Re\chi^{(3)}$ are one order of magnitude larger than those of $Im\chi^{(3)}$, which determine the magnitude of third-order nonlinear susceptibilities $\chi^{(3)}$. This indicated that the nonlinear refraction is predominant mechanism for the nonlinear optical response of three phthalocyanine polymers.

Furthermore, Researchers (Manas et al., 1997) have described the effect of intermacrocycle interactions on the second hyperpolarizabilities of phthalocyanine dimer and trimer. It was considered that stack form of phthalocyanine can induce molecular electronic interaction between neighbouring phthalocyanine rings and induce charge transfer between them. As shown in Figure 10, the cage structure of phthalocyanine polymer presents more possibilities of electronic interaction and distortion of electron cloud in three dimensions at intermolecular scale. This would elevate the spatial polarizability of molecular and induces large nonlinear coefficient. However, the trimer or binuclear phthalocyanine could only offer this intermolecular possibility from two dimensions.

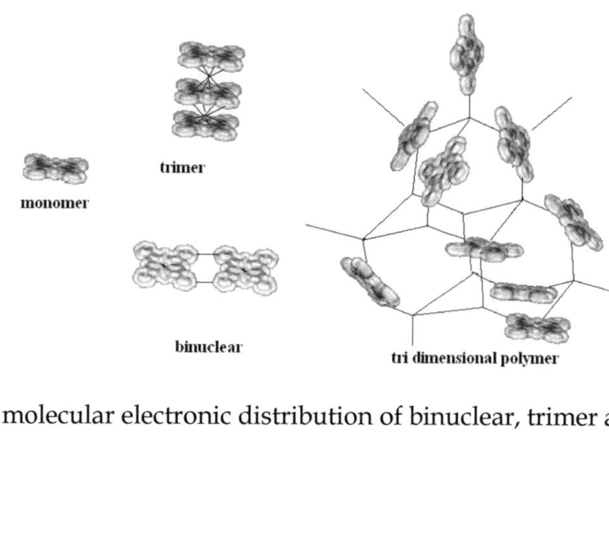

Fig. 10. Calculated molecular electronic distribution of binuclear, trimer and polymer of phthalocyanine.

5. Conclusions

Major advances have been made in the design and synthesis of the novel phthalocyanines and related compounds for third-order optical nonlinearity. They provide useful examples to illustrate the new features of the NLO response of the phthalocyanines and related compounds. This review summarizes our results recently obtained on the correlation between molecular structure and NLO response and offers some strategies for rendering new systems with improved NLO properties.

Owing to the extended π system, it is well-known that these Pc-related compounds exhibit a high aggregation tendency, and the aggregates usually display outstanding nonlinear optical properties. Moreover, by introducing peripheral substituents or by adding axial substituents, one may obtain three-dimensional (3D) structures with pyramidal shape and examine the effect of a third-dimension on the NLO response. Some of the four isoindole units can be formally replaced by other heterocyclic moieties, giving rise to different phthalocyanine analogues. All these chemical variations can alter the electronic structure of the macrocyclic core, and therefore, they allow the fine-tuning of the nonlinear response.

In conclusion, one could say that structural variations explored in phthalocyanines, such as metal insertion, introduction of functional groups into the periphery of the macrocycle, extension of conjugation, and variation of the main structure of the macrocycle, allow the tuning of the nonlinear responses. However, a detailed understanding of the factors affecting the nonlinear response is still necessary, and further work should be devoted to this objective.

6. Acknowledgments

We appreciate the help of Dr. Li Deng and Professor Zhenrong Sun, Key Laboratory of Optical and Magnetic Resonance Spectroscopy, East China Normal University, in carrying out the DFWM experiment. This work was supported by National Natural Science General Foundation of China (Grants 20773077, 20572059 and 20502013) and National Key Fundamental Research Program (2007CB808000).

7. References

Perry, J.W.; Mansour, K.; Mander, K.J.; Perry, K.J.; Alvarez, D. & Choong, I. (1994). Enhanced Reverse Saturable Absorption and Optical Limiting in Heavy-Atom-Substituted Phthalocyanines, *Optical Letters*, Vol.19,PP.625-627.

Perry, J.W.; Mansour, K.; Lee, I.-Y.S. & Wu, X.-L. (1996). Organic Optical Limiter with a Strong Nonlinear Absorptive Response, *Science*, Vol.273,PP.1533-1536.

Shirk, J.S.; Pong, R.F.S.; Flom, S.R.; Heckmann, H. & Hanack, M. (2000). Effect of Axial Substitution on the Optical Limiting Properties of Indium Phthalocyanines, *Journal of Physical Chemistry A*, Vol.104,PP.1438-1449.

Zyss, J. (Ed.). (1994). *Molecular Nonlinear Optics : Materials, phisics and Devices*, Academic Press, New York.

Ho, Z.Z.; Ju, C.Y. & Hetheringto, W.M. (1987). Third Harmonic Generation in Phthalocyanines, *Journal of Applied physics*, Vol.62, PP.716-718.

Shirk, J.S.; Lindle, F.J.; Bartoli, C.A.; Hoffman, C.A.; Kafafi, Z.H. & Snow, A.W. (1989). Off-Resonant Third-Order Optical Nonlinearities of Metal-Substituted Phthalocyanines, *Applied Phisics Letters*, Vol.55, PP.1287-1288.

Kambara, H.; Muruno, T.; Yamashita, A.; Matsumoto, S.; Hayashi, T.; Konami, H. & Tanaka, N. (1996). Third-Order Nonlinear Optical Properties of Phthalocyanine and Fullerene, *Journal of Applied Physics*, Vol.80, PP.3674-3682.

Ma, g.; He, J.; Kang, C. & Tang, S. (2003). Excited State Dynamics Studies of Iron(III) Phthalocyanine Using Femtosecond Pump-Probe Techniques, *Chemical Physics Letters*, Vol.370, PP.293-299.

Zhou, J.; Mi, J.; Zhu, R.; Li, B. & Qian, S. (2004). Ultrafast Excitation Relaxation in Titanylphthalocyanine Thin Film, *Optical Materials*, Vol.27, PP.377-382.

He, C.; Wu, Y.; Shi, G.; Duan, W.; Song, W. & Song, Y. (2007). Large Third-Order Optical Nonlinearites of Ultrathin Films Containing Octacarboxylic Copper Phthalocyanine, *Organic Electronics*, Vol.8, PP.198-205.

Hermann, J.P.; Ricard, D. & Ducuing, J. (1973). Optical Nonlinearities in Conjugated Systems: β-Carotene, *Applied Physics Letters*, Vol.23, PP.178-180.

Linder, R.C.; Steel, D.G. & Dunning, G.J. (1982). Phase Conjugation by Resonantly Enhanced Degenerate Four-Wave Mixing, *Optical Engineering*, Vol.21, PP.190-198.

Sheik-Bahae, M.; Said, A.A. & van Stryland, E.W. (1989). High-Sensitivity, Single-Beam n_2 Measurements, *Optical Letters*, Vol.14, PP.955-957.

Sheik-Bahae, M.; Said, A.A.; Wei, T.H.; Hagan, D.J. & van Stryland, E.W. (1990). Sensitive Measurement of Optical Nonlinearities Using a Single Beam, *IEEE Journal of Quantum Electronics*, Vol.26, PP.760-769.

Torre, G.; Vazquez, P.; Agullo-Lopez, F. & Torres, T. (2004). Role of Structure Facror in the Nonlinear Optical Properties of Phthalocyanines and Related Compounds, *Chemical Reviews*, Vol.104, PP.3723-3750.

Nalwa, H.S.; Kakuta, A. & Mukoh, A. (1993). Third-Order Optical Nonlinearities of Tetrakis-*n*-Pentoxy Carbonyl Metallo-Naphthalocyanines, *Chemical Physics Letters*, Vol.203, PP.109-113.

Nalwa, H.S.; Hanack, M.; Pawlowski, G. & Klaus, E.M. (1999). Third-Order Nonlinear Optical Properties of Porphyrazine, Phthalocyanine and Naphthalocyanine Germanium Derivatives: Demonstrating the Effect of π-Conjugation Length on Third-Order Optical Nonlinearity of Two-Dimensional Molecules, *Chemical Physics*, Vol.245, PP.17-26.

He, C.; Chen, Y.; Nie, Y. & Wang, D. (2007). Third Order Optical Nonlinearities of Eight-β-Octyloxy-Phthalocyanines, *Optics Communications*, Vol.271, PP.253-256.

Li, Z.; Chen, Z.; Xu, S.; Niu, L.; Zhang Z.; Zhang, F. & Kasatni, K. (2007). Off-Rsonant Third-Order Optical Nonlinearities of Novel Diarylethene-Phthalocyanine Dyads, *Chemical Physics Letters*, Vol.447, PP.110-114.

Chen, Z.; Zhong, C.; Zhang, Z.; Li, Z.; Niu, L.; Bin, Y. & Zhang, F. (2008). Photoresponsive J-Aggregation Behavior of a Novel Azobenzene-Phthalocyanine Dyad and Its Third-Order Optical Nonlinearity, *Journal of Physical Chemistry B*, Vol.112, PP.7387-7394.

Huang, X.; Zhao, F.; Li, Z.; Tang, Y.; Zhang, F. & Tung, C.-H. (2007). Self-Assembled Nanowire Networks of Aryloxy Zinc Phthalocyanines Based on Zn-O Coordination, *Langmuir*, Vol.23, PP.5167-5172.

Huang, X.; Zhao, F.; Li, Z.; Huang, L.; Tang, Y.; Zhang, F. & Tung, C.-H. (2007). A Novel Self-Aggregates of Phthalocyanine Based on Zn-O Coordination, *Chemistry Letters*, Vol.36, PP.108-109.

Chen, Z.; Dong, S.; Zhong, C.; Zhang, Z.; Niu, L.; Li, Z. & Zhang, F. (2009). Photoswitching of Third-Order Optical Properties of Azobenzene-Containing Phthalocyanines Based on Reversible Host-Guest Interactions, *Journal of Photochemistry and Photobiology A: Chemistry*, Vol.206, PP.213-219.

Breslow, R. & Dong, S. (1998). Biomimetic Reactions Catalyzed by Cyclodextrins and Their Derivatives, *Chemical Review*, Vol.98, PP.1997-2012.

Dugave, C. & Demange, L. (2003). Cis−Trans Isomerization of Organic Molecules and Biomolecules: Implications and Applications, *Chemical Review*, Vol.103, PP.2475-2532.

González-Cabello, A.; Claessens C, G.; Martin-Fuch, G.; Ledoux-Rack, I.; Vázquez, P.; Zyss, J.; Agulló-López, F. & Torres, T. (2003). Phthalocyanine-Ferrocene Dyads and Triads for Nonlinear Optics, *Synthsis Metal*, Vol. 137, PP.1487-1488.

Bin, Y.; Xu, S.; Li, Z.; Huang, L.; Zhang, Z. & Zhang, F. (2008). Large Third-Order Optical Nonlinearity of a Novel Copper Phthalocyanine Ferrocene, *Chinese Physics Letters*, Vol.25, PP.3257-3259.

Liu, J.; Zhang, F. ; Zhao, Y ; Tang, Y ; Song, X. & Yao, G. (1995). Complexation of Phosphorus (III) with a Novel Tetrapyrrolic Phthalocyanine-Like Macrocyclic Compound, *Journal of Photochemistry and Photobiology A: Chemistry*, Vol.91, PP.99-104.

Chen, Z.; Zhou, X.; Li, Z.; Niu, L.; Yi, J. & Zhang, F. (2011). The Third-Order Optical nonlinearities of Thiophene-Bearing Phthalocyanines Studied by Z-Scan Technique, *Journal of Photochemistry and Photobiology A: Chemistry*, Vol.211, PP.64-68.

Huang, L. Li, Z. ; Zhang, F.; Tung, C. & Kasatani, K. (2008). Off-Resonant Optical Nonlinearities of Phthalocyanine Analogues: Dihydroxy Phosphorus (V) Tetrabenzotriazacorroles, *Optics Communication*, Vol.281, PP.1275-1279.

Zhao, P.; Xu, S.; Li, Z. & Zhang, F. (2008). Nonlinear Optical Properties of Novel Polymeric Rare Earth Phthalocyanine Studied Using Picosecond Z-scan Technique, *Chinese Physics Letters*, Vol.6, PP.2058-2061.

Manas, E.S ; Spano, F.C & Chen, L.X. (1997). Nonlinear Optical Response of Cofacial Phthalocyanine Dimers and Trimers, *Journal of Chemical Physics*, Vol.107, PP.707-719.

14

Broadband Instability of Electromagnetic Waves in Nonlinear Media

Sergey Vlasov, Elena Koposova and Alexey Babin
Institute of Applied Physics, Russian Academy of Science
Russia

1. Introduction

Wave propagation in nonlinear media is usually accompanied by variation of their space-time spectra (V.I. Bespalov & Talanov, 1966; Lighthill, 1965; Litvak & Talanov, 1967; Benjamin & Feir, 1967; Zakharov & Ostrovsky, 2009; Bejot et al., 2011). The character of this variation is frequently explained in terms of modulation instability which occurs, in particular, for electromagnetic waves in cubic media, where polarization is approximately equal to the cube of electric field intensity. Modulation instability manifests itself in partitioning of the originally uniform packet into separate beams and pulses. It has been studied in ample detail for the perturbations whose frequencies and propagation directions slightly differ from those of intense waves (pump waves), i.e., temporal and spatial spectra of the waves participating in the process are rather narrow (paraxial approximation) (V.I. Bespalov & Talanov, 1966; Litvak & Talanov, 1967; Agraval, 1995; Talanov & Vlasov, 1997). In this case, modulation instability is described by the nonlinear parabolic equation for a wave train envelope.

The instability is quite different, if the nonlinearity is higher than the third order of magnitude. It was shown in (Talanov & Vlasov, 1994; Koposova & Vlasov, 2007) that the wave propagating in such a medium may be unstable relative to collinear perturbations at frequencies so high that not only wave packet envelope but the structure of each wave in the packet may change too. The parabolic equation does not hold for description of such phenomena; hence, the methods for solution of wave equations in a wide frequency band developed in (Talanov & Vlasov, 1995; Brabec & Krausz, 1997; V.G. Bespalov et al., 1999; Kolesic & Moloney, 2004; Ferrando et al., 2005; Koposova et al., 2006) are employed. One of such methods, namely, the technique of pseudodifferential operators (Koposova et al., 2006) is used in the current paper.

The first part of the paper that is an extension of (Talanov & Vlasov, 1994) is concerned with the instability of perturbation waves at combination frequencies noncollinear to pump for the case of nonlinear polarization represented as a polynomial of arbitrary but finite degree. Further, application of the theory to air, for which dielectric permittivity may be represented in the form of a polynomial, is addressed (Loriot et al., 2009). Finally, methods of finding amplitude and phase distribution of electric field in an ultrashort wave packet by analyzing signals from intensity autocorrelator traditionally used for measuring duration of ultrashort laser pulses are considered.

2. Plane wave instability in media with polynomial nonlinearity

Consider a linearly polarized wave packet propagating along the z-axis. We will make use of the equations for field E in this beam:

$$\frac{\partial E}{\partial z}+i\sqrt{-\frac{1}{c^2}\frac{\partial^2}{\partial t^2}n^2(-i\frac{\partial}{\partial t})+\Delta_\perp}\,E=i\frac{1}{2\sqrt{-\frac{1}{c^2}\frac{\partial^2}{\partial t^2}n^2(-i\frac{\partial}{\partial t})+\Delta_\perp}}\frac{4\pi}{c^2}\frac{\partial^2 P_{NL}}{\partial t^2}, \tag{1}$$

obtained in (Koposova et al., 2006). In Eq. (1), $n(-i\frac{\partial}{\partial t})$ is the operator describing linear dispersion of the medium; for the processes $\sim\exp[i\omega t]$ stationary in time, $n(\omega)$ is the index of refraction at circular frequency ω; $\Delta_\perp$ is Laplace operator in $\vec{r}_\perp$ coordinates transverse to the propagation direction; P_{NL} is nonlinear polarization; and c is the velocity of light. Assuming the angular difference between the directions of the interacting waves to be small, we will neglect dispersion of the nonlinear polarization coefficient and the Laplace operator in the right-hand side of (1) and suppose that

$$\frac{1}{\sqrt{-\frac{1}{c^2}\frac{\partial^2}{\partial t^2}n^2(-i\frac{\partial}{\partial t})+\Delta_\perp}}\approx\frac{c}{\omega\sqrt{\varepsilon(\omega)}},$$

where $\varepsilon(\omega)=n^2(\omega)$ is the dielectric permittivity at the same frequency.
We will describe the polarization by the polynomial dependence on the magnitude of electric field:

$$P_{NL}=\sum_{i=1}^{S}\chi^{(2N+1)}E^{2N+1}, \tag{2}$$

where $\chi^{(2N+1)}$ are $(2N+1)-th$ order susceptibilities, and $2S+1$ is the highest degree polarization taken into consideration.
To the first approximation of the asymptotic theory of nonlinear oscillations (Bogolyubov & Mitropol'skii, 1958), a steady-state plane wave

$$E(t,\vec{r})=\frac{A_0\exp[i\omega_0 t-ih_0 z]+A_0^*\exp[-i\omega_0 t+ih_0 z]}{2} \tag{3}$$

having amplitude A_0, frequency ω_0, and propagation constant $h_0=k_0(1+\frac{\delta\varepsilon_{NL}}{2})$ (where $k_0=\frac{\omega_0}{c}n(\omega_0)$ is the propagation constant of the wave of the same frequency in a linear medium and $\delta\varepsilon_{NL}$ is a nonlinear additive to permittivity) may propagate in the medium with permittivity (2). The nonlinear additive is represented as a sum $\delta\varepsilon_{NL}=\sum_{N=1}^{S}\delta\varepsilon_{2N+1,NL}$ of nonlinear additives

$$\delta\varepsilon_{2N+1,NL}=\frac{4\pi}{\varepsilon(\omega_0)}\frac{(2N+1)!}{N!(N+1)!2^{2N}}\chi^{(2N+1)}\left|A_0\right|^{2N}=\varepsilon'_{2N+1}\left|A_0\right|^{2N} \tag{4}$$

generated by each term of the series (2) and related to the nonlinearity of degree $2N+1$, the nonlinearity coefficient ε'_{2N+1} in (4) is of the $(2N+1)-th$ order of magnitude.

We will study stability of the solution relative to two-frequency perturbations of the form $u_1\exp[i(\omega_1 t-hz)-i\vec{k}_\perp\vec{r}_\perp]+u_2\exp[i(\omega_2 t-hz)+i\vec{k}_\perp\vec{r}_\perp]$, $\vec{k}_\perp$ being their transverse wave number. Wave frequencies ω_1 and ω_2 and their propagation constant h will be found from solution of dispersion equations for perturbations.

We will seek solution of Eq. (1) in the form

$$E=\frac{A_0\exp(i\varphi_0)+u_1\exp[i\varphi_1-i\vec{k}_\perp\vec{r}_\perp]+u_2\exp[i\varphi_2+i\vec{k}_\perp\vec{r}_\perp]+c.c.}{2}, \tag{5}$$

where $\varphi_0=\omega_0 t-h_0 z$, $\varphi_1=\omega_1 t-hz$, $\varphi_2=\omega_2 t-hz$. Among the polarization terms with the perturbation of the first degree u_1 and u_2^* we will select the terms of identical degrees A_0 and A_0^*:

$$\sim|A_0|^{2N}u_1\exp[i\varphi_1-i\vec{k}_\perp\vec{r}_\perp] \tag{6}$$

and the terms whose difference of degrees A_0 and A_0^* is equal to the even integer $2M\le 2N$:

$$\sim A_0^{N+M}A_0^{*N-M}\exp[i(N+M)\varphi_0-i(N-M)\varphi_0]u_2^*\exp[-i\varphi_2-i\vec{k}_\perp\vec{r}_\perp]\ . \tag{7}$$

If the conditions

$$2M\omega_0=\omega_1+\omega_2\,,\qquad Mh_0=h \tag{8}$$

are fulfilled, for definite values of $\omega_{1,2}$ and h the terms (6) and (7) will be synchronous, i.e., they will have identical frequencies and propagation constants. Sets of frequencies $\omega_{1,2}$ and propagation constants h will be different for $M=1$, $M=2$, and so on. In other words, solution of the form (5) may have M branches, with the characteristics of branch M depending on nonlinear polarization terms with indices $M\le N\le S$. For $M=1$, we have perturbations near carrier frequency - the well-known modulation instability describing variations of the wave packet envelope that are slow compared to the carrier.

Making use of the equality (8) we will express the frequencies $\omega_1=M\omega_0+\Omega,\ \ \omega_2=M\omega_0-\Omega$ through their difference $\Omega=\dfrac{\omega_1-\omega_2}{2}$ and frequency $M\omega_0$. Assuming that $k(\omega)=\dfrac{\omega n(\omega)}{c}$ is real in the frequency range under consideration, from Eq. (1) we obtain for the functions u_1 and u_2^* the following system of equations

$$\begin{aligned}\frac{\partial\{u_1\}}{\partial z}&=-i\beta_{M+}u_1-i\gamma_{M+}u_2^*\\ \frac{\partial\{u_2^*\}}{\partial z}&=i\beta_{M-}u_2^*+i\gamma_{M-}u_1,\end{aligned} \tag{9}$$

where

$$\beta_{M\pm}=\sqrt{k^2(M\omega_0\pm\Omega)-k_\perp^2}-Mh_0+k_0\frac{(M\omega_0\pm\Omega)n(\omega_0)}{\omega_0 n(M\omega_0\pm\Omega)}\sum_{N=1}^{S}\frac{N+1}{2}\delta\varepsilon_{2N+1,NL}\,,$$

$$\gamma_{M\pm} = k_0 \frac{(M\omega_0 \pm \Omega)n(\omega_0)}{\omega_0 n(M\omega_0 \pm \Omega)} \frac{A_0^{\pm M}}{A_0^{*\pm M}} \sum_{N=M}^{S} \frac{\delta\varepsilon_{NL,2N+1} N!(N+1)!}{2(N-M)!(N+M)!}.$$

Solutions to (9) are sought in the form

$$u_1 = A_1 \exp[-iH_1 z],$$

$$u_2^* = A_2^* \exp[-iH_1 z],$$

constants A_1 and A_2^* are found from the system of linear equations with constant coefficients the determinant of which is a characteristic equation for the corrections H_M to the propagation constants $Mh_0 = h$. The determinant has the form

$$\{H_M - \beta_{M+}\}\{H_M + \beta_{M-}\} + \gamma_{M+}\gamma_{M-} = 0. \tag{10}$$

For the corrections H_M to the propagation constants we have

$$H_M = \frac{\beta_{M+} - \beta_{M-}}{2} \pm \sqrt{\frac{(\beta_{M+} + \beta_{M-})^2}{4} - \gamma_{M+}\gamma_{M-}}. \tag{11}$$

By wave of illustration we present in Fig. 1 the diagram of wave vectors $\vec{k}(\omega_1)$, $\vec{k}(\omega_2)$, $\vec{k}(\omega_0)$ for the considered effects in the case $M = 2$, when the inequality $k(2\omega_0) = \frac{2\omega_0}{c} n(2\omega_0) > 2k(\omega_0) = \frac{2\omega_0}{c} n(\omega_0)$ or $n(2\omega_0) > n(\omega_0)$ is met.

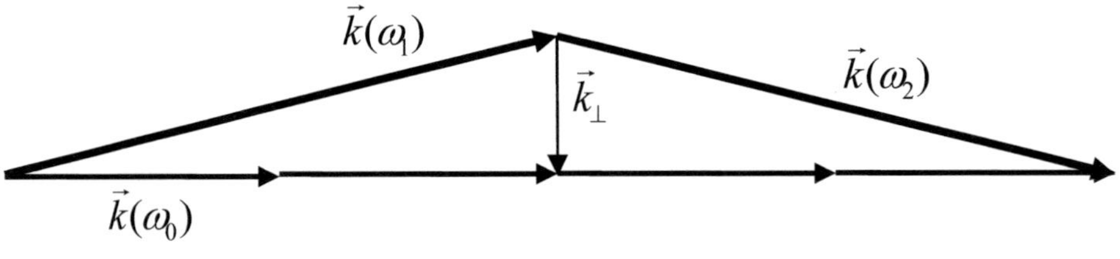

Fig. 1. Wave vector diagram of converting frequency ω_0 to the frequency close to the second harmonic for modulation instability and condition $|k(\omega_1)| > 2|k(\omega_0)|$.

The latter inequality is satisfied if the frequencies ω_0 and $2\omega_0$ are in one transparency band of the substance (Born & Volf, 1964). The wave vector diagram shows that in this case the waves at frequencies ω_0 and $2\omega_0$ may be synchronized during propagation in the direction of vector $\vec{k}_0$, which is accompanied by transformation of four vectors (quanta) of frequency ω_0 into two vectors (quanta) of frequency $2\omega_0$.

Let us study (12) in more detail in the quasiparaxial approximation, omitting Ω everywhere except the expression $\sqrt{k^2(M\omega_0 \pm \Omega) - k_\perp^2}$. Let us designate mismatches as

$$\Delta k_M = \frac{\sqrt{k^2(M\omega_0 + \Omega) - k_\perp^2} + \sqrt{k^2(M\omega_0 - \Omega) - k_\perp^2}}{2k_0} - M.$$

For $M=1$ (modulation instability) the radicand (11) is rewritten in the form

$$D=(\Delta k)^2+(\Delta k)\Delta\varepsilon\,. \tag{12}$$

In (12) we have

$$\Delta\varepsilon=\sum_{N=1}^{S}N\delta\varepsilon_{2N+1,NL}=\sum_{N=1}^{S}I\frac{d}{dI}\delta\varepsilon_{2N+1,NL}=I\frac{d}{dI}\delta\varepsilon_{NL}$$

that is equal to the product of the derivative of the nonlinear additive to intensity and intensity. In the case of cubic nonlinearity, this expression is equal to the additive.

The modulation instability occurs (Litvak & Talanov, 1967) at $D<0$. Its behavior changes as a function of the signs of Δk and $\Delta\varepsilon$. For $\Delta\varepsilon>0, \Delta k<0$, perturbations with spatial scale are more pronounced ("self-focusing" instability); whereas for $\Delta\varepsilon<0, \Delta k>0$, perturbations with temporal scales come to the forefront. Note that the modulation instability increments turn to zero at $(\Delta k)=0$ and

$$\Delta\varepsilon=I\frac{d}{dI}\delta\varepsilon_{NL}=\sum_{N=1}^{S}N\delta\varepsilon_{2N+1,NL}=0\,. \tag{13}$$

From the latter condition it follows that the increments produced by different terms in the expansion (2) may "obliterate" each other under certain conditions.

For $M\geq 2$, in the paraxial approximation we obtain

$$H_M=k_0(\Delta k_M\pm\sqrt{(\Delta k_M+\Delta\varepsilon)^2-(\Delta\overline{\varepsilon})^2}), \tag{14}$$

where

$$\Delta\overline{\varepsilon}=\frac{M}{2}\sum_{N=M}^{S}\frac{\delta\varepsilon_{NL,2N+1}N!(N-1)!}{(N-M)!(N+M)!}$$

The propagation constant H_M will be a complex one, and instability will occur, given $|\Delta k_M+\Delta\varepsilon|<|\Delta\overline{\varepsilon}|$. The increment reaches its maximum near zeros of the expression

$$\Delta k_M+\Delta\varepsilon=0\,, \tag{15}$$

with the increment being of order

$$H_M=k_0\Delta\overline{\varepsilon}\,. \tag{16}$$

The expression may become zero at an arbitrary sign of nonlinear additive due to appearance of a transverse component of the perturbation wave number on a certain curve

$$\Delta k_M+\Delta\varepsilon\approx M[\frac{n(M\omega_0)}{n(\omega_0)}-1]+\frac{\Omega^2}{2k_0}\frac{\partial^2 k}{\partial\omega^2}\mid_{\omega=M\omega_0}-\frac{k_\perp^2}{2k_0^2}+\Delta\varepsilon=0 \tag{17}$$

on the $\omega,k_\perp$-plane. At normal dispersion $\frac{\partial^2 k}{\partial\omega^2}\mid_{\omega=M\omega_0}>0$ and frequencies $M\omega_0$ and ω_0 located in one transparency band of the substance, the equality (15) is fulfilled for the hyperbolae (17). There exists in this case a minimal value of the transverse wave number

$$\frac{k_\perp}{k_0} \approx \sqrt{M[\frac{n(M\omega_0)}{n(\omega_0)} - 1]}, \quad M \geq 2 \tag{18}$$

at which instability occurs at arbitrary weak nonlinearity. Fulfillment of the equality (18) indicates the presence of conic radiation.

3. Intense plane wave instability in air

Consider the effects in air. The dependence at normal pressure of the nonlinear index of refraction $n_{NL} = \frac{\sqrt{\varepsilon_0}}{2} \sum_{N=1}^{S} \varepsilon'_{2N+1} |A_0|^{2N}$ on intensity $I = \frac{c}{8\pi} |A_0|^2$:

$$n_{NL} = \sum_{N=1}^{S} n_{NL,N} I^N \tag{19}$$

taken from (Loriot et al., 2009). is shown in Fig. 2. Curve 1 is plotted taking into consideration four (all known from (Loriot et al., 2009)) terms, curve 2 taking into consideration two terms, and curve 3 taking into consideration only the first term, when purely cubic nonlinearity occurs. Note that for curve 1 there exist unstable branches at $M = 1,2,3,4$, and for curve 2, despite its qualitative coincidence with curve 3 (one maximum), the instability branches exist only at $M = 1,2$; for curve 3 instability known as modulation (self-focusing) instability occurs at $M = 1$.

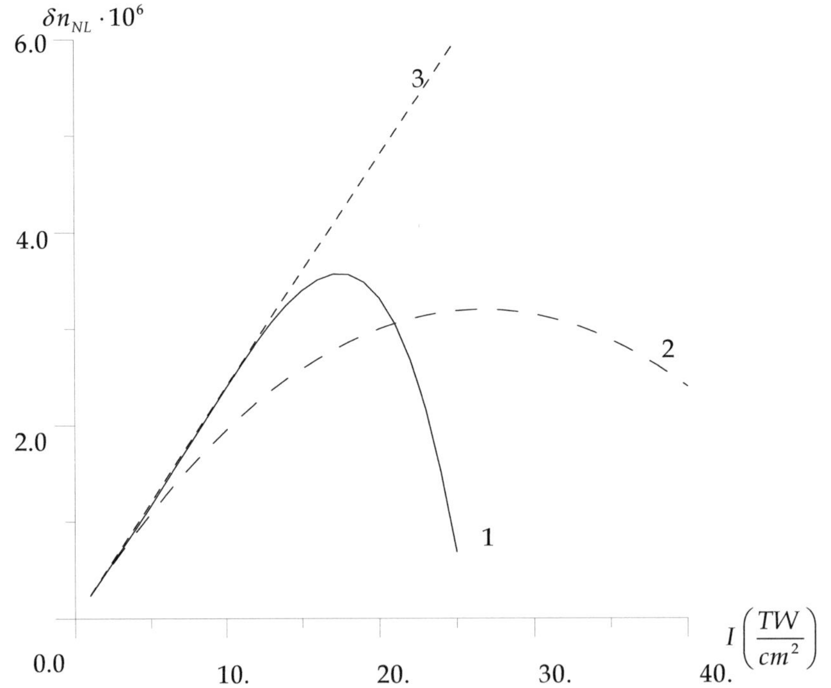

Fig. 2. Nonlinear additive to the index of refraction of air as a function of intensity 1– four-term approximation, 2– two-term approximation, 3– one-term approximation.

The instability regions on the plane of transverse wave numbers and frequencies within the 5-octave band for air, where the frequency dependence of refractive index is known (Grigor'ev & Melikhov, 1991), are shown in Fig. 3 for two values of intensity: $I = 15\,TW / cm^2$ and $I = 19\,TW / cm^2$ for $\omega_0 = 7.85 \cdot 10^{14} s^{-1}$, which corresponds to the radiation wavelength $\lambda_0 = \frac{2\pi}{k_0} \approx 0.8 \mu m$. The shadow density is proportional to the value of the normalized increment $\bar{H}_M = \frac{H_M}{k_0}$. In the first case, the intensity is smaller than its value at maximum nonlinear additive to permittivity and $\Delta\varepsilon > 0$. In the second case, the intensity is larger than its value at maximum nonlinear additive to permittivity and $\Delta\varepsilon < 0$.

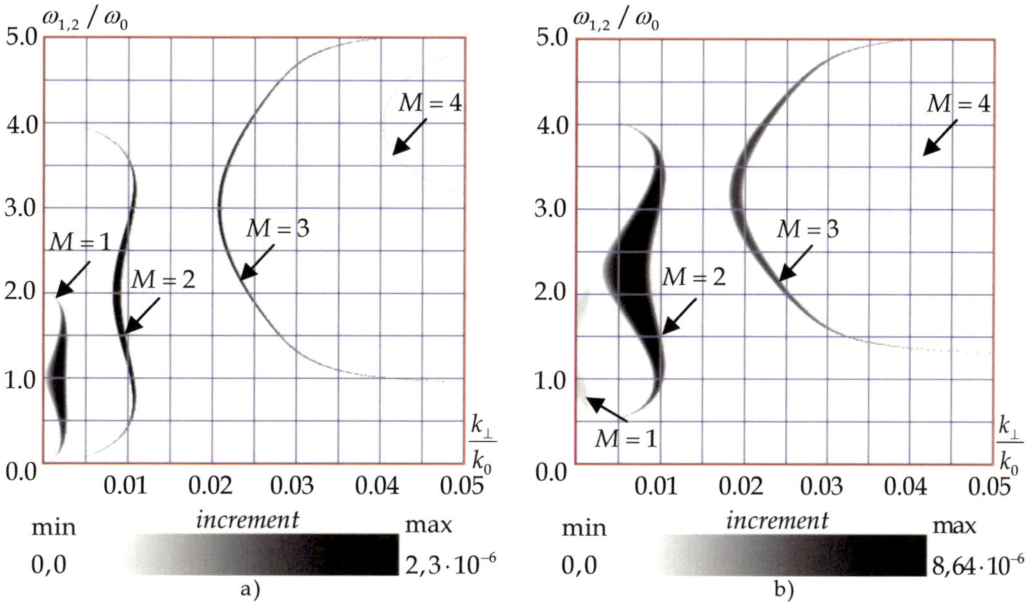

Fig. 3. Instability increment $\bar{H}_M$ in air on the plane of parameters $k_\perp / k_0, \Omega / \omega_0$ for $I = 15\,TW / cm^2$ (a) and $I = 19\,TW / cm^2$ (b) at $\lambda_0 = 0.8 \mu m$ for different instability branches $M = 1,2,3,4$.

For the values of intensities given above, the increments are different for all branches. At small intensities, $I < 15\,TW / cm^2$, the increment is largest for $M = 1$ (modulation instability); at large intensities it is largest for $M = 2$. The increment for perturbations with $M = 2,3$ attains its maximum near the frequencies $2\omega_0$ and $3\omega_0$, respectively, for the values of $\kappa_\perp$ satisfying the equality (15). It should be born in mind that accuracy of estimates becomes worse near the boundaries of the frequency interval $\omega \ll \omega_0$, $\omega \sim 5\omega_0$.

Fig. 4, that supplements Fig. 3., demonstrates maximal increments as a function of intensity, with the maximum attained for each branch and each value of intensity at definite values of frequency and transverse wave number that are also functions of intensity and number of the branch. For the branch $M = 1$, the increment becomes small when $\Delta\varepsilon$ vanishes to zero; for the other branches, the increments grow with increasing intensity in the $I > 10\,TW / cm^2$ region.

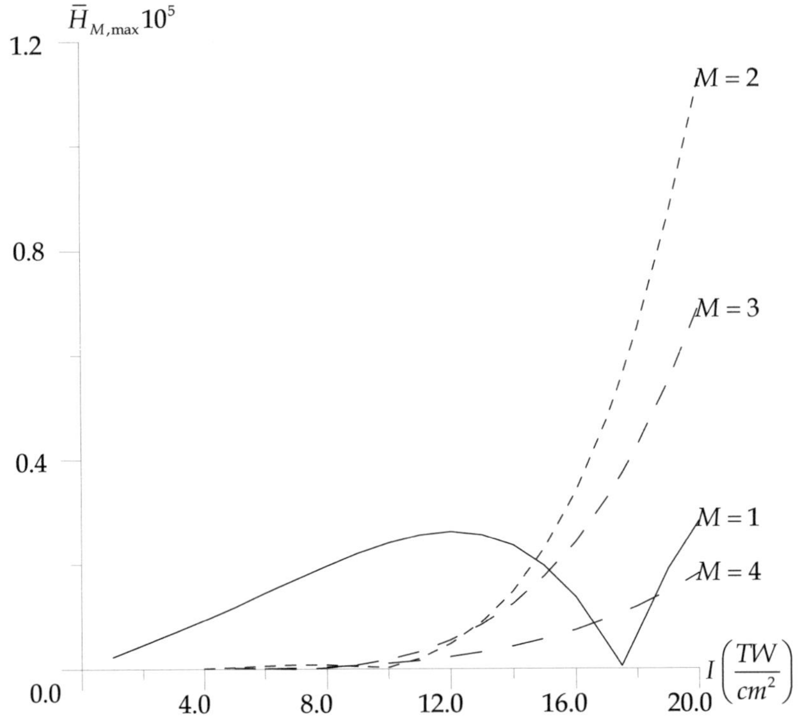

Fig. 4. Maximum value of increment $\bar{H}_{M,\max}$ versus intensity.

The studied instability may be one of the mechanisms of conic beam generation (Nibbering et al., 1996; Dormidontov et al., 2010).

4. Analysis of characteristics of ultrashort wave packet phase modulation by means of autocorrelation intensity function

It has been demonstrated above that the originally spectrum-limited powerful short wave packet is phase modulated during propagation in a nonlinear medium. In this Section we will show that some characteristics of the acquired phase modulation may be measured by analyzing the intensity autocorrelator signal usually used for measuring duration of ultrashort laser pulses. In addition, it is possible to qualitatively assess by the shape of this signal the presence of phase modulation in the studied light signal. Basically, solution of this problem allows retrieving amplitude and phase distribution of electric field in an ultrashort wave packet.

Methods for retrieving the total field of an optical pulse may conventionally be divided into three groups. The first group includes techniques based on interference measurements. They have a long history and, consequently, have been developed most comprehensively. It is clear that, based on the interference of an ultrashort pulse in the temporal or spectral domain, one can in principle derive information on the phase distribution of the studied field. A great number of publications, from which we cite only the key papers (Kuznetsova, 1968; Verevkin et al., 1971; Sala et al., 1980; Diels et al., 1985;

Naganuma et al., 1989; Iaconis & Walmsley, 1998), are devoted to interference measurements. Another group of fairly recent publications (Naganuma et al., 1989; Iaconis & Walmsley, 1998; Delong et al., 1996; Koumans & Yariv, 1999; Kane, 1999) take advantage of the mathematically proven fact that a multi-dimensional (noninterference) auto- or cross-correlation function allows retrieving the phase distribution of a short pulse. A two-dimensional signal distribution on the frequency-delay plane is usually measured in experiment, and then an iterative processing algorithm yields the desired parameters. The most advanced to-date technique of this group is referred to as FROG in the literature (Kane, 1999). The third group of methods (Naganuma et al., 1989; Kane, 1999; Peatross et al., 1998; Nicholson et al., 1999), that are based on the results of simultaneous spectral and correlation energy measurements and some iterative algorithm, also make it possible to determine phase and amplitude characteristics of an optical pulse.

The interference methods of retrieving ultrashort wave packet field parameters are simpler for implementation and in terms of processing than the methods of the second group. However, despite the fact that the speed of obtaining qualitative information for the latter techniques often does not meet the needs of experiment, such techniques are clearer, since the form of phase modulation of an optical pulse can qualitatively be deduced immediately from a two-dimensional distribution of the obtained signal. In this respect, methods of the third group are much faster, as the corresponding iterative retrieval procedure is based on operations with one-dimensional data files. However, the problem of accuracy of retrieval of the desired parameters remains open for such methods. In this part we demonstrate that in certain cases, the interference methods can also ensure clear, simple, and fast acquisition of information on the amplitude and phase of an optical pulse. In our opinion, the present part is methodical to a significant extent, although it has practical applications related to measuring dispersion characteristics of optical materials. It can easily be shown that the output signal from the photodetector of an interferometric intensity autocorrelator, whose scheme can be found in a number of publications (e.g., Krukov, 2008), has the form:

$$U(\tau) \approx \int_{-\infty}^{\infty} \left\{ \rho^2(t) + \rho^2(t+\tau) + 2\rho(t) * \rho(t+\tau) * \cos[\phi(t) - \phi(t+\tau)] \right\}^2 dt \tag{20}$$

where $\rho(t)$ and $\phi(t)$, respectively, are the slowly varying amplitude (envelope) and phase of the optical-pulse field

$$E(t) = \rho(t) * \exp[j\varphi(t)] + c.c. \tag{21}$$

Equation (20) is known (Akhmanov, Vysloukh & Chirkin, 1988) to be valid as long as the inequality $\omega_0 \tau_0 > 1$ is satisfied, i.e., if the envelope $\rho(t)$ contains at least a few optical cycles. Here, ω_0 and τ_0 are the central frequency and duration of the wave packet envelope, respectively. Note that, strictly speaking, the signal (20) contains the intensity autocorrelation function

$$G(\tau) = \int_{-\infty}^{\infty} \rho^2(t) * \rho^2(t+\tau) dt \tag{22}$$

but is not equal to it. In addition to an informative component, Eq. (20) for $U(\tau)$ comprises a constant background which usually impedes experimental measurement of temporal characteristics. This well-known fact was extensively discussed in the literature (Kuznetsova, 1968; Krukov, 2008). It is not difficult to design the scheme of the so-called background-free autocorrelator for which the output signal is free of this background. However, it follows from Eq. (20) that in this case, information on the phase structure of a pulse is completely lost. Nevertheless, such devices for measuring temporal parameters of ultrashort pulses are widely used in experiments, as they yield fairly reliable information on the wave packet envelope duration.

It follows from Eq. (20) that the signal $U(\tau)$ contains information on the time dependence of the envelope amplitude and phase. Typically of inverse problems, it is impossible to retrieve parameters of the field $E(t)$ in the general case, since the integral equation (20) is ill-posed. However, in certain particular cases of practical importance, $\rho(t)$ and $\phi(t)$ can be found from this expression. Consider this possibility in more detail.

Assume that the slowly varying envelope $\rho(t)$ has a Gaussian profile, while the phase $\phi(t)$ can be described by a cubic polynomial:

$$\rho(t) = \rho_0 * \exp(-\frac{t^2}{2\tau_0^2}); \qquad \varphi(t) = \omega_0 * t + t^2 * \frac{\alpha}{2} + t^3 * \frac{\beta}{3}. \tag{23}$$

where is the coefficients α and β refer to the linear and quadratic frequency chirps, respectively. Within the framework of this assumption, the temporal distribution of the wave packet envelope is determined by one parameter, the envelope duration τ_0, and the phase distribution by two coefficients α and β. Representation of the phase in the form given by Eq. (23) corresponds to a Taylor expansion in which any higher-order term is much less than the previous lower-order one. It usually suffices to use such a phase expansion for an ultrashort wave packet, for which Eq. (21) is valid, propagated in a substance with weak dispersion, i.e., in the transparency band. Let us firstly put $\beta = 0$, i.e., allow for only a linear chirp of an input optical pulse. Then Eq. (20) for the autocorrelator output signal $U(\tau)$ takes the form

$$U(\tau) \approx 1 + 2G(\tau) + [G(\tau)]^{L^2} \cos 2\omega_0\tau + 4[G(\tau)]^{(2+L^2)/4} \cos\omega_0\tau * \cos\left[\sqrt{L^2-1} * \tau^2 / 4\tau_0^2\right], \tag{24}$$

Here, L is the temporal-compression ratio of an initial phase modulated pulse with envelope duration τ_0 due to compensation for its quadratic phase (Akhmanov, Vysloukh & Chirkin, 1988):

$$L^2 = (\frac{\Delta\omega}{\Delta\omega_0})^2 = 1 + (\alpha\tau_0^2)^2, \tag{25}$$

$\Delta\omega$ is the spectrum width of an input optical signal, and $\Delta\omega_0 = 1/\tau_0$ is the spectrum width of a transform limited pulse for which $\alpha = 0$ and, therefore, $L = 1$. Note that Eq. (24) was derived in (Sala, Kenney-Wallace & Hall, 1980; Diels, Fontane, McMichel & Simoni, 1985). However, the authors of (Sala, Kenney-Wallace & Hall, 1980). did not reduce it to such a clear form, which seemingly impeded its further analysis, while the approximation of the upper and lower branches of the envelope of the signal $U(\tau)$ used in (Diels, Fontane, McMichel & Simoni, 1985) is not sufficiently accurate for retrieval of the parameter a, especially for an analysis of few-optical-cycle pulses.

Let us analyze the obtained expression. In fact, the signal $U(\tau)$ is not equal to the intensity autocorrelation function (22) even for a transform limited optical signal and contains several characteristic temporal scales. The largest one, $\tau_1 \sim \sqrt{2}\,\tau_0$, is related to the duration of the envelope $\rho(t)$ and is determined by the function $G(\tau) = \exp[-\tau^2 / (2\tau_0{}^2)]$. The next, shorter-term scales

$$\tau_2 \sim \frac{2\tau_0}{(L^2-1)^{1/4}}; \qquad \tau_3 \sim \frac{2\sqrt{2}\tau_0}{(2+L^2)^{1/2}}; \qquad \tau_4 \sim \frac{2\tau_0}{L}. \tag{26}$$

are determined by the third and fourth terms on the right-hand side of Eq. (24). Finally, the minimum temporal scale $\tau_5 \sim 1/\omega_0$ in Eq. (24) is determined by the optical cycle of a pulse. If $L > 1$, then the hierarchy of these temporal scales is as follows: $\tau_1 > \tau_2 > \tau_3 > \tau_4 > \tau_5$. This means that the signal $U(\tau)$ in the presence of a linear chirp should have an oscillatory component near $\tau = 0$, which is primarily determined by the last term on the right-hand side of Eq. (24), and smooth wings with shape determined by the function $G(\tau)$. In this case, the larger the value of L, i.e., the greater deviations of an optical pulse from a transform limited one, the more prominent the localization of the oscillatory component of such a signal in the vicinity of $\tau = 0$. If the averaging over the shortest scale τ_5 is performed in the operation of a detecting system, then the signal $U(\tau) \sim 1 + 2G(\tau)$ does not contain information on the optical pulse phase. This is also a well-known fact (Diels, Fontane, McMichel & Simoni, 1985 ; Krukov, 2008).

The function $U(\tau)$ calculated using Eq. (24) for various values of the pulse duration τ_0 and the parameter L is plotted in Fig. 5. It follows from analysis of these plots that, despite the fact that it is impossible to obtain the actual autocorrelation function from the signal of an interferometric intensity autocorrelator, the duration of a transform limited optical pulse can be determined with experimentally plausible accuracy using the upper branch $(7G(\tau)+1)$ of the oscillatory-component envelope of the signal. This fact is quite pleasant, as namely such a technique for measuring durations of ultrashort optical pulses is used by virtually all researchers. The appearance of smooth signal wings without any periodic modulation is indicative of a quadratic phase modulation of the studied wave packet. Hence, the form of the measured function $U(\tau)$ provides information on the presence of a linear frequency chirp in the optical band. In follows from Eq. (25) that for finding the numerical value of the coefficient a one should determine L and τ_0. In what follows, when discussing the experimental verification of the calculations, we describe in detail a procedure for determining these quantities. Here, we only note that the sign of the coefficient α, as is seen from Eq. (25), cannot be specified by this method; its determination requires either some *a priori* information or an additional experiment.

Let us now turn to analysis of the effect of the nonlinear frequency chirp, i.e., the cubic additive to the phase in Eq. (23) on the signal profile $U(\tau)$. For this, we put $\alpha = 0$ in Eq. (23). This can be done experimentally, if the quadratic phase is pre- compensated using, e.g., a prism dispersion compensator. With allowance for this assumption, we can also find an analytical formula for the output signal of an interferometric intensity autocorrelator

$$V(\tau) \approx 1 + 2G(\tau) + G(\tau) * \sqrt[-4]{4F^2(\tau) - 3}\cos\Phi(\tau) + 4[G(\tau)]^{1-1/4F^2(\tau)} * \cos\Psi(\tau) * 1/\sqrt{F(\tau)} \tag{27}$$

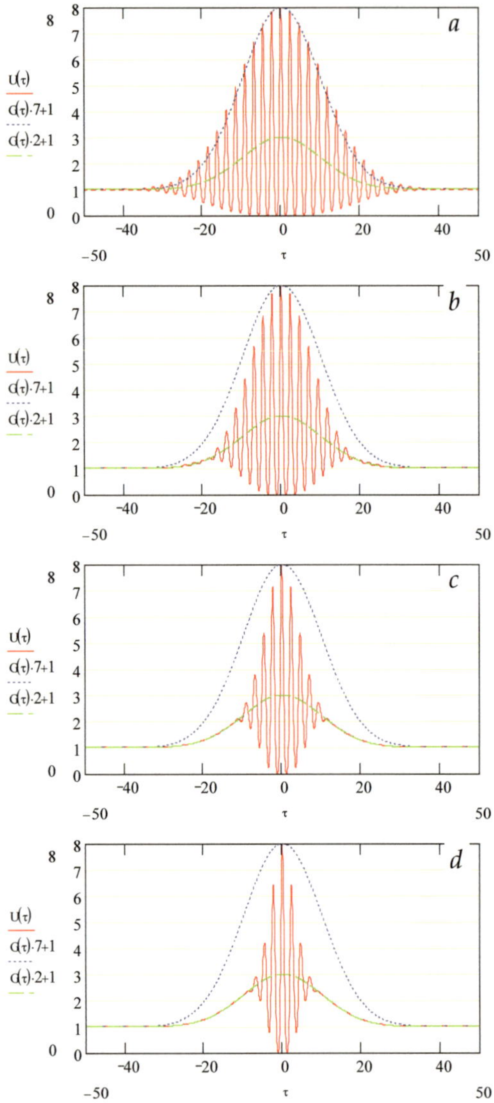

Fig. 5. Profile of the signal $U(\tau)$ for $L=1$ (*a*), $L=2$ (*b*), $L=4$ (*c*), and $L=6$ (*d*) in the case of a pulse of duration $\tau_0 = 10\,fs$ with quadratic phase modulation.

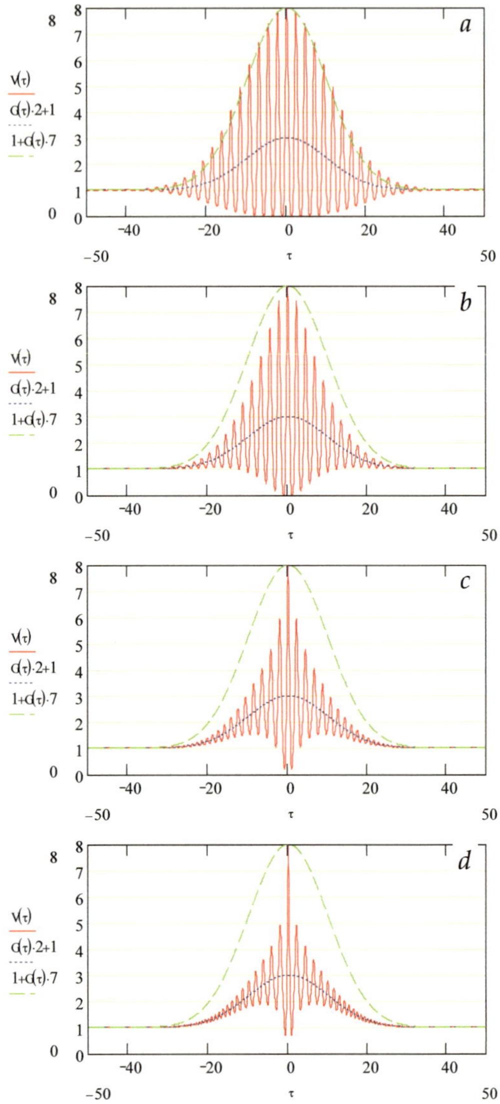

Fig. 6. Profile of the signal $V(\tau)$ for $K=1$ (*a*), $K=5$ (*b*), $K=20$ (*c*), and $K=50$ (*d*) in the case of a pulse of duration $\tau_0 = 10\,fs$ with cubic phase modulation.

where

$$F(\tau) = \sqrt{1+(K^2-1)*(\tau/2\tau_0)^2}\ , \tag{28a}$$

$$\Phi(\tau) = 2\omega_0\tau + 0.5a\tan\left[(\tau/\tau_0)\sqrt{K^2-1}\right] + (\tau/\tau_0)^3(\sqrt{K^2-1})/6 \tag{28b}$$

$$\Psi(\tau) = \omega_0\tau + 0.5\,\mathrm{a}\tan\left((\tau / 2\tau_0)\sqrt{K^2-1}\right) + \left(1/16F^2(\tau) + 1/12\right)(\tau/\tau_0)^3\sqrt{K^2-1} \quad (28c)$$

$$K^2 = 1 + (\beta\tau_0^3)^2. \quad (28d)$$

It follows from Eq. (27) that the formula for $V(\tau)$ is analogous to Eq. (24). The first two terms in these formulas are identical, while the third and fourth terms are oscillatory additives. The physical meaning of the parameter K is similar to that of the parameter L specified above. Note that in the considered case, localization of the oscillatory component of a signal near $\tau = 0$ is less prominent than for a quadratic phase since the exponent of the function $G(\tau)$ in the last term on the right-hand side of Eq. (27), which determines this localization, tends to unity with increasing K. Moreover, it follows from Eqs. (27) and (28) that the output signal modulation becomes much more aperiodic than in the case of a linear frequency chirp, and the aperiodicity of the observed oscillations is the stronger, the larger the parameter K Therefore, the presence of a linear frequency chirp in a signal can be determined from broadening of the profiles of the harmonics of the signal $V(\tau)$ at the frequencies ω_0 and $2\omega_0$.

The signal profiles $V(\tau)$ for various K are plotted in Fig. 6. Analysis of these plots shows that, as the behavior of the curves $V(\tau)$, especially for small K impedes finding any features entirely determined by τ_0 or β, the procedure of determining the parameters τ_0 and β is more difficult and complicated in this case than in the case of a linear frequency chirp. If the phase modulation is relatively weak ($K < 5$), then, in contrast to the case of a quadratic phase, the obtained dependences are close to the signal $V(\tau)$ for $K = 1$. Of course, this makes retrieval of the desired quantities from experimental data more difficult. The difference in the functions $V(\tau)$ for different K is quite measurable for $N \geq 5$, so that the above mentioned features or specific components can already be pointed out. Therefore, the parameters τ_0 or β can quite easily be determined in this case. However, such large values of K can hardly be realized in practice. In the limiting case of $K \gg 1$, the signal $V(\tau)$ has a smooth shape, determined by the first two terms in Eq. (27), with a very narrow, the so-called coherence peak in the vicinity of $\tau = 0$, because $V(0) = 8$ in any case. Determination of the sign of β, as in the case of a quadratic phase modulation, is impossible. This is explained by the physical principle of operation of this correlator. Therefore, the possibility of experimental measuring of a cubic phase of an ultrashort optical pulse using an ordinary interferometric intensity autocorrelator seems very problematic, except probably for some special cases.

If the studied optical signal comprises both quadratic and cubic phase modulation, then the output signal of an interferometric intensity autocorrelator can also be found analytically assuming a Gaussian signal envelope. The resulting formula is similar to the functions $U(\tau)$ and $V(\tau)$ obtained above, but is much more cumbersome and, correspondingly, far less illustrative than the formula for $U(\tau)$.

To check the results of calculation, we performed an experiment in which the signal from an interferometric intensity autocorrelator was obtained for a phase-modulated ultrashort optical pulse. The femtosecond ring laser described in (Babin, Kiselev, Kirsanov & Stepanov, 2002) was used as the radiation source. An external prism dispersion compensator mounted immediately after the output mirror was used to compensate for the dispersion of the substrate material of this mirror to ensure that a transform limited wave packet is input to the autocorrelator. The total duration $2\tau_0$ of the compensated pulse measured using its interference autocorrelation function (Fig. 7.) amounted to about 20 fs at the e^{-1}-folding level of the maximum intensity. The measured value of the product $\Delta\omega\tau_0$ was equal to 1.3, which is 30% larger than the corresponding value for a transform limited Gaussian pulse. This is known (Rousseau, McCarthy & Piche, 2000) to be related to a slight deviation of the generated spectrum from the Gaussian one, as is the case for the considered experiment. Note that hardware averaging of the autocorrelation function shown in Fig. 7 over fast oscillations yields the same value of τ_0, which, according to Eqs. (24) and (27), is indicative of absence of phase modulation of the optical pulse at the compensator output.

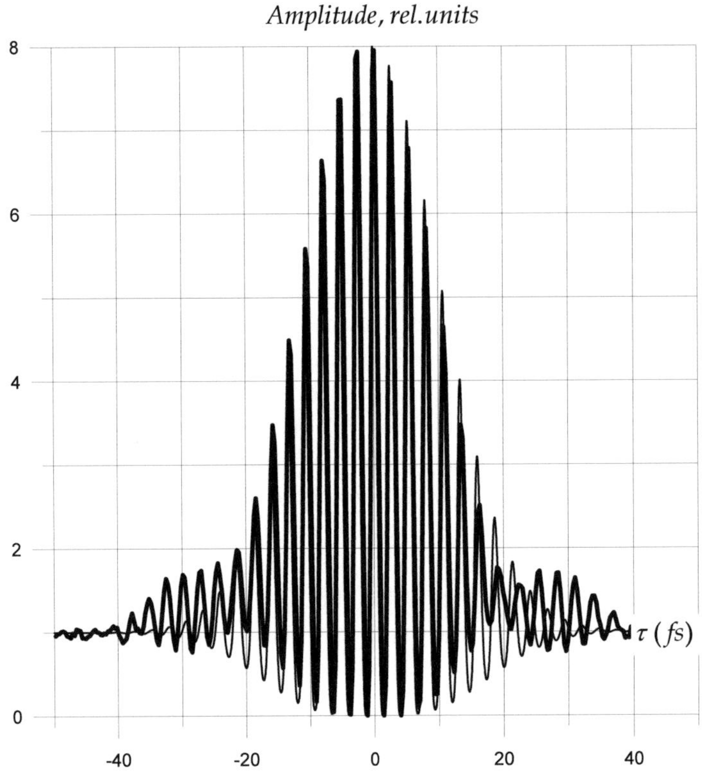

Fig. 7. Output signal of interferometric intensity autocorrelator for a transform limited wave packet. The bold and fine curves are for the experiment and the calculation using Eq. (24) for $L = 1$, respectively.

The idea of the experiment is fairly simple: to introduce a chirp in an initial transform limited laser pulse and to check the resulting output signal of the autocorrelator. It is known (Akhmanov, Vysloukh & Chirkin, 1988) that an ultrashort optical pulse can easily be phase modulated upon propagation through a linear dispersive medium. To realize this experimentally, we mount plane-parallel plates made of various materials immediately before the autocorrelator input.

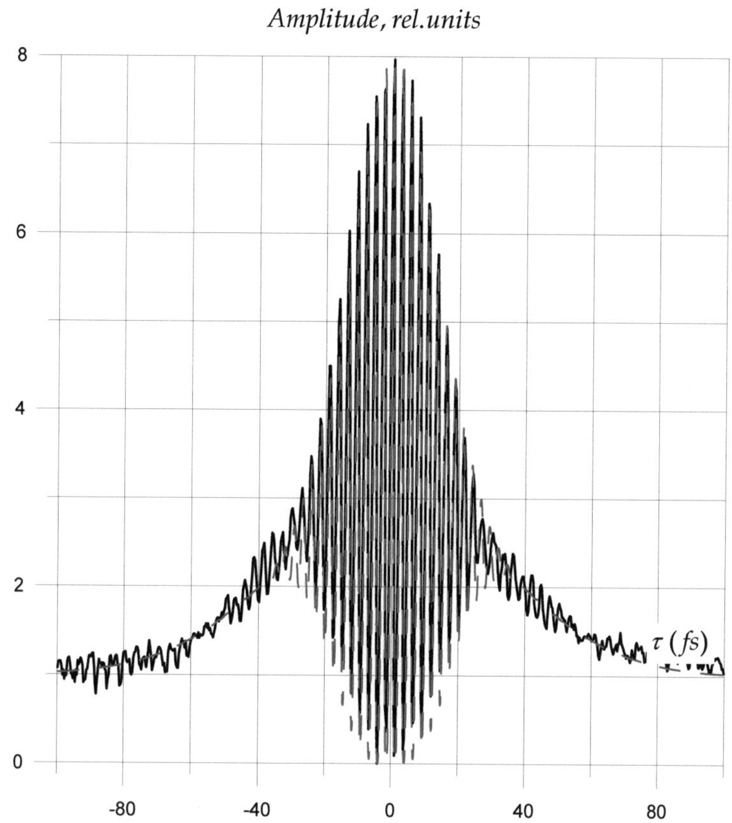

Fig. 8. Output signal of interferometric intensity autocorrelator for a phase modulated wave packet. The initial transform limited signal passed through a 3.25 mm thick ZnSe plate. The solid and dashed curves are, respectively, for the experiment and the calculation using Eq. (24) for $\tau_0 = 33.5\,fs$ and $L = 4.47$.

Fig. 8. shows a typical profile of the output signal obtained in this case. It follows from this figure that the observed picture fully agrees with the theoretical analysis for the case of a quadratic phase. Note that a similar profile was obtained in (Sala, Kenney-Wallace & Hall, 1980) for a linearly chirped pulse of about 13 ps in duration. Excess noise in the signal is due to a nonoptimal frequency band of the amplifier used in our experiment. Let us find the parameters of a phase modulated optical pulse input to the correlator assuming that a substance gives only a quadratic additive to the phase. The envelope duration τ_0 can most easily be derived from the profile of the wings of the signal $U(\tau)$. According to Eq. (24), the shape of the signal wings is described by the formula $U(\tau) \sim 1 + 2G(\tau)$. Therefore, the parameter τ_0 is readily determined for a wave packet with Gaussian envelope. The next step is to appropriately choose the parameter L to ensure the best agreement between the oscillating parts of experimental and theoretical functions $U(\tau)$. Note that one more parameter, ω_0, entering Eq. (24), can easily be found either from the period of oscillations in Fig. 7 or from the measured average frequency of the spectrum of the analyzed pulse. Hence, the formulated problem is solved completely in this approximation, i.e., we have measured the parameter τ_0 and the coefficient α of the quadratic phase using the output signal of an interferometric intensity autocorrelator. A cubic phase additive is known to be much less than the quadratic one for transparent optical materials. Therefore, in our case, this additive has almost no impact on the increase in the envelope duration τ_0. This means that the parameter K is close to unity and, according to the above theoretical analysis, it is impossible to measure the parameter β in the experiment. In this experiment, using a scanning intensity autocorrelator, we actually determined the averaged parameters of an ultrashort wave packet, as the analyzed signal had the shape of a femtosecond pulse train with a repetition rate of about 100 MHz. In principle, a similar interference autocorrelator can also be realized for rarely repeated or even single optical pulses if, as was proposed in (Brun, Georges & LeSaux, 1991), the standard scheme of a single-pulse correlator (Brun, Georges & LeSaux, 1991) including a nonlinear crystal with tangential (superwide angular) synchronism is used.

5. Conclusion

To conclude we enumerate the principal results of the work. It was shown that in media with nonlinearity described by a finite-degree polynomial, instability may develop at frequencies greatly exceeding the carrier frequency. Increments of these frequencies are found. Methods of measuring temporal characteristics of femtosecond pulses were analyzed. Retrieval of amplitude and phase modulated wave packets by means of interference intensity autocorrelator were demonstrated theoretically and experimentally.

6. References

Bespalov, V.I. & Talanov, V.I. (1966). About filamentation of beams of light in a nonlinear liquid. *JETP lett.*, Vol.3, No.12, pp.307-310

Lighthill, M.J. (1965). Contribution to the theory of waves in Non-linear Dispersion Systems. *J. Ins. Math. Aplic.*,Vol.1, No.3, p.269-306

Litvak, A.G. & Talanov, V.I. (1967). A parabolic equation for calculatioting the fields in dispersive nonlinear media. *Radiophys. Quant. Electron.*, Vol.10, No.4, pp.296-302

Benjamin, T.B. & Feir, J.E. (1967). The disintegration of wave train on deep water, Part I, Theory. *J. Fluid Mech.*,Vol.27, No.3, p.417-430

Zakharov, V.E. & Ostrovsky, L.A. (2009). Modulation instability: The beginning. *Physica D*, 2009, Vol.238, pp. 540-548

Bejot, P.; Kibler, B.; Hertz, E.; Lavorel, B.; & Fisher, O. (2011). General approach to spatiotemporal modulational instability processes, *Physical Review A, Vol.* 83, No.1, pp. 013830

Agraval, G.P. (1995). *Nonlinear fiber optics*, Academic Press, San Diego- Boston- New York- London- Sydney- Tokyo- Toronto, Second edition

Talanov, V.I. & Vlasov, S.N. (1997). *Wave Self-Focusing*, [in Russian], IAP RAS, N.Novgorod

Talanov, V.I. & Vlasov, S.N. (1994). Modulation instability at sum and difference frequencies in media with high-order nonlinearity. *JETP lett.* Vol.60, No. 9, pp. 639-642

Koposova, E.V. & Vlasov, S.N.(2007). Wideband modulation instability at combination frequencies in wave beams. *Radiophys. Quant. Electron.*, Vol.50, No.6, pp.477-486

Talanov, V.I. & Vlasov, S.N. (1995). The parabolic equation in the theory of wave propagation (on the 50th anniversary of its publication). *Radiophys. Quant. Electron.*, Vol.38, No.1-2, pp. 1-12

Brabec, Th. & Krausz, F. (1997). Nonlinear optical pulse propagation in the single-cycle regime. *Phys. Rev. Letts.*, Vol. 78, No.17, pp.3282-3285

Bespalov, V.G.; Oukrainski, A.O.; Kozlov, S.A. et al. (1999) Spectral evolution of propagation extremely short pulses. *Physics of Vibration*, Vol.7, No.1, pp.19-27

Kolesic, M. & Moloney, J.V. (2004). Nonlinear optical pulse propagation simulation: from Maxwell's to unidirectional equations. *Phys. Rev.* E , Vol.70, No.3, pp.036604

Ferrando, A.; Zacares, M.; de Cordoba, P.F. et al. (2005). Forward-backward equation for nonlinear propagation in axialy invariant optical systems. *Phys. Rev.* E, Vol. 71, No.1, , pp. 016601

Koposova, E.V.; Talanov, V.I. & Vlasov, S.N. (2006). Use of decomposition of the wave equations and pseudo-differential operators for the description of nonparaxial beams and broadband packets, *Radiophys. Quant. Electron.*, Vol.49, No.4, pp.258-301

Loriot ,V.; Hertz, E.; Faucher, O. et al.(2009). Measurement of high order Kerr refractive index of major air components. *Optics express*, Vol.17, No. 16, p.13429 - 13434

Bogolyubov, N.N., & Mitropol'skii Yu. A., Asimptotic method in the theory nonlinear vibration, [in Russian], Fizmatgiz, Moscow (1958)

Born, M. & Volf E, Principle of optics, Pergamon Press, Oxford-London-Edinburg-New-York-Paris-Frankfurt, 1964

Grigor'ev, I.S. & Melikhov, E.Z. (Eds). (1991). *Physical Quantities* [in Russian], Energoatomizdat, Moscow

Nibbering, E.T.J.; Curley, P.F.; Grillon, G. et al. (1996). Conical emission from self-guided femtosecond pulses in air. *Optics Letters*, Vol.21, No 1, pp. 62-65

Dormidontov, A.E.; Kandidov, V.P.; Kompanets, V.O. & Chekalin, S.V. (2010). Interference effects in the conical emission of femtosecond filament in fused silica. JETP lett., Vol.91, No .8, pp. 373-377

Kuznetsova, T. (1968). To the problem of recording ultrashort pulses *Zh. Eksp .Teor. Fiz.*. Vol. 55, No.6(12), , pp..2453-2458

Verevkin, Yu.; Daume, E.; Makarov, A.; Novikov, M. &. Khizhnyak, A.I. (1971). To the question about measurement time duration for ultrashort pulses. *Radiophys.Quantum Electron.*. Vol..14, No.6, pp..840-844

Sala, K.; Kenney-Wallace, G. & Hall, G. (1980). CW autocorrelation measurements picosecond laser pulse., *IEEE QE*, 1980, Vol,16, 1980, pp..990-996

Diels, M.; Fontane, J.; McMichel, I. & Simoni, F. (1985). Control and measurement of ultrashort pulse shape in amplitude and phase. *Appl.Opt.* Vol.24, No.9, pp.1270-1282

Naganuma, K.; Mogi, K. & Yamada, H. (1989). General method for ultrashort light pulse chirp measurement. *IEEE, QE*, Vol..25, No.6, pp.1225-1233

Iaconis, C. & Walmsley, I. (1998). Spectral phase interferometry for direct electric-field reconstruction of ultrashort optical pulses. *Opt.Lett.* Vol.23, No.19, pp.792 -796

DeLong, K.; Fittinghoff, D. & Trebino, R. (1996). Practical issues in ultrashort laser pulse measurement using frequency-resolved optical gating. *IEEE QE* Vol.32, No.7, pp.1253-1263

Koumans, R. & Yariv, A. (2000). Time-resolved optical gating based on dispersive propagation: a new method to characterize optical pulses. *IEEE QE*, Vol.36, No.2, pp.137-144

Kane, D. (1999). Recent progress toward real-time measurement of ultrashort laser pulses. *IEEE QE*, Vol.35, No.4, pp.421-431

Peatross, J. & Rundquist, A. (1998).Temporal decorrelation of ultrashort laser pulses. *J.Opt.Soc.Am.B*, Vol.15, No.1, pp.216-222

Nicholson, J.; .Jasapara, J.; Rudolph, W.; Omenetto, F. & Taylor A. (1999). A.Full-field characterization of femtosecond pulses by spectrum and cross-correlation measurements. *Opt.Lett.* Vol.24, No.23, pp.1774-1776

Krukov, P. (2008). *Femtosecond pulses. Introduction to the new field of laser physics.*Physmathlit, Moscow

Akhmanov, S.; Vysloukh, V. & Chirkin, A.(1988). *Optics of femtosecond pulses.* Nauka, Moscow

Babin, A.; Kiselev, A.; Kirsanov, A. & Stepanov, A.(2002). A 10-fs Ti:sapphire laser with folded ring resonator. *Quant.Electr.*, Vol.32, No.5, pp.401-403

Rousseau, G.; McCarthy, N. & Piche, M. (2000). Generation and characterization of sub-10 femtosecond laser pulses. *Proc. SPIE,* Vol.4087, pp.910-920

Brun, A.; Georges, P. & LeSaux, G. (1991). Single-short characterization of ultrashort laser pulses. *J.Phys.D:Appl.Phys* Vol.24, pp.1225-1233

Part 4

Metrology

15

Quantification of Laser Polarization by Position Dependent Refractive Indices

Yong Woon Parc and In Soo Ko
Department of Physics, Pohang University of Science and Technology, Pohang Korea

1. Introduction

The electro-optic (EO) crystals used in the generation of tera-hertz radiation [Shan et al, 2000; Wen & Lindenberg, 2009; Shen et al., 2008] can be applied to non-invasive measurements of the electron bunch in the accelerator. The bunch length and arrival time of the electron bunch with respect to the laser can be measured with this technique. The electric field generated by the electron bunch changes the optical properties of the EO crystal. The laser passing through the EO crystal will thus experience a modulation in the polarization state. The amount of optical power modulation, which can be detected by detectors such as photodiodes and cameras, can be calculated by Jone's matrices [Jones, 1941; Hecht, 2002].

In recent years, the EO technique has been implemented to measure femtosecond electron bunches with high energy at several facilities such as the free-electron laser for infrared experiments (FELIX), the sub-picosecond pulse source (SPPS), and the free-electron laser in Hamburg (FLASH) [Yan et al, 2000; Wilke et al., 2002; Berden et al., 2004; Cavalieri et al., 2005; Casalbuoni et al., 2008; Steffen et al., 2009]. In these applications, the electric field from the electron bunch makes a finite angle with the crystallographic axis, because the electric field is generated radially from the electron bunch.

In a real diagnostics setup, the principal axes at the passing position of the laser in the EO crystal are no longer 45° with respect to the crystallographic axes. However, the polarization vector of the laser is usually fixed as parallel to an crystallographic axis of the EO crystal in the measurement. When an electric field is applied to the EO crystal, it shows a birefringence which is described by an ellipsoid equation of refractive indices. If the ellipsoid equation can be expressed without any cross term between the three axes in a coordinate system, they are called the principal axes. If a linearly polarized laser is passing through the EO crystal with an external electric field applied to it, the polarization of the laser is modulated elliptically or even circularly. Conventional theories to describe the measurement result of an electron bunch with EO crystal have only used the principal axes and the principal refractive indices. To quantify the polarization state, we need to define amplitudes and phases of the electric field components of the laser. If the amplitudes of the decomposed electric field are the same, we need the phase information only to quantify the polarization state. To decompose the electric field with the same amplitude, we need to choose decomposition axes at 45° with respect to the polarization vector. The spatial decoding method uses a laser propagated to the EO crystal at a certain angle with respect to the surface of the crystal to measure the electron bunch [Cavalieri et al., 2005; Parc et al., 2009a]. In this situation, we need to know refractive indices in the plane perpendicular to the propagation direction.

The principal axes are not in the plane in the spatial decoding method. Thus, the principal axes are not a good choice for the decomposition of the polarization vector. Even though this problem exists in the application of the EO technique to the accelerator, simulations for the EO measurement of the electron bunch have been conducted without consideration of this problem [Casalbuoni et al., 2008]. Refractive indices along the 45° axes with respect to the polarization vector are introduced by the authors in Ref. [Parc et al., 2009a] to compare a measurement result with the electron bunch and the simulation result.

This article presents the detailed derivation process of the refractive indices along the 45° axes with respect to the polarization vector in the plane perpendicular to the laser propagation direction. The optical power modulation of the laser is calculated by the matrix multication method invented by Jones with newly derived refractive indices. These refractive indices can be used to directly express experimental and numerical study results of the polarization state in the laser by the relative phase shift. The variation of the relative phase shifts will be presented when the distance between the electron bunch and EO crystal are changed. The paper is organized as follow: The theory of EO effect is reviewed in Section 2. The theory of detection of the optical power modulation of the laser is developed in Section 3. An experiment conducted in FLASH is analyzed in Section 4. Simulation results are shown in Section 5, and the conclusion and discussion are provided in Section 6.

2. Theory of refractive indices in electro-optic crystal

There are two methods to get the principal refractive indices. The two methods are reviewed in next two subsections, respectively. The principal refractive indices calclated by the two methods are revealed as the same and new form of the principal refractive indices is also obtained in this review.

2.1 Principal refractive indices obtained by solving the eigenvalue equation

In this subsection, the method introduced in Ref. [Casalbuoni et al., 2008] is reviewed to compare the principal refractive indices calculated in the reference with the other results which will be introduced in next subsection. The refractive indices of EO crystal such as GaP are obtained from an ellipsoid equation [Yariv & Yeh, 2003], written as,

$$\frac{1}{n_0^2}\left(x^2+y^2+z^2\right)+2r_{41}\left(E_x yz+E_y zy+E_z xy\right)=1 \tag{1}$$

where n_0 is the initial refractive index, r_{41} is the electro-optic constant, and E_i ($i = x, y, z$) is the electric field component applied to the crystal along the corresponding axis as shown in Fig. 1(a). If the ellipsoid equation can be expressed without any cross term in a certain coordinate system, the coordinates are called the principal axes. The X, Y, Z axes in Fig. 1(b) are defined as the crystallographic axes of the EO crystal in this study.

The optical property of isotropic EO crystal depends on the direction of the applied electric field [Berden et al., 2004; Parc et al., 2009a]. Thus, the directions of the principal axes of the isotropic crystal are changed along the direction of applied electric field. A correlation

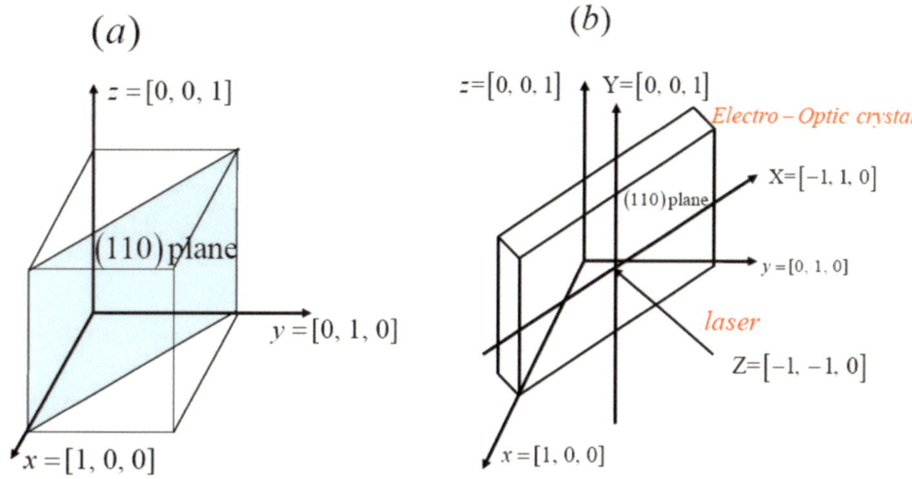

Fig. 1. (a) Right angle coordinate system of the crystal axis and (110) plane. The electro-optic crystal is cut in (110) plane. (b) The coordinate system X, Y, Z is shown with the electro-optic crystal. Laser will be passing into the Z direction in the experiment.

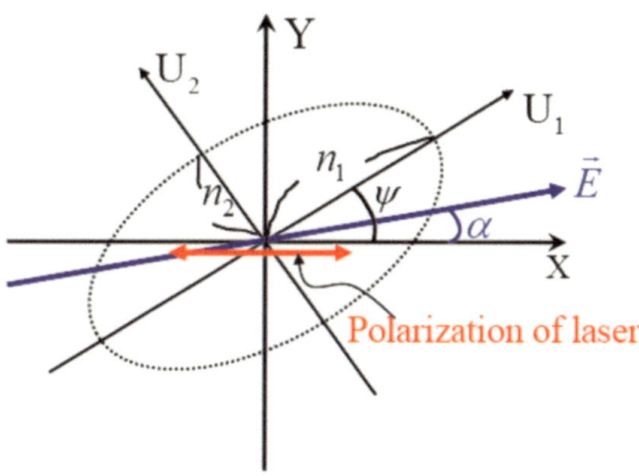

Fig. 2. The index ellipse is shown with the principal axis U_1 and U_2. a represents the angle between the crystal axis X and the applied electric field E. The ψ is the angle between the principal axis and the crystal axis. The probe laser is linearly polarized as parallel to the X axis.

between the direction of the principal axes and the external applied electric field can be calculated from the constant energy surface in the electric displacement vector space and the impermeable tensor that is linear to the electric field [Yariv & Yeh, 2003; Casalbuoni et al. 2008]. The electric field vector from the electron beam with an angle a with respect to X = [-1, 1, 0] axis in Fig. 2 can be represented as

$$\vec{E}=E\cdot\begin{pmatrix} -\cos\alpha/\sqrt{2} \\ \cos\alpha/\sqrt{2} \\ \sin\alpha \end{pmatrix}=\begin{pmatrix} E_x \\ E_y \\ E_z \end{pmatrix} \tag{2}$$

in the (x, y, z) coordinate system, where E is the amplitude of the electric field. The impermeability tensor in the (x, y, z) coordinate system with the same electric field vector is defined as

$$\eta = \frac{1}{n_0^2}\begin{pmatrix} 1 & 0 & 0 \\ 0 & 1 & 0 \\ 0 & 0 & 1 \end{pmatrix} + r_{41} \cdot E \cdot \begin{pmatrix} 0 & \sin\alpha & \cos\alpha/\sqrt{2} \\ \sin\alpha & 0 & -\cos\alpha/\sqrt{2} \\ \cos\alpha/\sqrt{2} & -\cos\alpha/\sqrt{2} & 0 \end{pmatrix}. \tag{3}$$

The eigenvalue equation is given by

$$\eta X = \xi X \text{ or } |\eta - \xi I| = 0. \tag{4}$$

Principal axes of the refractive index ellipsoid equation and the eigenvalues can be known by solving Eq. (4) with the impermeable tensor. The eigenvalues of the impermeability tensor are, then,

$$\begin{aligned} \xi_1 &= \frac{1}{n_0^2} - \frac{r_{41}E}{2}\left(\sin\alpha + \sqrt{1+3\cos^2\alpha}\right), \\ \xi_2 &= \frac{1}{n_0^2} - \frac{r_{41}E}{2}\left(\sin\alpha - \sqrt{1+3\cos^2\alpha}\right), \\ \xi_3 &= \frac{1}{n_0^2} + r_{41}E\sin\alpha. \end{aligned} \tag{5}$$

The eigenvectors U_1, U_2 can be obtained as shown in Fig. 2. The refractive index ellipsoid equation in Eq. (1) is now transformed into the form

$$\frac{U_1^2}{\xi_1^{-1}} + \frac{U_2^2}{\xi_2^{-1}} + \frac{U_3^2}{\xi_3^{-1}} = 1 \text{ or } \frac{U_1^2}{n_1^2} + \frac{U_2^2}{n_2^2} + \frac{U_3^2}{n_3^2} = 1 . \tag{6}$$

The principal refractive indices n_i can, then, be following:

$$n_i^2 = \xi_i^{-1}. \tag{7}$$

If the following relation is available in Eq. (5)

$$r_{41}E \ll 1/n_0^2, \tag{8}$$

the indices of refraction of the crystal along the principal axes in Fig. 2 are given as [Casalbuoni et al. 2008],

$$\begin{aligned} n_1 &= n_0 + \frac{n_0^3 r_{41}E}{4}\left(\sin\alpha + \sqrt{1+3\cos^2\alpha}\right), \\ n_2 &= n_0 + \frac{n_0^3 r_{41}E}{4}\left(\sin\alpha - \sqrt{1+3\cos^2\alpha}\right),. \\ n_3 &= n_0 - \frac{n_0^3 r_{41}E}{2}\sin\alpha \end{aligned} \tag{9}$$

where n_0 is the initial refractive index, r_{41} is the electro-optic constant, E is the electric field applied to the crystal, and α is the angle between the crystal axis and the electric-field. The relation of the angle ψ between the principal axis and the X axis and the angle α between the electric field and the X axis is given by [Casalbuoni et al. 2008; Parc et al., 2008a]

$$\cos 2\psi = \frac{\sin\alpha}{\sqrt{1+3\cos^2\alpha}}.. \tag{10}$$

2.2 Principal refractive indices obtained by coordinate transformation

In Ref. [Planken et al., 2001], another method is introduced to get the principal axes and the principal refractive indices. By transformation of the axes, the principal axes and the principal refractive indices can be obtained. The first transformation of the crystallographic axes x, y, z to new axes X, Y, Z in Fig. 1(b) is a rotation of 45° around the z axis,

$$\begin{aligned} x &= \frac{1}{\sqrt{2}}Z - \frac{1}{\sqrt{2}}X, \\ y &= \frac{1}{\sqrt{2}}Z + \frac{1}{\sqrt{2}}X, \\ z &= Y. \end{aligned} \tag{11}$$

The ellipsoid equation of Eq. (1) is now transformed into the new form in the new coordinated system X, Y, Z as

$$X^2\left(\frac{1}{n_o^2} - E_z r_{41}\right) + \frac{Y^2}{n_o^2} + 2\sqrt{2}E_x r_{41} XY + Z^2\left(\frac{1}{n_o^2} + E_z r_{41}\right) = 1. \tag{12}$$

To eliminate the remaining cross term between X and Y, we need one more rotation around Z axis as follow:

$$\begin{aligned} X &= U_1\cos\psi - U_2\sin\psi, \\ Y &= U_1\sin\psi + U_2\cos\psi, \\ Z &= U_3. \end{aligned} \tag{13}$$

Using the rotation relation Eq. (13), Eq. (12) is now expressed as

$$\begin{aligned} &U_3^2\left(\frac{1}{n_o^2} + E_z r_{41}\right) + U_1^2\left(\cos^2\psi\left(\frac{1}{n_o^2} - E_z r_{41}\right) + \frac{1}{n_o^2}\sin^2\psi + 2\sqrt{2}E_x r_{41}\cos\psi\sin\psi\right) \\ &+ U_2^2\left(\sin^2\psi\left(\frac{1}{n_o^2} - E_z r_{41}\right) + \frac{1}{n_o^2}\cos^2\psi - 2\sqrt{2}E_x r_{41}\sin\psi\cos\psi\right) \\ &+ U_1U_2\left(-2\cos\psi\sin\psi\left(\frac{1}{n_o^2} - E_z r_{41}\right) + \frac{(2\sin\psi\cos\psi)}{n_o^2} + 2\sqrt{2}E_x r_{41}\left(-\sin^2\psi + \cos^2\psi\right)\right) = 1. \end{aligned} \tag{14}$$

There is a cross term between U_1 and U_2. If the cross term is eliminated, we can get the principal axes and the principal refractive indices. Thus, the condition is

$$-2\cos\psi\sin\psi\left(\frac{1}{n_o^2} - E_z r_{41}\right) + \frac{(2\sin\psi\cos\psi)}{n_o^2} + 2\sqrt{2}E_x r_{41}\left(-\sin^2\psi + \cos^2\psi\right) = 0. \tag{15}$$

Using Eq. (2), Eq. (15) gives us next relation:

$$\tan 2\psi \tan\alpha = 2\ . \tag{16}$$

It is easy to show that Eq. (16) is a different expression of Eq. (10). Using Eq. (2) and after an algebra, Eq. (14) is now shown as more compact form

$$\begin{aligned} &U_3^2\left(\frac{1}{n_o^2}+Er_{41}\sin\alpha\right)+U_1^2\left(\frac{1}{n_o^2}-Er_{41}\left(\sin\alpha\cos^2\psi+\cos\alpha\sin 2\psi\right)\right) \\ &+U_2^2\left(\frac{1}{n_o^2}-Er_{41}\left(\sin\alpha\sin^2\psi-\cos\alpha\sin 2\psi\right)\right)=1. \end{aligned} \tag{17}$$

The principal refractive indices are now obtained with the same condition in Eq. (8),

$$\begin{aligned} n_1 &= n_0+\frac{1}{2}n_0^3 r_{41}E\left(\sin\alpha\cos^2\psi+\cos\alpha\sin 2\psi\right), \\ &= n_0+\frac{1}{4}n_0^3 r_{41}E\left(\sin\alpha+\sqrt{1+3\cos^2\alpha}\right), \\ &= n_0+\frac{1}{2}n_0^3 r_{41}E\left(\sin\alpha\sin^2\psi+\sin(\alpha+2\psi)\right). \end{aligned} \tag{18}$$

$$\begin{aligned} n_2 &= n_0+\frac{1}{2}n_0^3 r_{41}E\left(\sin\alpha\sin^2\psi-\cos\alpha\sin 2\psi\right), \\ &= n_0+\frac{1}{4}n_0^3 r_{41}E\left(\sin\alpha-\sqrt{1+3\cos^2\alpha}\right), \\ &= n_0+\frac{1}{2}n_0^3 r_{41}E\left(\sin\alpha\cos^2\psi-\sin(\alpha+2\psi)\right). \end{aligned} \tag{19}$$

The first expressions of n_1 and n_2 are the newly derived form in this review. The second expressions of n_1 and n_2 are derived in section A. The third expressions are shown to compare our results with Eq. (7) in Ref. [15Planken]. The second and the third expressions in n_1 and n_2 can be obtained also from the first expression by using the relation Eq. (16). Note that the angle definition a of the electric field in this review is equivalent with a+90° in Ref. [Planken et al., 2001]. The ψ in this study is expressed as θ in Ref. [Planken et al., 2001].

2.3 Refractive indices in the plane perpendicular to the laser propagation direction

The configuration for the measurement of the electron beam with EO crystal is shown in Fig. 3. A method called the spatial decoding is used to measure the bunch length and the timing jitter of the electron beam as shown in Fig. 3 [Cavalieri et al., 2005; Azima et al., 2006]. In the spatial decoding method, the laser pulse is propagated through the EO crystal with the incident angle φ_0, and the timing information of the electron beam is converted to the spatial information of the image measured by an ICCD camera as shown in Fig. 3 [Cavalieri et al., 2005; Azima et al., 2006]. The modulation of the polarization of the laser can be measured by a prism. The prism in Fig. 3 splits the laser as horizontal and vertical components.

When an electron beam is propagating beside the EO crystal, the detail configuration around EO crystal is shown in Fig. 4(a). The red line in Fig. 4(a) represents the laser which is shaped like a thin pencil. The principal axes U_1, U_2 of the crystal is also shown in Fig. 4(a). The U_1, U_2, X, and Y axes are all in the same plane (110) which is described in Fig. 2. The front view of Fig. 4(a) is shown in Fig. 4(b). The angle between the crystallographic axis X and the electric field E from the electron beam is represented by α in Fig. 4(b). The y axis in Fig. 4(b) is the trace of the laser on the EO crystal. R is the distance between the electron beam and the laser passing position when the electron beam is arrived at the EO crystal. The perpendicular distance from the electron beam and the y axis is denoted by r_o. The side view of Fig. 4(a) is drawn in Fig. 4(c). The pencil shape laser makes an incident angle φ_o with EO crystal in the spatial decoding method [Cavalieri et al., 2005]. $\vec{k}$ is the wave vector of the laser inside of EO crystal. The laser is propagating to the $\vec{k}$ direction drawn as a dotted arrow. The angle φ_L is the angle between the laser and EO crystal inside of the crystal.

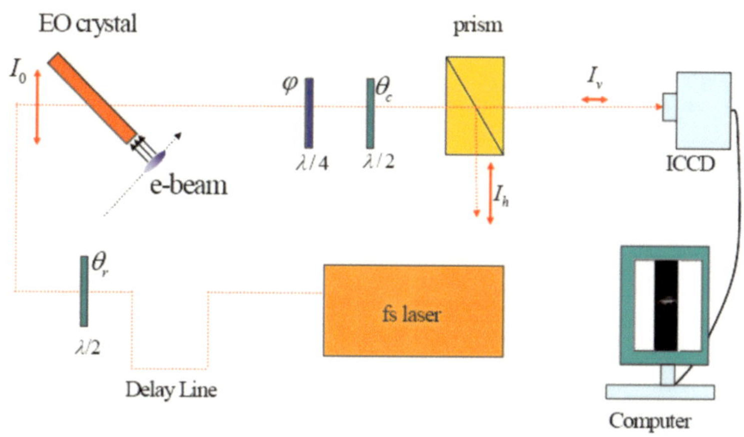

Fig. 3. Layout of the EO sampling experiment of electron beam in TEO setup at FLASH in DESY.

The linear polarization of the laser can be expressed as a vector which will be called as the polarization vector denoted as Ψ_L in Fig. 5(a). The inner parts of all axes in the ellipsoid are denoted by dotted lines in Fig. 5(a). $\vec{k}$ is the wave vector of the laser shown in Fig. 4(c). A plane perpendicular to the wave vector $\vec{k}$ will cut the ellipsoid, thus we can see an ellipse as shown in Fig. 5(a). The 45° axes with respect to the polarization vector are denoted by A and B in this study as shown in Fig. 5(a). A and B axes are in the same plane with the ellipse. Note that U_1 and U_2 axes are not in the same plane. The polarization vector Ψ_L can be also decomposed as two vectors denoted by Ψ_A and Ψ_B along the A and B axes, respectively, as shown in Fig. 5(a). The amplitude of the vector represents the electric field amplitude of the laser. In Fig. 5(a), Ψ_A is defined as a decomposed vector along A axis which makes 45° with the polarization vector Ψ_L of the laser. These vectors are all in the same plane. The ellipse and the A axis makes an intersection point M as shown in Fig. 5(a). The distance from the origin of the ellipsoid to the intersection point M will give new refractive index n_A along the A axis as shown in Fig. 5(b). Similarly, the refractive index along the B axis is denoted by n_B in Fig. 5(b).

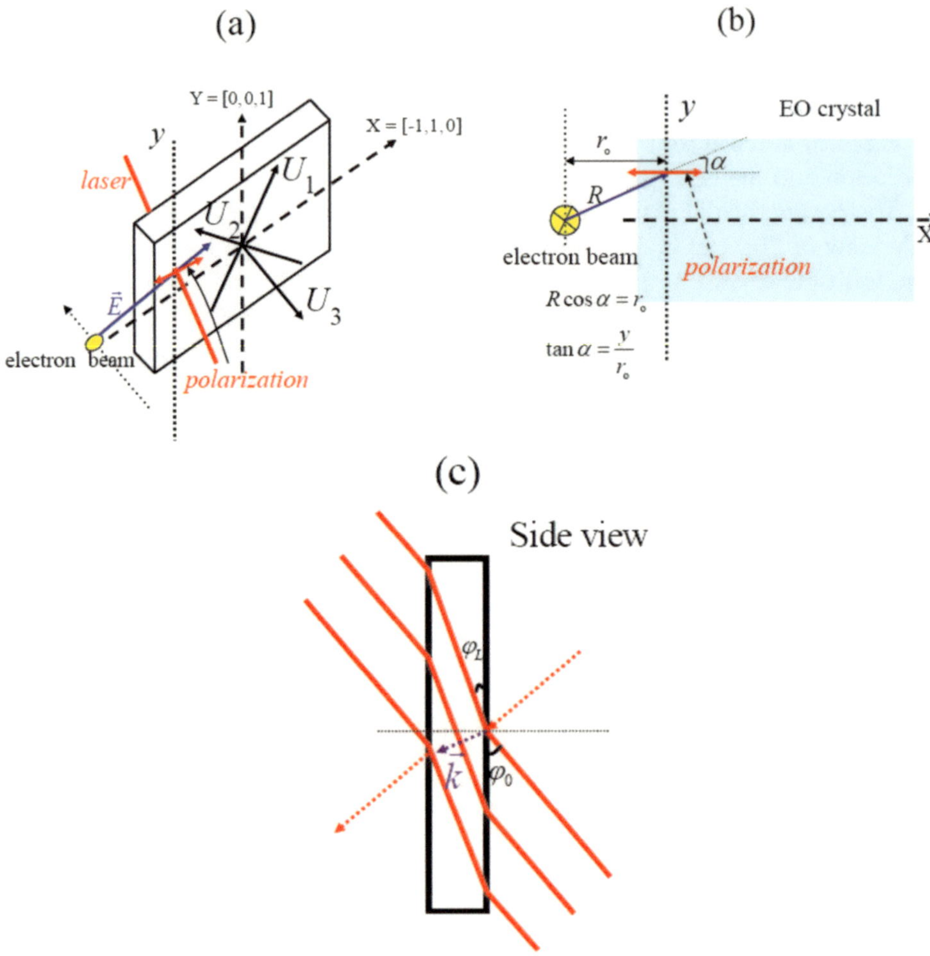

Fig. 4. Spatial decoding method in EO measurement of electron beam is shown with the principal axis

The angle of the prism in Fig. 3 determines the horizontal H and vertical V axis in Fig. 5(b). Because the polarization vector is usually fixed as parallel to X axis, we should control the prism angle such that the horizontal axis H to be matched with X axis as shown in Fig. 5(b). The polarization vector Ψ_L can be decomposed as two vectors denoted by Ψ_h and Ψ_v along the horizontal H and the vertical V axis, respectively. There is no vertical component of the polarization in Fig. 5(b). However, by the modulation of the polarization by passing through the EO crystal, the vertical component will be created and we can measure the vertical component by a prism as shown in Fig. 3. This is the key factor to measure the electron beam properties by EO crystal nondestructively. The G point in Fig. 5(b) is the projection point of the polarization vector Ψ_L to A axis.
The 3D configuration for the vector decomposition with the principal axes is shown in Fig. 6. The A axis in Fig. 5 is in the same direction with the OG line in Fig. 6. Note that OG line and X axis forms 45° angle as described in Fig. 5(b). For a given wave vector $\vec{k}$ of the laser

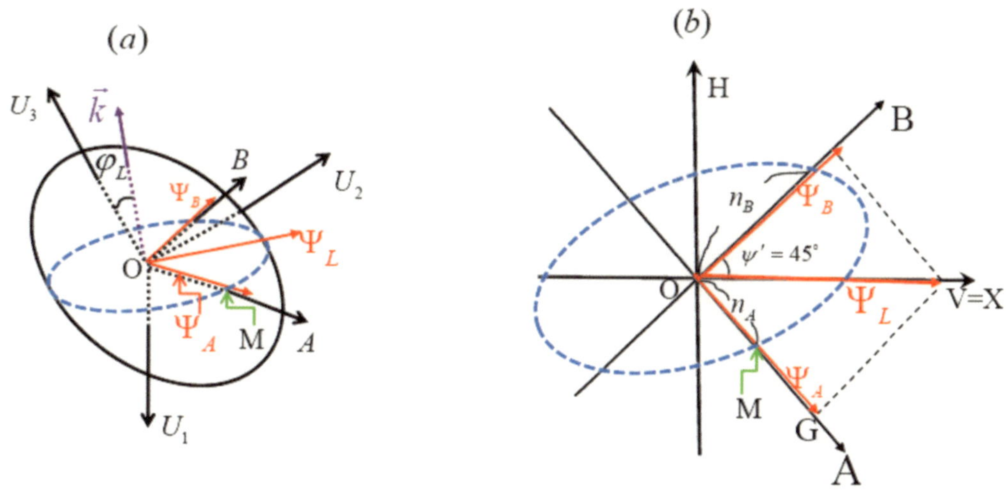

Fig. 5. (a) The index ellipsoid is shown with the principal axes U_1, U_2, and U_3. (b) The 2D configuration for the decomposition of the laser polarization vector Ψ_L. A and B axes make 45° with respect to the polarization vector Ψ_L. H and V axes are the horizontal and vertical axes in the AB plane. The polarization vector Ψ_L is fixed as parallel to the X axis. n_A and n_B are the refractive indices along A and B axes, respectively.

as shown in Fig. 5(a), the index n_A can be derived from the index ellipsoid equation and the OG line. The OG line will make an intersection point M with the ellipse as shown in Fig. 5(b). Our goal is to get an expression of the G point in the principal axes system to calculate the index n_A.

Lengths of OG, GF and FH in Fig. 6 are calculated as:

$$\begin{aligned} \mathrm{OG} &= \Psi_L \cos(45^\circ) = \frac{1}{\sqrt{2}}\Psi_L = \Psi_A \\ \mathrm{GF} &= \mathrm{OG}\sin(45^\circ) = \frac{1}{2}\Psi_L \\ \mathrm{FH} &= \mathrm{GF}\cos(\varphi_L) = \frac{1}{2}\Psi_L\cos(\varphi_L) \end{aligned} \quad . \tag{20}$$

We need to know the G point in the principal axes U_1, U_2, U_3 system. Each component of the G point can be calculated as the projection of the OG line to each axis as shown:

$$\begin{aligned} \mathrm{GH} &= \mathrm{GF}\sin(\varphi_L) = \frac{1}{2}\Psi_L\sin(\varphi_L) = G_3 \\ \mathrm{OH} &= \sqrt{\mathrm{OF}^2 + \mathrm{FH}^2} = \frac{1}{2}\Psi_L\sqrt{1+\cos^2\varphi_L} \\ \mathrm{OK} &= \mathrm{OH}\cos(\psi - \theta_L) = G_1 \\ \mathrm{HK} &= \mathrm{OH}\sin(\psi - \theta_L) = -G_2 \end{aligned} \quad . \tag{21}$$

The G point in the principal axes is now shown as:

$$\begin{aligned} G &= (G_1, G_2, G_3) \\ &= (\mathrm{OK}, -\mathrm{HK}, \mathrm{GH}) \\ &= \left(\mathrm{OH}\cos(\psi - \theta_L), \ -\mathrm{OH}\sin(\psi - \theta_L), \ \frac{1}{2} E_L \sin\varphi_L \right) \\ &= \frac{1}{2}\Psi_L \left(\cos(\psi - \theta_L)\sqrt{1+\cos^2\varphi_L}, \ -\sin(\psi - \theta_L)\sqrt{1+\cos^2\varphi_L}, \ \sin\varphi_L \right) \end{aligned} \quad (22)$$

Next relation can be calculated easily from Eq. (22)

$$G_1^2 + G_2^2 + G_3^2 = \frac{1}{2}. \quad (23)$$

The projection line equations of the OG line to the planes generated by the principal axes U_1, U_2, U_3 are given by

$$\begin{aligned} G_2 U_1 - G_1 U_2 &= 0 \\ G_3 U_2 - G_2 U_3 &= 0 \\ G_1 U_3 - G_3 U_1 &= 0 \end{aligned} \quad (24)$$

Using Eqs. (24) and (7), next relation can be derived

$$\left[\left(\frac{G_1}{n_1}\right)^2 + \left(\frac{G_2}{n_2}\right)^2 + \left(\frac{G_3}{n_3}\right)^2 \right] U_i^2 = G_i^2, \quad (25)$$

where i is the index for the principal axes U_1, U_2, U_3. The refractive indices n_A for the A axis is derived by the distance from the origin of the ellipsoid to the intersection point M as shown in Fig. 5(a). Thus, the refractive index n_A along the A axis is defined as

$$n_A = \sqrt{U_1^2 + U_2^2 + U_3^2}. \quad (26)$$

The calculation of n_B for B axis is similar with n_A, not shown in this study. The two indices are [Parc et al., 2009a],

$$\begin{aligned} n_A &= \sqrt{\frac{2}{\left(\frac{\sin^2(\psi - \theta_L)(1+\cos^2\varphi_L)}{n_1^2}\right) + \left(\frac{\cos^2(\psi - \theta_L)(1+\cos^2\varphi_L)}{n_2^2}\right) + \left(\frac{\sin^2\varphi_L}{n_3^2}\right)}} \\ n_B &= \sqrt{\frac{2}{\left(\frac{\cos^2(\psi - \theta_L)(1+\cos^2\varphi_L)}{n_1^2}\right) + \left(\frac{\sin^2(\psi - \theta_L)(1+\cos^2\varphi_L)}{n_2^2}\right) + \left(\frac{\sin^2\varphi_L}{n_3^2}\right)}} \end{aligned} \quad (27)$$

We can also get the relation between the angle φ_L and θ_L.

$$\begin{aligned} OF &= OH\cos(\theta_L) \\ (1+\cos^2\varphi_L)\cos^2\theta_L &= 1 \end{aligned}, \quad (28)$$

where φ_L is the angle between the wave vector $\vec{k}$ and U_3 axis, θ_L is the angle between the line OH and X axis in Fig. 6(b). The refractive indices in Eq. (27) are converted into the principal refractive indices n_1 and n_2 if the angle φ_L is 0° and ψ is 45°. The formula of the relative phase shift Γ in this review is now given by

$$\Gamma = \frac{\omega_0 d}{c}(n_A - n_B)\,. \tag{29}$$

where d is the crystal thickness, ω_0 is the center angular frequency of the laser pulse, and c is the speed of light.

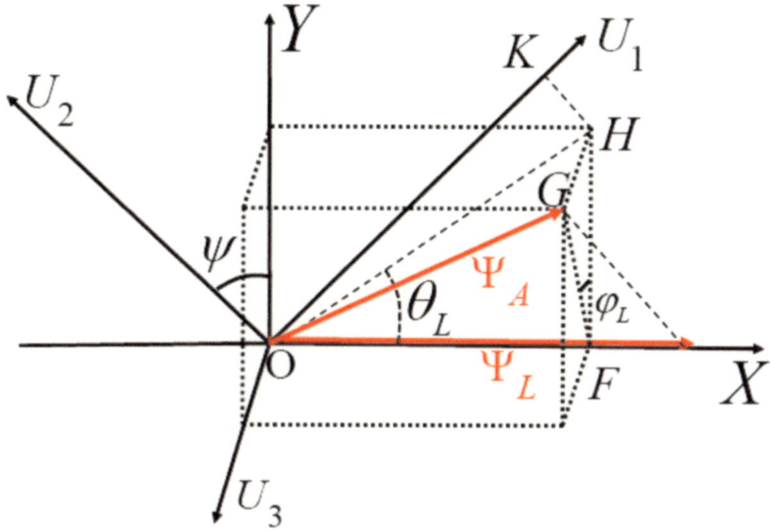

Fig. 6. The 3D configuration for the decomposition of the laser polarization vector Ψ_L. Ψ_A is the electric field component of the laser along the A axis. ψ is the angle between the principal axis and the crystal axis X. φ_L is the angel between the wave vector $\vec{k}$ of the laser and the U_3 axis. θ_L is the angle between OH and X axis. The probe laser is linearly polarized as parallel to the X axis. ψ' is the angle between the A axis and the crystal axis X.

3. Theory for the detection of the modulation of the polarization

The polarization of the laser can be rotated by a half wave plate at the front of the EO crystal in a certain experimental situation as shown Fig. 3. This half wave plate will be named as 'Rotator' in this study. To increase the signal to noise ratio of the measurement result, a half wave plate and a quarter wave plate can be used in the downstream of the EO crystal [Steffen et al., 2009, Parc et al., 2008a]. The half wave plate in the downstream of EO crystal will be named as 'Compensator' in this study. In the next matrices and equations, Γ is the phase difference between the two decomposed vectors of the laser along A and B axes given by Eq. (29), θ_r is the angle of the rotator, θ_c is the angle of the compensator, φ is the angle of quarter wave plate, and ψ' is the angle between the A axis and X axis in the crystal as shown in Fig. 5(a). For the general purpose, the angle ψ' is set as arbitrary value in the

theory. To use the theory in the analysis of the experimental result, this angle is set to be 45° because the polarization vector is set as parallel with X axis as shown in Fig. 5(b).
The rotation matrix R, Jones matrices for the half (H) wave plate, quarter (Q) wave plate, and EO crystal (EO) are given by, respectively,

$$\Gamma R = \begin{pmatrix} \cos\zeta & -\sin\zeta \\ \sin\zeta & \cos\zeta \end{pmatrix}$$

$$EO(\Gamma) = \begin{pmatrix} e^{-i\Gamma/2} & 0 \\ 0 & e^{i\Gamma/2} \end{pmatrix}$$

$$Q = \begin{pmatrix} e^{-i\pi/4} & 0 \\ 0 & e^{i\pi/4} \end{pmatrix}$$

$$H = \begin{pmatrix} e^{-i\pi/2} & 0 \\ 0 & e^{i\pi/2} \end{pmatrix}$$

where ζ represents the angle of each wave plate with respect to the laser polarization direction in the measurement. The initial polarization vector in front of the rotator in Fig. 3 can be represented by 2×1 matrix such as

$$\Psi_L = \begin{pmatrix} 1 \\ 0 \end{pmatrix}.$$

All quantities of the laser such as the amplitude of the electric field and the intensity are normalized by its maximum value in all formulae for the convenience of derivation.
The horizontal electric field Ψ_h of the laser after passing through all optical components in Fig. 3 can be calculated by the product of matrices:

$$\Psi_h = (0 \quad 1)\cdot R[-\theta_c]\cdot H\cdot R[\theta_c]\cdot R[-\varphi]\cdot Q\cdot R[\varphi]\cdot R[-\psi']\cdot EO(\Gamma)\cdot R[\psi']\cdot R[-\theta_r]\cdot H\cdot R[\theta_r]\cdot\begin{pmatrix} 1 \\ 0 \end{pmatrix}. \quad (30)$$

The intensity I_h of the horizontal component is proportional to the square of the horizontal electric field, and the normalized expression is given by:

$$\begin{aligned} I_h(\theta_c,\varphi,\psi',\theta_r:\Gamma) &= \Psi_h^2 \\ &= \frac{1}{2}\begin{bmatrix} \sin(\Gamma)\sin(4\theta_c-2\varphi)\sin(4\theta_r-2\psi') \\ +\left\{\sin^2(2(\theta_c-\theta_r))+\sin^2(2(\theta_c+\theta_r-\varphi))\right\}\cos^2(\Gamma/2) \\ +\left\{\sin^2(2(\theta_c+\theta_r-\psi'))+\sin^2(2(\theta_c-\theta_r-\varphi+\psi'))\right\}\sin^2(\Gamma/2) \end{bmatrix}. \end{aligned} \quad (31)$$

The vertical electric field Ψ_v can be obtained by the similar calculation:

$$\Psi_v = (1 \quad 0)\cdot R[-\theta_c]\cdot H\cdot R[\theta_c]\cdot R[-\varphi]\cdot Q\cdot R[\varphi]\cdot R[-\psi']\cdot EO(\Gamma)\cdot R[\psi']\cdot R[-\theta_r]\cdot H\cdot R[\theta_r]\cdot\begin{pmatrix} 1 \\ 0 \end{pmatrix}. \quad (32)$$

The intensity I_v of the vertical component is proportional to the square of the vertical electric field, and the normalized expression is also shown as:

$$I_v(\theta_c, \varphi, \psi', \theta_r : \Gamma) = \Psi_v^2$$
$$= \frac{1}{2}\begin{bmatrix} -\sin(\Gamma)\sin(4\theta_c - 2\varphi)\sin(4\theta_r - 2\psi') \\ +\left\{\cos^2(2(\theta_c - \theta_r)) + \cos^2(2(\theta_c + \theta_r - \varphi))\right\}\cos^2(\Gamma/2) \\ +\left\{\cos^2(2(\theta_c + \theta_r - \psi')) + \cos^2(2(\theta_c - \theta_r - \varphi + \psi'))\right\}\sin^2(\Gamma/2) \end{bmatrix}. \quad (33)$$

The intensity of the horizontal part of the laser through the EO crystal and wave plates can be derived from Eq. (31). If we set the angle ψ' to be 45° as shown in Fig. 5(b) and the rotator angle θ_r is 0°, we can obtain the intensity function for the horizontal part of the laser and is given by [Steffen et al., 2009, Parc et al., 2008a]:

$$I_h\left(\theta_c, \varphi, \psi' = 45^\circ, \theta_r = 0^\circ : \Gamma\right) = \left[1 - \cos(\Gamma)\cos(4\theta_c - 2\varphi)\cos(2\varphi) - \sin(\Gamma)\sin(4\theta_c - 2\varphi)\right]/2 \quad (34)$$

This kind of detection scheme is called as 'Near Crossed Polarizer' scheme [Steffen et al., 2009]. The angles of the quarter wave plate and compensator is usually set as a few degrees in the experiment [Parc et al., 2008a]. The intensity of the vertical part of the laser is also calculated as:

$$I_v\left(\theta_c, \varphi, \psi' = 45^\circ, \theta_r = 0^\circ : \Gamma\right) = \left[1 + \cos(\Gamma)\cos(4\theta_c - 2\varphi)\cos(2\varphi) + \sin(\Gamma)\sin(4\theta_c - 2\varphi)\right]/2 . \quad (35)$$

The difference between two components is easily calculated as

$$\Delta I \equiv I_v - I_h = \cos(\Gamma)\cos(4\theta_c - 2\varphi)\cos(2\varphi) + \sin(\Gamma)\sin(4\theta_c - 2\varphi) \quad (36)$$

For the balanced detection, the compensator angle θ_c should be set as 0°, the quarter wave plate angle φ should be set as 45° and the result is given by

$$\Delta I \equiv I_v - I_h = \sin(\Gamma) . \quad (37)$$

This is well known formula derived with principal refractive indices [Casalbuoni et al., 2008]. Note that the relative phase Γ in Eq. (37) is defined by Eq. (29) with refractive indices given by Eq. (27).

4. Experiment

A test experiment was conducted to see the peak change in the CCD image with the change of laser arrival time with respect to the electron bunch at FLASH facility in DESY with the configuration in Fig. 3. The GaP crystal in this experiment was 180 μm thick. The radial distances between the electron bunch and the crystal were 3, 2.5 and 2 mm. The total charge Q of the electron bunch in the measurement was 0.6 nC and the beam energy was 682 MeV. Arrival timing jitter of the electron bunch can be measured with an EO crystal due to the change of the crossing position of the laser as the electron bunch arrives [Cavalieri et al., 2005]. The arrival timing information can be extracted from the spatial information of the laser intensity measured by ICCD camera. When the arrival timing of electron bunch with respect to the laser is changed, the modulation of polarization of the laser will also be changed. To investigate the modulation, the laser delay was controlled with a certain delay time with respect to the electron beam and the vertical component of the laser was measured by ICCD camera. Changing the laser arrival time control the crossing positions of the laser with the EO crystal when the electron beam arrived at the EO crystal. The movement of the signal peak in the ICCD image was observed due to the change of crossing position of the laser as the electron bunch arrives at the EO crystal.

The plots in Figure 7 indicate the arrival time of the electron bunch with respect to the laser beam as observed by the ICCD camera. The time axis can be converted to the transverse length of the laser pulse [Azima et al., 2006]. A background measurement without an electron bunch is shown in Fig. 7 (a). A measurement result with the electron bunch is shown in Fig. 7(b). The background is subtracted from the result in Fig. 7(c). It shows a clear signal peak around 9 ps. Several measurement results with different arrival times of the laser are overlapped in Fig. 7(d). The peaks are moving by controlling of the laser arrival time, and the peak heights are also changed.

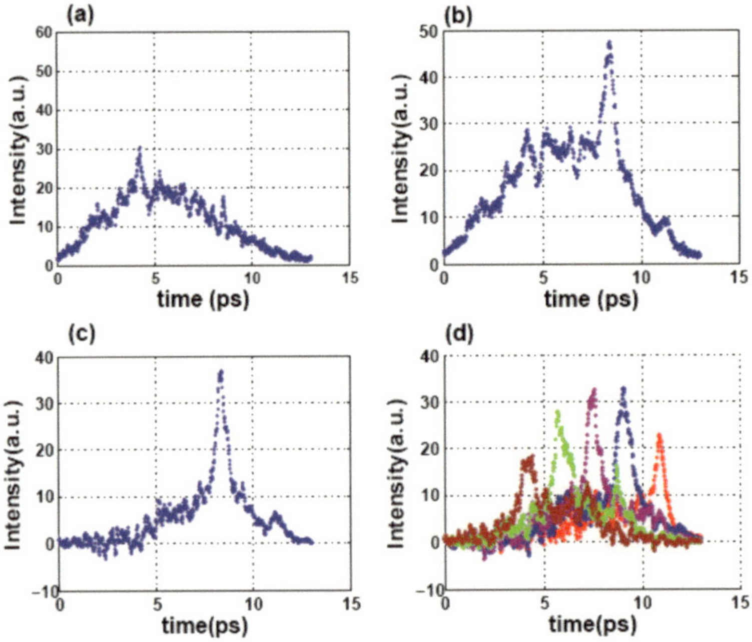

Fig. 7. Electron beam measurement by TEO setup at FLASH. The GaP crystal thickness is 180 μm. The distance between the electron beam and the laser at the crystal is 3 mm on average. The half wave plate angle is 0° and the quarter wave plate angle is -2°. X-axis represent the arrival time of the electron beam with respect to the laser beam. (a) The background measurement without electron beam. (b) The measurement result of the electron beam. (c) The difference between (a) and (b) is shown in (c). (d) Several measurement results with the change of the laser arrival time.

The measurement result must be expressed in terms of the relative phase shift Γ in Eq. (29) by using refractive indices in Eq. (27) to compare it with the simulation result. The measurement result can be analyzed in terms of the relative phase shift Γ with Jones matrices of the each wave plates as mentioned in Section 3 [Parc et al., 2008a, 2009b]. The analysis of the measured data in terms of the relative phase shift in Eq. (29) is shown in Fig.

8. The y axis in Fig. 8 represents the same axis shown in Fig. 4. The origin of the y axis is determined as the place where the relative phase shift reaches its maximum. Measurements are done with three r_o values defined in Fig. 4(b). The dots are the result with $r_o = 3$ mm, the squares are with $r_o = 2.5$ mm, and the triangles are with $r_o = 2$ mm. The relative phase shifts with the same condition of the electron beam show rapid change in spite of small changes of y as shown in Fig. 8. The smallest r_o has a higher relative phase shift at $y = 0$ mm. The peak value in Fig. 8 is about 12°. We define the degree of polarization as 0% for the linear polarization and 100% for the circular polarization. In our experimental result, we can conclude that the laser is 13% elliptically polarized by the electron beam with respect to the linear polarization.

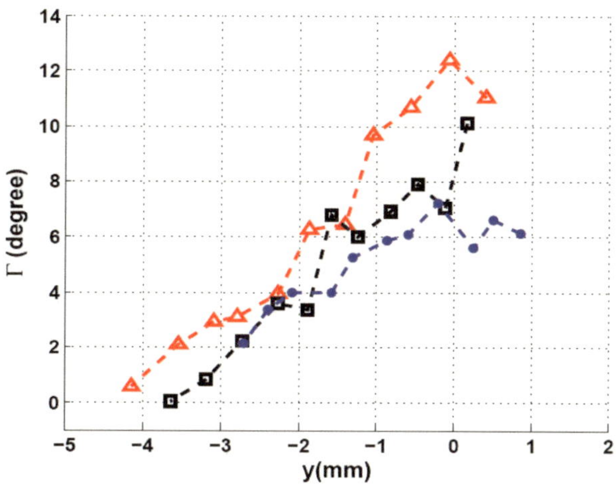

Fig. 8. The analysis results of the experiment in Fig. 6 in terms of the relative phase shift with new refractive indices for different distance r_o at $y = 0$ mm ($a = 0°$) case are shown. The dots are the result with $r_o = 2$ mm, the squares are with $r_o = 2.5$ mm, and the triangles are with $r_o = 3$ mm. y is the laser passing position in the crystal along the angle a in Fig. 4. The total beam charge Q measured in the experiment is 0.6 nC.

5. Simulation

The measurement of an electron bunch with a EO crystal is also simulated. The GaP crystal has a frequency dependent refractive index [Casalbuoni et al., 2008]. The electric field profile from the electron bunch is expanded to a Fourier series to know the response of the Fourier component inside the crystal [Parc et al., 2007, 2008b]. Each Fourier component propagates at a different speed according to the real part of the complex refractive index [Casalbuoni et al., 2008]. The speed difference between the two decomposed vectors of the laser polarization makes the polarization modulation of the laser.

The simulation carefully considered the important aspects of the propagation of the electric field in the EO crystal [Parc et al., 2008b, 2009a,]. The electric field affecting the laser is not constant owing to the difference of the group velocity between the laser and the electric field. In this simulation, the crystal was divided into N sections to calculate the total relative phase shift of the laser during its propagation. The total relative phase shift with new refractive indices is the sum of each relative phase shift at each section as shown by

$$\Gamma = \sum_{j=1}^{N} \frac{d}{\lambda N}\left(n_A(E_j) - n_B(E_j)\right), \tag{38}$$

where all the parameters are same as those of Eq. (14) except that the index j is used to denote the jth section of the crystal, and N is the number of sections to be summed. The thickness of GaP was 180 μm, which is the same as the experiment. The electron bunch was assumed to have a Gaussian distribution. The beam charge q for the electron beam was 0.1 nC. The electron beam charge in the experiment is larger than the charge used in the simulation because the shape of the electron beam was not Gaussian [Parc et al., 2009b]. The configuration in the simulation is the same as in the experiment.

A simulated signal output with the electron bunch expressed by the relative phase shift Γ is shown in Fig. 9. In this figure, τ is the time delay between the probing laser pulse and the electric field at the surface of EO crystal in the simulation [Casalbuoni et al., 2008]. Peak values of the phase shift Γ in Fig. 9 for different distances r_o along the y axis are presented in Fig. 10. The range of y axis is determined as the same with Fig. 8. The r_o values were the same as in Fig. 8. The dots are the results with r_o = 3 mm, the squares with r_o = 2.5 mm, and the triangles with r_o = 2 mm. These results are similar to the experimental results in Fig. 8. The relative phase shifts in the simulation are also within 12°. Both experimental and simulation found higher relative phase shifts around y = 0 mm at the smaller r_o value. For the large absolute values of y, the relative phase shifts decreased for all cases. The angle a between the line of electric field and polarization vector of the laser increased when the y value increased. For larger a, the difference between the refractive indices in Eq. (29) decreased, as shown in Fig. 11. With the shorter distance r_o, the relative phase shift decreased faster than with the longer r_o. At around y = –2 mm in Fig. 10, the three lines representing the relative phase shifts for the three cases cross. This result is very similar to the one obtained by the experiment in Fig. 8. This result can be understood from the fact that the angle a is larger for the shorter distance r_o at the same position of y as shown in Fig. 4.

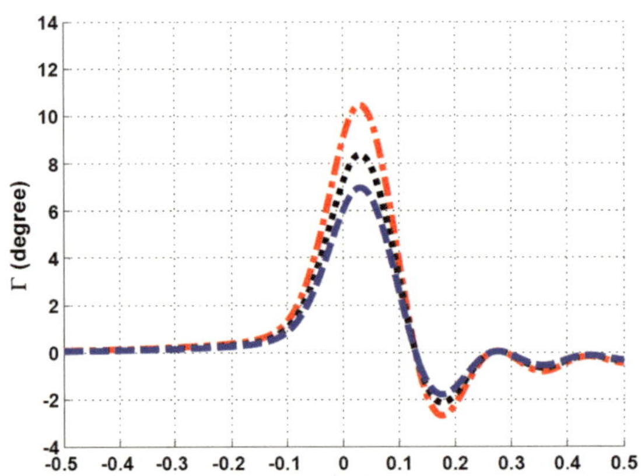

Fig. 9. Simulation result for the different distance r_o at $y = 0$ mm (a = 0°) case. The dash-dotted line is for r_o = 2 mm, the dotted line for r_o = 2.5 mm, and the dashed line for r_o = 3 mm. τ is the delay time of the electron beam with respect to the laser arrival time to sample the electric field from the electron beam in the simulation. The beam charge q for the bunched part in the electron beam for the simulation is 0.1 nC.

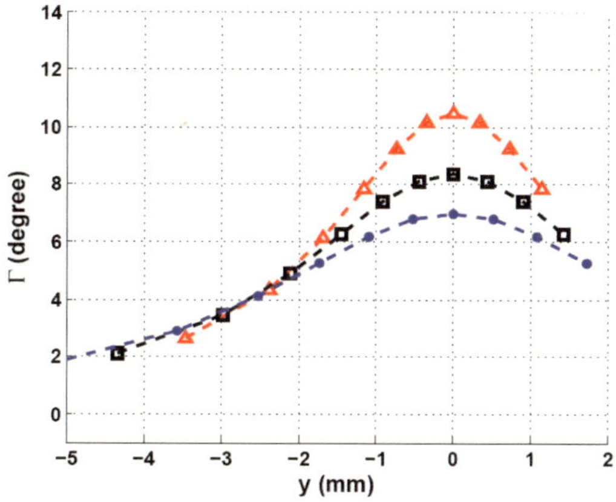

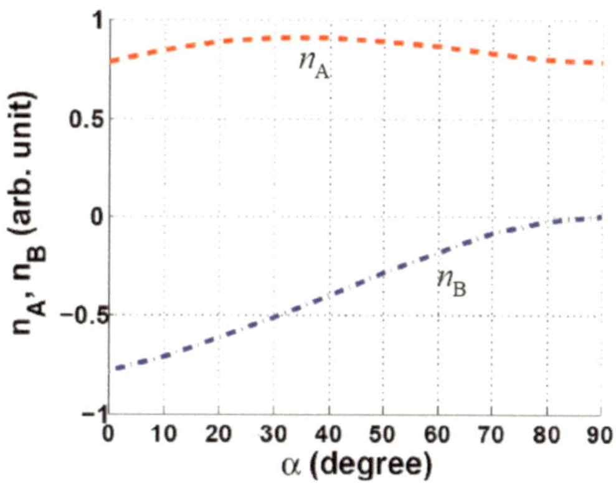

Fig. 10. Maximum relative phase shifts in Fig. 8 for different distances r_o with respect to the y are presented. The dots are the result with r_o = 2 mm, the squares with r_o = 2.5 mm, and the triangles with r_o = 3 mm. y is the laser passing position in the crystal along the angle a in Fig. 3. The beam charge q for the bunched part in the electron beam for the simulation is 0.1 nC.

Fig. 11. The refractive indices in Eq. (27) vs a.

6. Conclusion and discussion

We have derived refractive indices with the 45° decomposition axes with respect to the laser polarization vector in the plane perpendicular to the laser propagation direction with 3D configuration to decompose the laser polarization vector. We have also developed a theory for the detection of the optical power modulation. The experimental results are analyzed by the newly derived refractive indices. It is revealed that the laser is elliptically polarized from the fact that the relative phase shift is 12° at maximum.

We have also developed the simulation to understand experimental results correctly. The simulation shows that the maximum phase shift is about 12°, which is well agreed with the experimental result. The decreasing behavior of the relative phase shift in terms of larger $|y|$ is observed in the simulation result. While the relative phase shift for three different r_o cases is decreasing with different ratio, the phase shifts near y = -2 mm are crossed for three r_o cases. This also agrees well with the experimental result.

With the refractive indices derived in this study, we can open the way to express the polarization state of the laser in the electron beam measurement with EO crystal with a single quantity, i.e. the relative phase shift in Eq. (29). The relative phase shift shows a dependency on the incident angle of the laser to the EO crystal. To analyze the experimental result precisely, the newly derived refractive indices in this study will be a good tool in the diagnostics of the electron beam with EO crystal. We believe that the refractive indices in Eq. (29) must also be used in the analysis of the THz measurement result shown in Ref. [Shan et al., 2000].

7. Acknowledgements

This work was supported by the Korea Science and Engineering Foundation (KOSEF) grant funded by the Korea government (Ministry of Education Science and Technology) (grant

No. 2008-0059842). One of authors (Y. W. Parc) extends his thanks to Deutsches Elektronen Synchrotron (DESY) for the hospitality during his stay (2006. Nov. ~ 2007. Apr.). Specially, we want to give great thanks to A. Azima for many discussions and for experimental help.

8. References

[Azima et al., 2006] A. Azima, S. Dusterer, M. Huning, E. -A. Knabbe, M. Roehrs, V. Rybnikov, H. Schlarb, B. Schmidt, B. Steffen, P. Schmuser, A. Winter, M. Ross, Proceedings of European Particle Accelerator Conference 2006, Edinburg, Scotland, 1049 (2006).

[Berden et al., 2004] G. Berden, S. P. Jamison, A. M. MacLeod, W. A. Gillespie, B. Redlich, A. F. G. van der Meer, Phys. Rev. Lett., 93, 114802 (2004).

[Casalbuoni et al., 2008] S. Casalbuoni, H. Schlarb, B. Schmidt, P. Schmuser, B. Steffen, A. Winter, Phy. Rev. ST Accel. Beams, 11, 072802 (2008).

[Cavalieri et al., 2005] A. L. Cavalieri, D. M. Fritz, S. H. Lee, J. B. Hastings, et al., Phys. Rev. Lett., 94, 114801 (2005).

[Hecht, 2002] E. Hecht: *Optics* (Addison-Wesley, New York 2002) 4th ed., p. 336.

[Jones, 1941] R. C. Jones, J. Opt. Soc. Am. 31, 488 (1941).

[Par et al., 2007] Y. W. Parc, J. H. Park, C. B. Kim, J. Y. Huang, T. H. Joo, and I. S. Ko, J. Korean Phys. Soc., 50, 1390 (2007).

[Parc et al., 2008a] Y. W. Parc, C. B. Kim, J. Y. Huang, and I. S. Ko, Nucl. Insrt. And Meth. A, 586, 452 (2008).

[Parc et al., 2008b] Y. W. Parc, J. H. Hong, C. B. Kim, J. Y. Huang, S. H. Lee, and I. S. Ko, Jap. J. Appl. Phy., 47, 342 (2008).

[Parc et al., 2009a] Y. W. Parc, I. S. Ko, J. Appl. Opt. A: Pure Appl. Opt., 11, 105704 (2009).

[Parc et al., 2009b] Y. W. Parc, I. S. Ko, J. Korean Phys. Soc., 54, 1481 (2009).

[Planken et al., 2001] P. C. M. Planken, H. K. Nienhuys, H. J. Bakker, and T. Wenckebach, J. Opt. Soc. Am. B, 18, 313 (2001).

[Shan et al., 2000] J. Shan, A. S. Weling, E. Knoesel, L. Bartels, M. Bonn, A. Nahata, G. A. Reider, T. F. Heinz, Opt. Lett. 25, 426 (2000).

[Shen et al., 2008] Y. Shen, G. L. Carr, J. B. Murphy, T. Y. Tsang, X. Wang, and X. Yang, Phys. Rev. A. 78, 043813 (2008).

[Steffen et al., 2009] B. Steffen, V. Arsov, G. Berden, W. A. Gillespie, S. P. Jamison, A. M. MacLeod, A. F. G. van der Meer, P. J. Phillips, H. Schlarb, B. Schmidt, and P. Schmuser, Phy. Rev. ST Accel. Beams, 12, 032802 (2009).

[Wen & Lindenberg, 2009] H. Wen and A. M. Lindenberg, Phys. Rev. Lett., 103, 023902 (2009).

[Wilke et al., 2002] I. Wilke, A. M. MacLeod, W. A. Gillespie, G. Berden, G. M. H. Knippels, A. F. G. van der Meer, Phys. Rev. Lett., 88, 124801 (2002).

[Yan et al., 2000] X. Yan, A. M. MacLeod, W. A. Gillespie, Phys. Rev. Lett., 85, 3404 (2000).

[Yariv & Yeh, 2003] A. Yariv and P. Yeh, *Optical Waves in Crystals: propagation and Control of Laser Radiation* (Wiley, New York, 2003), p. 229.